V. V. Nikulin I. R. Shafarevich

Geometries and Groups

Translated from the Russian
by M. Reid

D0103264

With 159 Figures

Springer-Verlag
Berlin Heidelberg New York
London Paris Tokyo

Viacheslav V. Nikulin Igor R. Shafarevich
Steklov Mathematical Institute
ul. Vavilova 42, 117966 Moscow, USSR

Miles Reid
Mathematics Institute, University of Warwick
Coventry CV4 7AL, Great Britain

QA445
N5513
1987

Title of the Russian original edition: *Geometrii i gruppy*
Publisher Nauka, Moscow 1983

This volume is part of the *Springer Series in Soviet Mathematics*
Advisers: L. D. Faddeev (Leningrad), R. V. Gamkrelidze (Moscow)

Mathematics Subject Classification (1980): Primary 51-01
Secondary 00A05, 20H15, 51H20, 51M05, 51M10

ISBN 3-540-15281-4 Springer-Verlag Berlin Heidelberg New York
ISBN 0-387-15281-4 Springer-Verlag New York Berlin Heidelberg

Libary of Congress Cataloging-in-Publication Data
Nikulin, V. V. (Viacheslav Valentinovich)
Geometries and groups.
(Universitext) (Springer series in Soviet mathematics)
Translation of: Geometrii i gruppy.
Bibliography: p. Includes index.
1. Geometry. 2. Groups, Theory of.
I. Shafarevich, I. R. (Igor' Rostislavovich), 1923–
II. Title. III. Series: Springer series in Soviet mathematics.
QA445.N5513 1987 516 87-12798
ISBN 0-387-15281-4 (U.S.)

This work is subject to copyright. All rights are reserved, whether the whole or part of the material is concerned, specifically the rights of translation, reprinting, reuse of illustrations, recitation, broadcasting, reproduction on microfilms or in other ways, and storage in data banks. Duplication of this publication or parts thereof is only permitted under the provisions of the German Copyright Law of September 9, 1965, in its version of June 24, 1985, and a copyright fee must always be paid. Violations fall under the prosecution act of the German Copyright Law.

© Springer-Verlag Berlin Heidelberg 1987
Printed in Germany

Printing and binding: Druckhaus Beltz, Hemsbach
2141/3140-543210

Universitext

Preface

This book is devoted to the theory of geometries which are locally Euclidean, in the sense that in small regions they are identical to the geometry of the Euclidean plane or Euclidean 3-space. Starting from the simplest examples, we proceed to develop a general theory of such geometries, based on their relation with discrete groups of motions of the Euclidean plane or 3-space; we also consider the relation between discrete groups of motions and crystallography. The description of locally Euclidean geometries of one type shows that these geometries are themselves naturally represented as the points of a new geometry. The systematic study of this new geometry leads us to 2-dimensional Lobachevsky geometry (also called non-Euclidean or hyperbolic geometry) which, following the logic of our study, is constructed starting from the properties of its group of motions. Thus in this book we would like to introduce the reader to a theory of geometries which are different from the usual Euclidean geometry of the plane and 3-space, in terms of examples which are accessible to a concrete and intuitive study. The basic method of study is the use of groups of motions, both discrete groups and the groups of motions of geometries.

The book does not presuppose on the part of the reader any preliminary knowledge outside the limits of a school geometry course. We have in mind a wide circle of possible readers: students of mathematics and physics in universities and technical colleges, high school teachers and students in the upper classes of high school... We hope that reading this book will enable even a reader with no professional mathematical training to become acquainted with one of the most attractive aspects of mathematics: that many of its problems are solved using methods and concepts which at first sight have nothing whatever to do with the original problem. It is only after fairly lengthy development that these methods lead to a solution of the problem which led to their appearance, and they often open up before the investigator a completely new field of study. This internal development of mathematics, whereby the needs of one area lead to the creation of new areas of research is complemented by the extraordinary phenomenon of its unity: theories created for various ends and developing in different directions unexpectedly turn out to be closely related. Naturally, in order to get a feeling for these special features of mathematical research, the reader must be ready to spend both time and effort overcoming the difficulties which he may face in reading the book; the difficulties are not caused by the use of complex techniques - the reader will be able to get by with a school geometry course - but by the need to get used to more complex and longer mathematical arguments.

DE 15 '89

The book divides into four chapters. It is our hope that the first can be read by a high school student with interest in mathematics. In it we treat the basic examples, which will allow the reader to gain some geometrical intuition in the new subject. Reading this chapter only will already provide some first impressions of the subject matter of the book.

The second chapter is the heart of the book: in it we introduce the new method, with which the problem stated at the end of Chapter I can be solved. This chapter is distinctly harder than the first; it contains the proofs of a number of theorems, some of which are not so easy even for a professional mathematician. It is, however, our hope that while working on Chapter I, the reader will have acquired certain abilities which will help him to overcome the difficulties of Chapter II. The reader who has worked his way through the first two chapters will already have a complete impression of the theory which forms the subject of the book. The remainder of the book relates this theory to other questions.

The reader will probably find Chapter III rather easier. Its first section §11 generalises the theory, constructed so far in two dimensions, to the 3-dimensional case; the second section §12 deals with the relation of this theory to crystallography. In the final fourth chapter of the book the logic of the development of our theory leads naturally to a completely new fundamental concept of geometry, Lobachevsky geometry.

Almost each section ends with some exercises. They do not aim to introduce the reader to new facts; in the main, they are intended to help him check the extent to which he has grasped the preceding text. For this reason they are as a rule very simple.

Some references for further reading are included in the text. The reader who would like to pursue the questions treated in this book in more depth is recommended the following books:

H. Cartan, Geometry of Riemannian spaces, Gauthier-Villars, Paris, 1951 and Math. Sci. Press, Brookline, 1983;

J.A. Wolf, Spaces of constant curvature, McGraw-Hill, New York, 1967;

S. Helgason, Differential geometry and symmetric spaces, Academic Press, New York, 1967;

L.R. Ford, Automorphic functions, Chelsea, New York 1951;

M. Klemm, Symmetrien von Ornamenten und Kristallen, Springer, New York - Berlin, 1982;

together with the survey article:

P. Scott, The geometries of 3-manifolds, Bulletin of London Math. Soc. **15** (1983), p. 401-487.

The authors would like to thank the translator M. Reid for a number of suggestions which have improved the book.

Table of contents

Chapter I
Forming geometrical intuition; statement of the main problem

§1. Formulating the problem

Many of the most outstanding achievements of geometry are based on the surprising circumstance that geometrical intuition, formed in some areas, turns out to be applicable in other, sometimes quite dissimilar, areas. In other words, we can imagine different worlds, in which the laws of geometry are different from ours, almost as well as if we lived in them. The aim of this book is to tell of one line of geometrical investigation in which this phenomenon manifests itself particular vividly.

The question with which we will be concerned is purely geometrical, but to formulate and explain it, it is simplest to begin by discussing it in physical terminology. In its widest form, the question is as follows: how can we convince ourselves that the theorems of geometry are actually valid in the space around us? Starting from this general formulation, we will gradually narrow the problem down until we get to the more concrete question which is the subject matter of this book.

First of all, when we speak of geometry, we have in mind only the part of this science which is well-known to all and covered in a school course. Even in this restricted sense, our question loses none of its interest.

As a first answer to the question, we could say that any theorem from a geometry textbook can be tested experimentally: for example, to test the theorem that the angles of a triangle add up to $180°$ one has to measure the angles of a genuine physical triangle. For this purpose we will ignore the fact that all physical measurements are only approximate, and suppose that the measurements can be made to any degree of accuracy.

The question arises: are such measurements always possible? The length of a reasonable-sized interval can be measured by comparing it with some unit of length; however, how are we to test if some theorem or other is valid for extremely large figures? For example, how can we test the theorem on the sum of the angles in the case of a triangle having as its vertices a point on the earth and two satellites? In this case all measurements we need can be carried out using a telescope. The same method can even be used to measure the distance from the earth to the stars. At this point, however, we run into the fact that our possibilities are limited, in this case by the effective radius of our telescopes. Of course this limit may change as telescopes are perfected, but at any given time it can be represented as a perfectly definite number. The actual value of this limit is not essential for our purposes: let us denote it by r.

Thus we have seen that there exists a magnitude (the effective radius of modern telescopes), such that all measurements which we are in a position to carry out refer only to figures all of whose points are at distance not more than r from us; in other words, to figures contained within a ball of radius r centred on the earth. Suppose that measurements showed that within this ball, all the theorems of geometry are valid. Can we then assert the same for the whole of space? The answer to this, of course, is negative: since all our information refers to measurements inside one ball, we cannot say anything about the results of measurements outside it.

To arrive at a question of more substance, we will make another assumption, which can never be tested by experience, but which seems natural from the logical point of view: suppose that the earth does not occupy any special position in space, but that on the contrary, all points of space have equal status. More precisely, we assume that all results which could be obtained by measurements within a ball of radius r centred at the earth would also be valid if we could carry out measurements in a ball of the same radius r centred at any point of space. And therefore, we suppose that all the theorems of geometry are satisfied by all figures contained in a ball of radius r centred at any point of space. This is of course an extremely substantial restriction of the general statement of the problem from which we started.

Thus we finally arrive at the following question: suppose that there exists some magnitude r such that all the theorems of geometry hold for all figures contained in a ball of radius r, centred at any point of space. Will the theorems then hold for any figures?

This question is in essence purely geometrical; in what follows we will study it as a problem in geometry.

The answer turns out completely unexpectedly. Firstly, the answer is negative. We will construct other geometries, distinct from the familiar geometry of school and of our real-life experience, and which are nevertheless indistinguishable from our 'ordinary' geometry if we restrict ourselves to considering their properties within any ball of some radius r. But on the other hand, we will see that there are not so many of these geometries; there are just a few types of them, and we will list them all.

Moreover, when we attempt to treat the set of geometries of one type, we will see that these geometries themselves are naturally represented by the points of a certain new geometry. In the most interesting case we thus arrive at another celebrated geometry, Lobachevsky geometry. And finally, yet another surprise: although our studies are apparently motivated by a purely theoretical problem, methods turn up in the course of them which relate to 'ordinary' geometry, and which are indispensable tools in many areas of geometry and physics, for example in such concrete areas of physics as crystallography.

We thus encounter a completely new phenomenon which is of great importance for science, the existence of different geometries. The trouble is, a Euclidean geometry course gives the impression that only one geometry is possible, with predetermined rules, and that the only task is to discover these. Before developing this new point of view, we must of course explain what it means to construct a geometry. A precise definition will be given in §6. We begin with examples, restricting ourselves to the 2-dimensional case for simplicity. As a first and extremely intuitive example of a geometry other than the well-known geometry of the Euclidean plane, we talk about so-called spherical geometry.

§2. Spherical geometry

In this section we are concerned with geometry on the sphere, that is, the surface of a ball. Clearly, this example is thoroughly realistic, since we ourselves are living on the surface of the earth, which is approximately a sphere.

What do we understand by the distance between two points on earth, for example between the north and south poles? There are two different possible answers to this question. We could either forget that our points are on earth, and measure the distance between them in the ambient cosmic space. Then for example, the distance between the north and south poles would be equal to the diameter of the earth. This distance is equal to the time needed to go from one point to the other, travelling at a constant unit speed, assuming that motion through the earth is possible. Alternatively, we could measure the distance as the minimum time taken to get from one point to the other, travelling at unit speed along the surface of the earth. This distance would be equal to the length of the shortest possible curve joining our two points and lying on the surface of the earth. In this case the distance from the north to the south pole would be equal to one half the length of a meridian. For terrestrial matters we generally adopt the second method of measuring distances.

It is this second method of measurement which leads us to spherical geometry. The word 'geometry' itself (from the Greek γεωμετρια) comes from 'Ge': the earth, and 'metreo': I measure. The points of this geometry are just the points of some sphere Σ. The distance between two points P and Q of Σ is taken to be the length of the shortest curve lying entirely on Σ and joining P and Q.

We now consider some of the simplest properties of spherical geometry.

Theorem 1. A curve of shortest length lying on Σ and joining two points P and Q of Σ is an arc of a great circle of Σ passing through P and Q.

Before starting on the proof of the theorem, we recall that a *great circle* of a sphere is its intersection with a plane through the centre. With the exception of *antipodal* (or diametrically opposite) points, there is a unique great circle passing through two points P and Q of the sphere, and they divide it into two arcs; when we speak of the arc of great circle joining P and Q, we will mean the shorter of the two. If P and Q are antipodal, then they are joined by an infinite number of arcs of great circles; all of these have the same length, equal to half the length of a meridian of Σ.

Proof of Theorem 1. We proceed by contradiction. Let ℓ be the arc of great circle joining P and Q, and $\ell^{\sim}$ be another curve lying on Σ and joining P and Q; suppose that $\ell^{\sim}$ is shorter than ℓ. We are not in this book going to go into the finer points of the definition of the length of a curve; the only thing which is important for our purposes is that this length can be obtained to any degree of accuracy by measuring the length of broken lines inscribed in it and having sufficiently small links. Choose points $P, P_1, P_2, ..., P_n, Q$ on $\ell^{\sim}$, and join each adjacent pair by an arc of great circle (Figure 2.1); this gives us a spherical broken line ℓ'. Choosing

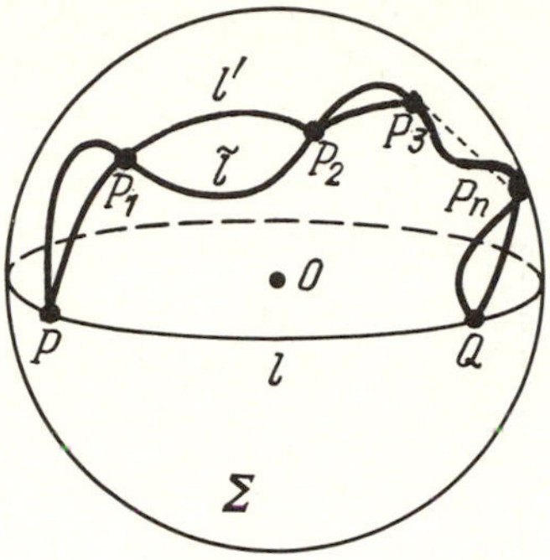

Figure 2.1

points $P_1, ..., P_n$ on $\ell^{\sim}$ sufficiently close together, we can arrange that ℓ' lies close to $\ell^{\sim}$, and that its length differs very little from that of $\ell^{\sim}$; in particular, ℓ' will also be shorter than ℓ.

Now we are going to delete the vertices $P_1, ..., P_n$ of ℓ' one by one. Deleting P_1, say, we join $P, P_2, P_3, ..., P_n, Q$ by arcs of great circles; we obtain a new broken line ℓ''. The theorem will be proved if we can show that ℓ'' is not longer than ℓ'. Indeed, since ℓ' was shorter than ℓ, it will follow from this that ℓ'' is even shorter. Repeating the same n times, we delete all the vertices of ℓ', arriving at a 'broken' line $\ell^{(n)}$ consisting just of the single arc of great circle joining P and Q, which will again be shorter than ℓ. But this is a contradiction, since ℓ and $\ell^{(n)}$ have the same length (in fact ℓ and $\ell^{(n)}$ are identical, provided that P and Q are not antipodal).

It remains to prove that ℓ'' is not longer than ℓ'. But all the links in these two broken lines are the same, except that ℓ'' has an arc of great circle joining P and P_2, whereas ℓ' has two arcs joining P to P_1 then P_1 to P_2. Let a, b and c denote the lengths of these three arcs. We thus need to prove the inequality

$$a \le b + c. \tag{1}$$

For this, we pass three planes through the centre O of Σ which cut out the great circles joining P to P_1, P_1 to P_2, and P to P_2. These form a solid angle OPP_1P_2 (Figure 2.2). Since a, b and c are arc lengths in circles having the same radius, they are proportional to the plane angles α, β and γ of the corresponding faces of the solid angle OPP_1P_2, so that (1) is equivalent to the inequality

$$\alpha \leq \beta + \gamma \tag{2}$$

for these angles; this inequality is well-known as one of the elementary properties of the face angles of a solid angle (and is taught in the standard geometry course in Soviet schools). It can easily be proved, for example by dropping a perpendicular from P_1 onto the plane OPP_2 of α.

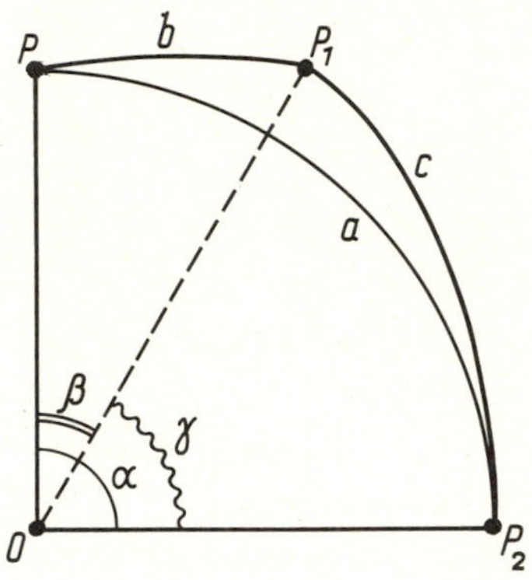

Figure 2.2

We have thus proved that the distance between two points in spherical geometry is equal to the length of the arc of a great circle joining them. In this sense, great circles play the same role in spherical geometry as lines in plane geometry. For this reason, in spherical geometry we define an *angle* CAB^ to be the portion of the sphere bounded by arcs of the great circles ABA′ and ACA′ passing through A, B and A, C respectively, and the magnitude of the angle CAB^ to be the magnitude of the plane angle formed by the half-planes ABA′ and ACA′ along the edge AA′ (Figure 2.3).

Using this definition, the simple well-known properties of angles in plane geometry are preserved: the angle formed by two adjacent angles whose magnitudes add up to less than 2π is an angle having magnitude the sum of the two; the full angle at any point is equal to four right angles, or 2π.

A *spherical triangle* is the figure formed by three points, assumed not to lie on a great circle, and the three arcs of great circles joining them. Many of the known theorems of plane geometry also hold for spherical triangles. For example, inequality (1) is the theorem that the length of a side of a triangle does not exceed the sum of the lengths of the other two sides. The property that any two points of the sphere can be joined by an arc of great circle corresponds to the property of plane geometry that any two points on the plane can be joined by a straight line.

However, there are a number of differences between spherical and plane

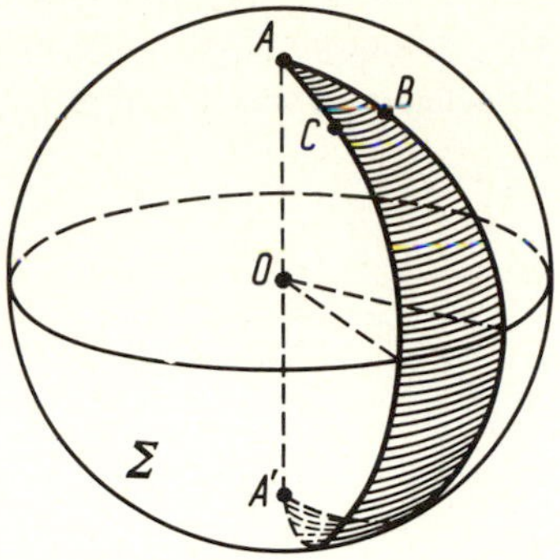

Figure 2.3

geometries. For example, the line passing through any two distinct points of the plane is unique, whereas the great circles passing through a pair of antipodal points of the sphere are infinite in number. In contrast to lines in the plane, any two great circles always meet: there is no analogue of parallel lines. The theorem on the sum of angles in a triangle also fails: three mutually perpendicular planes through the centre of a sphere intersect the sphere in a triangle with three right angles! Perhaps the most striking divergence from plane geometry is the fact that, in contrast to lines, great circles have finite length - if we continue round a great circle, we eventually arrive back to our starting point (on the surface of the earth this property was first tested by a ship of Magellan's expedition in 1522).

All the differences between spherical and plane geometry we have listed can be detected on sufficiently large pieces of the sphere. Thus, any two great circles meet in a pair of antipodal points, so that if we only consider a sufficiently small region of the sphere that does not contain such a pair of points, then we will not discover their second intersection. It might therefore be suspected that in sufficiently small regions, spherical geometry does not differ from plane geometry - this would be an example of the phenomenon which was discussed as a possibility in §1. We show that this is not the case, giving two examples of properties of spherical geometry which distinguish it from plane geometry, and which can be detected in any region of the sphere, no matter how small.

The first property concerns the perimeter of a circle. In spherical geometry, a circle is the set of all points whose distance from some point is a fixed length ρ, called the radius; the same definition as in plane geometry. If the sphere Σ has radius R, then $\rho = R\varphi$, where φ is the angle in radians between the rays from the centre of the sphere to the centre of the circle and to one point of the circle (Figure 2.4). From the point of view of the ordinary geometry of the ambient space, we have a circle whose radius ρ' is given by $\rho' = R \sin \varphi = R \sin(\rho/R)$, as can be seen from Figure 2.4. Therefore the perimeter of this circle equals $2\pi\rho'$, that is

$$2\pi R \sin(\rho/R). \tag{3}$$

We see from this that in spherical geometry, a circle of radius ρ has a different perimeter from that in plane geometry, $2\pi R \sin(\rho/R)$ instead of $2\pi\rho$. This difference can of course be detected on arbitrarily small circles.

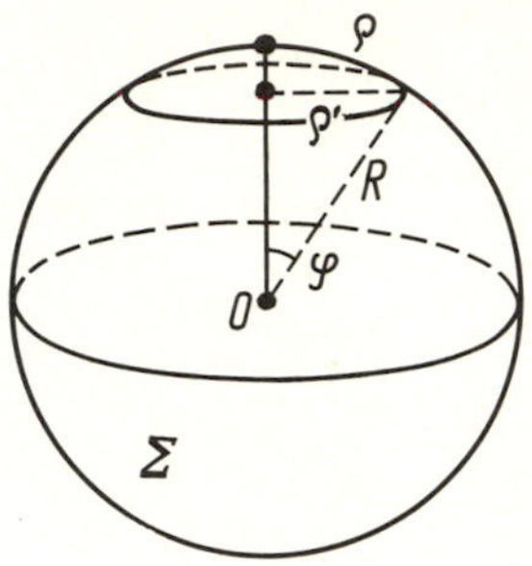

Figure 2.4

The second distinction concerns the sum of angles in a triangle.

Theorem 2. In spherical geometry, the sum of the angles of any triangle is greater than two right angles, that is, greater than π.

According to the definition of the magnitude of an angle in spherical geometry, this is equivalent to saying that the sum of the plane angles in any 3--dimensional angle is greater than 2π. One could give a proof of this as an assertion in 3-dimensional Euclidean geometry, with no mention of the sphere. But it would be more satisfying to give a proof in which substantial use is made of the geometry of the sphere. We now do this.

Proof. Extend the sides AB, BC and CA of the spherical triangle ABC to great circles (AB), (BC) and (CA) (see Figure 2.5, (a)). Let's find the areas of the segments of sphere Σ_A, Σ_B and Σ_C bounded by the two great circles out of each of the three angles A, B and C of the triangle; (Figure 2.5, (b) shows one of these segments, Σ_A).

Write $\hat{A}$ for the magnitude of the angle $\widehat{BAC}$. The area of Σ_A as a proportion of the whole sphere is obviously the same as $2\hat{A}$ as a proportion of the angle 2π. Since the area of the sphere is $4\pi R^2$, the area of Σ_A is

$$4\pi R^2 \cdot \frac{2\hat{A}}{2\pi} = 4r^2\hat{A};$$

similarly, the areas of Σ_B and Σ_C are $4R^2\hat{B}$ and $4R^2\hat{C}$ respectively.

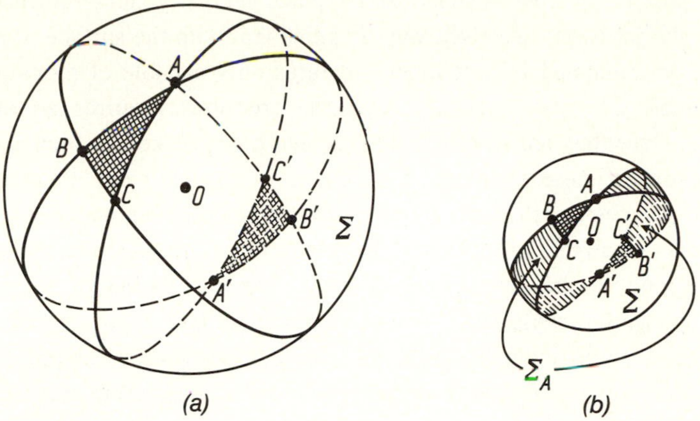

Figure 2.5

Note that Σ_A, Σ_B and Σ_C together cover Σ, and they overlap in the spherical triangle ABC and the antipodal triangle A′B′C′ (see Figure 2.5, (a)), and hence

$$4\pi R^2 \hat{A} + 4\pi R^2 \hat{B} + 4\pi R^2 \hat{C} = 4\pi R^2 + 2S_{ABC} + 2S_{A'B'C'} = 4\pi R^2 + 4S_{ABC};$$

therefore

$$\hat{A} + \hat{B} + \hat{C} - \pi = \frac{S_{ABC}}{\pi R^2} > 0, \tag{4}$$

which proves the theorem.

The discrepancies we have discovered between spherical and plane geometries show that it is impossible to make an exact map of any region of the sphere, however small, since a map is drawn in the plane! Here by an exact map we have in mind one in which the distance between any two points is equal to the distance between their representations on the map. In practice, maps of the surface of the earth are only usable because rather than exact maps, one is satisfied with maps whose distortion is sufficiently small. It can be shown that the smaller a region of the sphere is compared to the radius of the sphere, the more accurately can it be represented on a map. This phenomenon is already illustrated by formula (3) for the perimeter of a circle on the sphere, and by formula (4) for the difference between the sum of angles of a triangle and π. As is well-known, $\sin\alpha/\alpha$ tends to 1 as α tends to 0, so that the smaller the angle α, the nearer $\sin\alpha$ is to α. Therefore when ρ is very small compared to R, the value of $2\pi R$

sin (ρ/R) is very close to 2πρ, that is, to the perimeter of a circle in plane geometry. The corresponding assertion for the sum of angles is even more obvious: (4) shows that the smaller the area of a triangle compared with the radius of the earth, the closer is the sum of its angles to 2π.

The difference between spherical and plane geometry can be illustrated for example in the following idealised way. Imagine that both the surface of a sphere and a plane are inhabited by intelligent creatures only capable of moving within these surfaces. Of course, visualising these creatures requires an effort of imagination, since they are strictly 2-dimensional beings, like ink-blots spreading out. Suppose that these creatures are capable of carrying out measurements, studying the geometry of their environment, and communicating with each other by radio. Then an inhabitant of the plane could enquire of an inhabitant of the sphere whether the formula 2πρ for the perimeter of a circle of radius ρ holds, and on receiving a negative reply, they would understand that their geometries are different; they could even come to this conclusion without stepping out of arbitrarily small regions of their respective geometries, considering only perimeters of sufficiently small circles.

The reader can consult M. Berger, Geometry II, Springer, 1987 for a detailed introduction to spherical geometry. A classical reference is J. Hadamard, Leçons de géométrie élémentaire, vol.2, Géométrie dans l'espace, A. Colin, Paris, 1949.

We have spent some time studying spherical geometry merely in order to give the simplest example of a geometry distinct from the well-known geometry of the Euclidean plane. We will now turn our attention to an example of a geometry which answers the question discussed in §1: this geometry is distinct from plane geometry, but is identical with it in sufficiently small regions.

Exercises

1. Say what the well-known properties of angles in triangles and polygons give for spherical geometry.
2. Prove the analogue of Pythagoras' theorem in spherical geometry: cos (c/R) = cos (a/R) cos (b/R), where c is the length of the hypothenuse, and b, c are the lengths of the other two sides of a right-angled spherical triangle. Prove that this approaches the usual Pythagoras' theorem for sufficiently small triangles. [Hint: $\cos^2 \alpha = 1 - \sin^2\alpha$, and therefore for small α, $\cos \alpha \approxeq 1 - \alpha^2$.]
3. Prove that the angles of a convex spherical polygon satisfy

$$A_1^\wedge + A_2^\wedge + \dots + A_n^\wedge - \pi(n-2) = S/R^2,$$

where $A_1^\wedge, A_2^\wedge, \dots, A_n^\wedge$ are the angles, and S the area of the polygon.
4. Prove that a convex polyhedron satisfies the Euler formula

$$V - R + E = 2,$$

where V is the number of vertices, R the number of edges, and E the number of faces of the polygon. [Hint: project from a point O inside the polyhedron onto a sphere centred at O, and express the area of the sphere as the sum of the areas of the spherical polygons mapped to by faces of the polyhedron.]

§3. Geometry on a cylinder

3.1. First acquaintance. Consider a cylindrical surface Σ obtained by moving a generating line (or *generator*) ℓ' parallel to a fixed direction around some closed plane curve with no self-intersections. We define geometry on Σ in the same way as we did on the sphere: we take the points of the geometry to be points of Σ, and define the distance between two points p and q to be the length of the shortest curve on Σ joining p and q.

To study the geometry of Σ we make use of a representation of Σ by unrolling it, or developing it onto the plane. Write c' to denote a cross-section of Σ perpendicular to the generator ℓ'; this c' is sometimes called a *directrix* of Σ. In the plane, consider a strip bounded by two parallel lines m_0 and m_1 and of width d equal to the length of c'; join m_0 and m_1 by a perpendicular line segment $[P_0P_1]$. Since the length of $[P_0P_1]$ is equal to that of c', it can be wrapped around c' without stretching, starting with P_0 at some point p of c', and going all the way round c' with P_1 ending up at the same point p. Now wrap the whole

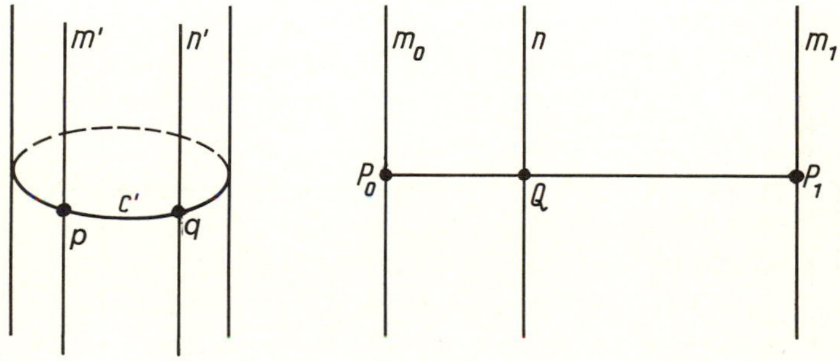

Figure 3.1

strip around Σ in such a way that a line n through a point Q of $[P_0P_1]$ parallel to the boundaries m_0 and m_1 goes to the generator n' of Σ through the corresponding point q of c' (see Figure 3.1).

This operation can be visualised as starting from a strip of sheet metal, which is rolled up until the edges m_0 and m_1 coincide, in such a way that the line segment $[P_0P_1]$ takes the shape of c'; the strip then becomes the cylindrical surface Σ.

In this representation, the strip covers the whole of Σ, and each point of Σ

is covered by exactly one point, with the only exception of points of the generator through p: each of these is covered by two points of the strip, one on m_0 and one on m_1. To avoid this, we will from now on allow as points of the strip only the points of m_0, and not those of m_1. Then each point of Σ is covered by a single point of the strip. The inverse correspondence of passing from a point of Σ to the point of the strip covering it will be called developing the cylinder onto the plane.

We will determine how to measure the distance between two points of Σ if we know the corresponding points of the strip. Here is the first step in this direction.

Theorem 1. As the strip covers Σ, any curve contained in the strip goes into a curve of Σ of the same length.

The theorem is intuitively convincing if we visualise Σ as above as obtained by rolling up a strip of sheet metal: since the metal neither stretches nor shrinks, the length of a curve is not altered.

For a formal proof, consider the case that the curve c' is a polygon, so that Σ is a polygonal cylinder (Figure 3.2). Then the edges $m', n', o', p', \ldots, t'$ of Σ

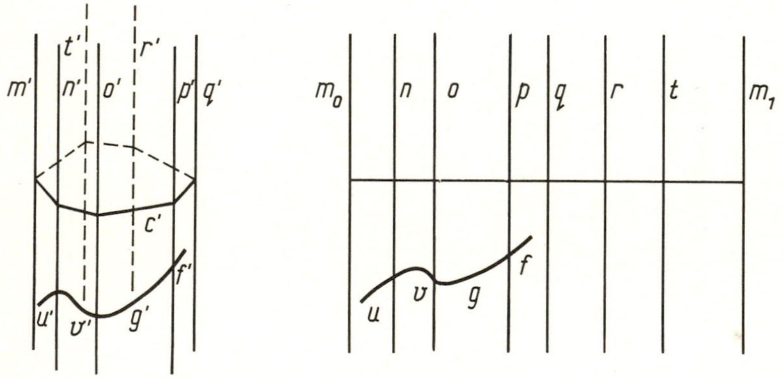

Figure 3.2

develop out to lines $m_0, n, o, p, \ldots, t$ of the strip; these lines divide the strip up into smaller strips m_0n, no, ..., tm_1, each of which covers a side of Σ by a rigid motion. Hence a curve lying in one of these smaller strips goes to an equal curve. Now any curve f is divided up by the lines n, o, p, ... into pieces u, v, ... But by what we have just said, the curves u, v, ... go over to equal curves u′, v′, ...; now since f is composed of the pieces u, v, ..., and f′ of u′, v′, ..., it follows that f and f′ have equal length.

For an arbitrary cylindrical surface Σ we must represent Σ as a limit of inscribed polygonal cylinders Σ_n, and a given curve f' on Σ as the limit of the curves f'_n obtained by projecting f' to Σ_n. After this, we can use the assertion already proved for the polygonal cyclinder Σ_n and the curves f'_n. The detailed proof is not difficult, but we do not give it here, since for the construction of cylindrical geometry which we are at present concerned with, we could perfectly well restrict attention to the case of a polygonal cylinder Σ.

In spite of the equality of lengths of curves established in Theorem 1, it is definitely not true that the distance between two points of the strip is equal to the distance between the points of Σ which they cover. The reason for this is that under the covering, the lines m_0 and m_1 of the strip are glued together into one line of Σ; because of this, certain pairs of points (Q_0, Q_1) of these lines are identified as single points q of Σ. This happens for points Q_0 of m_0 and Q_1 of m_1 such that the line segment $[Q_0Q_1]$ is perpendicular to m_0 and m_1. In what follows such points will be called *opposite* . Of course, you may object that we have agreed to exclude points of m_1 from the strip. This is true, but even excluding them, we see that the same phenomenon reappears in a different guise: two points S_0 and S_1 lying within the strip and close to Q_0 and Q_1 will be far apart on the strip (their distance is approximately the width of the strip), but will go into points of Σ which are close together (see Figure 3.3).

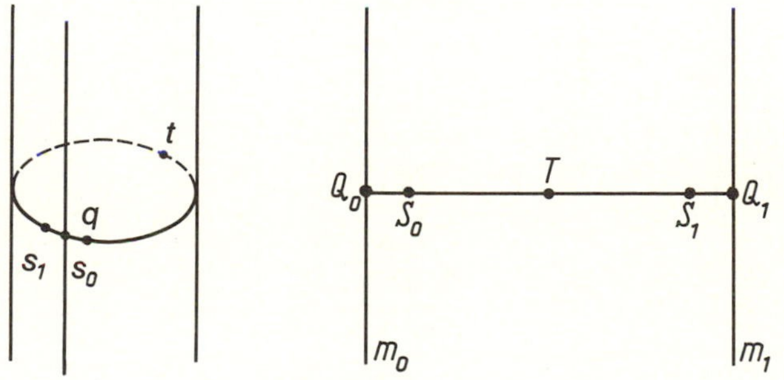

Figure 3.3

We have reached a seemingly self-contradictory situation: both in the plane and in Σ, the distance between two points is the length of the shortest curve joining them. The length of a curve in the strip is equal to the length of the curve in Σ to which it maps, but the distance between the endpoints joined by these curves may be different! To understand the reason behind this, notice that two points s_0 and s_1 of Σ can be joined by either of two arcs of c': a 'short' arc s_0qs_1 and a 'long' arc s_0ts_1. Of these, the 'long' arc develops into the line segment $[S_0TS_1]$ joining

S_0 and S_1 in the strip; what does the 'short' arc develop to? Although obvious, the answer is unexpected: it develops into the two intervals $[Q_0S_0]$ and $[S_1Q_1]$ which are disconnected in the strip, but join together to form a single curve in Σ.

At this point we need to make our terminology more precise. Up to now we have used the word *curve* to mean a curve which we can draw (in the plane or in Σ) continuously, that is, without lifting the pencil off the surface. We will continue to use the term in the same way in what follows. A figure made up of several disconnected curves (for instance $[Q_0S_0]$ and $[S_1Q_1]$) will be called a *track* . Now we can say that the reason for our apparent contradiction is that in Theorem 1 we were not dealing with arbitrary curves in Σ; in fact, curves in the strip do not cover just any curves in Σ, but only those curves which do not cross the generator m'. However, in the definition of distance in Σ we must consider all curves, including those which cross m', and to which Theorem 1 therefore does not apply.

To solve this problem as to how distances in Σ are measured in terms of its representation in the strip, we therefore have to determine how any curve in Σ develops in the plane, including curves crossing m'. The answer is obvious (see Figure 3.4): each time a curve f' in Σ crosses the generator m' in a point q, its

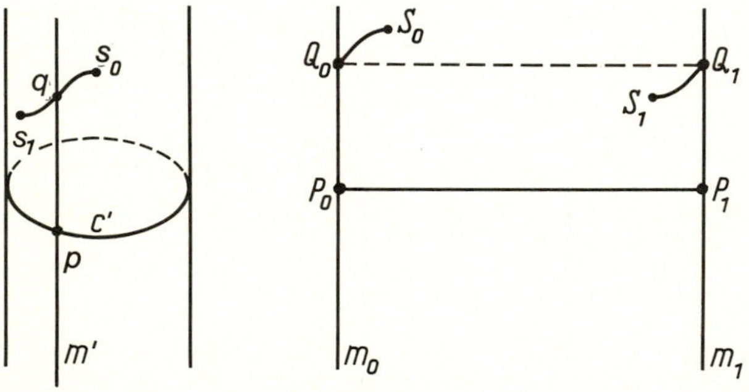

Figure 3.4

development in the strip splits into two curves S_1Q_1 and Q_0S_0, of which S_1Q_1 ends at Q_1, and Q_0S_0 starts at the opposite point Q_0. Thus any curve in Σ develops into a track in the strip which can consist of several curves, with each curve starting at the point of the strip opposite the endpoint of its predecessor. Since Theorem 1 is applicable to each curve, the length of a curve in Σ is equal to the length of the track in the strip to which it develops.

We thus get the following answer to the question we asked:

Theorem 2. The distance between points s and t of a cylindrical surface is equal to the length of the shortest track f in the strip m_0m_1 with the following properties: f consists of curves $f_1, f_2, \ldots, f_n$; f_1 starts at the point S covering s, and f_n ends at the point T covering t; and for each $i = 1, 2, \ldots, n-1$, f_{i+1} starts at the point opposite the endpoint of its predecessor f_i.

Theorem 2 can be interpreted by imagining that an instantaneous jet service operates between opposite points of the strip, so that arriving at a point of m_0, one can instantaneously transfer to the opposite point of m_1, and conversely. An inhabitant of the strip can move about the strip with unit speed, and make free use of the jet service. The distance in Σ between s and t is equal to the minimum time which is needed to travel from S to T.

This is not yet the definitive answer, since we have not indicated how to find the shortest of all possible paths joining S and T; but at least we have reduced the study of geometry on Σ to a certain problem in plane geometry.

Exercises

1. Prove that in the definition of distance between points of Σ given in Theorem 2, it is sufficient to consider only tracks f for which each curve f_i is a line segment.
2. Suppose that s and t are points of Σ covered by points S and T of the strip which lie on a line m parallel to the sides m_0 and m_1 of the strip. Prove that the distance from s to t is equal to the distance from S to T.
3. Suppose that s and t are points of Σ covered by points S and T of the strip which lie on a line segment perpendicular to the sides m_0 and m_1. How do you express the distance from s to t in terms of the distance from S to T and the width d of the strip?

3.2. How to measure distances. In the rule for measuring distances formulated in Theorem 2, we can get rid of the at first sight unfamiliar use of disconnected tracks (or the 'jet service'). Suppose for example that a track f breaks up as two curves f_1 and f_2, with f_1 ending at a point R of m_1, and f_2 starting at the opposite point of m_0 (see Figure 3.5, (a)). Make a translation of the strip bounded by m_0 and m_1 taking the points of m_0 into the points of m_1 opposite them (see Figure 3.5, (b)). Suppose that this translation takes m_1 into the line m_2, and that T goes into a point $T^\sim$ in the strip bounded by m_1 and m_2, and f_2 into a new curve $f_2^\sim$ in the same strip. Then the curves f_1 and $f_2^\sim$ form a single curve $f^\sim$, because now f_1 ends, and $f_2^\sim$ starts at the same point R. Obviously f and $f^\sim$ have the same length, since they break up into equal pieces. Thus, we can replace the track f by a curve $f^\sim$ of the same length, only the endpoint of the curve $f^\sim$ we obtain is not T, but the point $T^\sim$ into which T goes under the translation of the strip.

Now we describe a similar construction in the general case, when f consists

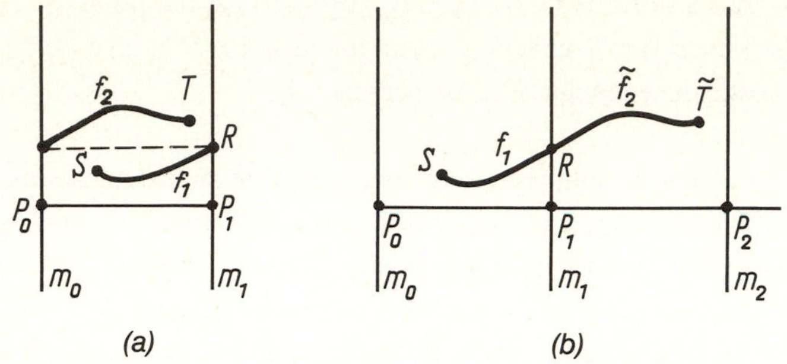

Figure 3.5

of any number of curves, rather than just two. In addition to the lines m_0 and m_1, we draw a further infinite number of parallel lines m_2, m_3, ... to the right of m_1, and m_{-1}, m_{-2}, m_{-3},... to the left of m_0 in such a way that each of the distances from m_i to m_{i+1} is equal to the width d of the strip (Figure 3.6). Then the whole plane is divided up into an infinite number of equal strips; suppose that the line (P_0P_1) meets each m_i in a point P_i.

Each of the strips m_im_{i+1} can be laid over the cylinder Σ in the same way

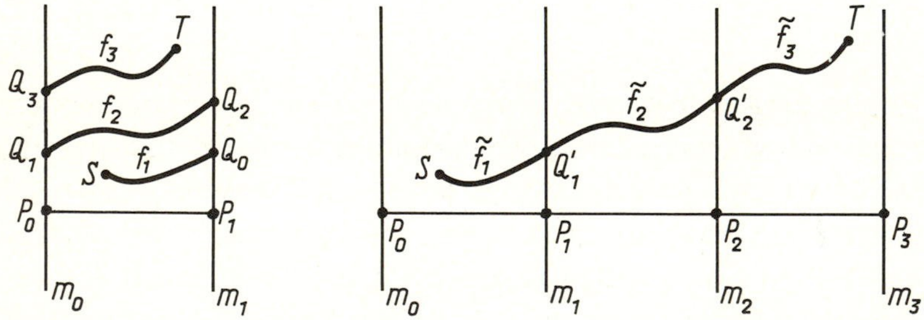

Figure 3.6

as the original strip m_0m_1, taking P_i to the point p of Σ. Since all the strips together cover the whole plane, this rolls up the whole plane onto Σ; this process can be visualised as rolling up a sheet of paper around a cylindrical stick. Under it, each point of Σ is covered by one point from each of the strips, and hence by an infinite number of points: these are the points which occupy the same position in different strips, that is, which will coincide if we move the plane by a translation parallel to (P_0P_1) to lay one strip over another. All such points lie on one line parallel to (P_0P_1) and form a series with adjacent points at a distance d apart, where d is the width of the strip. In other words, these points can all be obtained from one of them by translations in vectors $k \cdot \vec{P_0P_1}$ which are integer multiples of the vector $\vec{P_0P_1}$. We will say that all the points belonging to such a series (that is, all the points which cover one point of Σ) are *equivalent*.

The inverse process of unrolling the surface Σ onto the plane can be visualised just as directly (for simplicity we assume that the curve c' is convex): lay the cylinder Σ on the plane so the the generator m' coincides with the line m_0, and start to roll Σ without slipping along the plane; when Σ has performed one full turn, m' will coincide with m_1, and during this turn each point of Σ will come into contact with the point of the strip to which it develops. If we continue rolling Σ along the plane, or roll it in the opposite direction, then m' will coincide in succession with the lines $m_1, m_2, ..., m_{-1}, m_{-2}, ...,$ and a point p of Σ will coincide in turn with points of the various strips m_im_{i+1} which are all equivalent to one another.

Suppose that in the strip, f is a track of the type described in Theorem 2 (Figure 3.6, (a) illustrates such a track made up of 3 curves). We can represent f on the plane, translating f_2 parallel to (P_0P_1) from m_0m_1 into m_1m_2 (so that Q_1 goes to $Q'_1 = Q_0$), then f_3 into m_2m_3 (so that Q_3 goes to Q'_2, the point which Q_2 has moved to), and so on. As a result, we get a curve $\tilde{f}$ consisting of pieces $\tilde{f}_1, \tilde{f}_2, \tilde{f}_3, ...,$ equal to $f_1, f_2, f_3, ...$ (Figure 3.6, (b)), and therefore having the same length as f. Conversely, any curve $\tilde{f}$ going from a point of the strip to any point of the plane is broken up by the lines m_i into pieces $\tilde{f}_i$ lying within one strip; we can translate each of the $\tilde{f}_i$ parallel to (P_0P_1) to lie in the strip m_0m_1, and we thus get in this strip curves f_i which together make up a track f of exactly the type described in Theorem 2.

Thus according to Theorem 2, any curve f' in Σ joining points s and t can be represented as a curve $\tilde{f}$ of the same length in the plane; $\tilde{f}$ starts at the point S of the strip m_0m_1 which covers s, but its endpoint $\tilde{T}$ may be contained in some different strip m_im_{i+1}; however, if we translate m_im_{i+1} parallel to (P_0P_1) to lay it onto m_0m_1, then $\tilde{T}$ goes to the point T of m_0m_1 which covers t. In other words, $\tilde{T}$ is equivalent to T.

Thus we have a complete description of all curves on Σ joining s and t if we are given the points S and T of m_0m_1 which cover them. Since the distance on Σ is defined as the length of the shortest such curve, we see that the distance from s to t equals the length of the shortest plane curve joining S to some point $\tilde{T}$ equivalent to T. But in the plane, the length of a curve joining S and $\tilde{T}$ will be a minimum if this curve is a line segment joining S and $\tilde{T}$, and its length will simply be the distance from S to $\tilde{T}$. This leads us to the final very simple conclusion, which we state as a theorem.

Theorem 3. Let s and t be points of Σ corresponding to points S and T of the strip m_0m_1. The distance in Σ from s to t is equal to the minimum of the distances from S to one of the points $T^{\sim}$ equivalent to T; (see Figure 3.7, in which the $T^{\sim} = T_i$, for $i = 0, \pm 1, \pm 2, \ldots$).

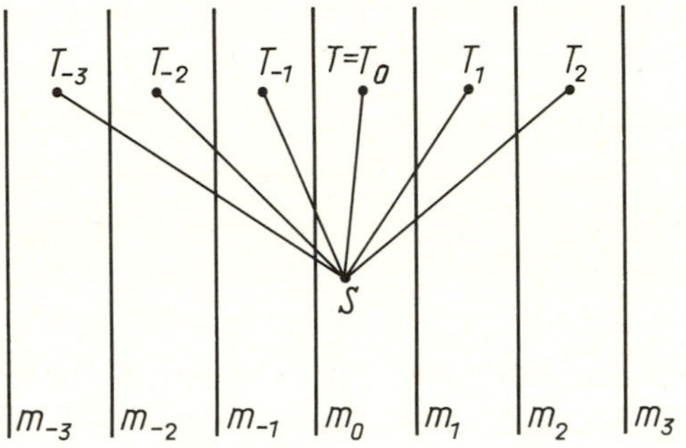

Figure 3.7

In Theorem 3, s and t play different roles: we represent s by a point S, always contained in the strip m_0m_1, whereas $T^{\sim}$ is allowed to be any point equivalent to T. It is easy to see that we could also replace S by any point $S^{\sim}$ equivalent to S without affecting the answer. In fact if we translate the whole plane to take the strip containing $S^{\sim}$ into m_0m_1, the series of points T_i equivalent to T goes into itself; hence the set of all distances of S to the T_i is just the same thing as the set of all distances of $S^{\sim}$ to the T_i. Using this, from now on we will not necessarily restrict S to lie in m_0m_1.

According to Theorem 3, to compute the distance from s to t in Σ we need to determine which is the shortest of the distances from S to the points T_i equivalent to T. The T_i all lie on one line ℓ perpendicular to m_0, and the distance between adjacent points T_i and T_{i+1} is the same, being equal to the width of the strip m_0m_1, which is equal to the length of the directrix c' of Σ; we write d for this width. The orthogonal projection of S onto the line ℓ falls into one of the intervals of ℓ divided by the T_i (Figure 3.8).

Suppose that this projection is contained between T_k and T_{k+1}. Then obviously the shortest distance from S to one of the T_i will be the distance from S to T_k or T_{k+1}, depending on which end of the interval $[T_k, T_{k+1}]$ the

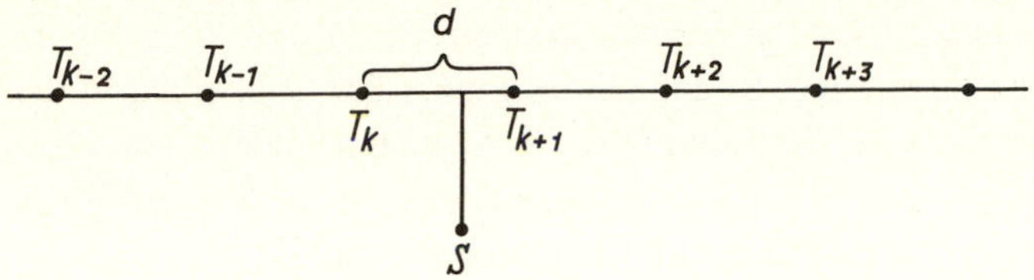

Figure 3.8

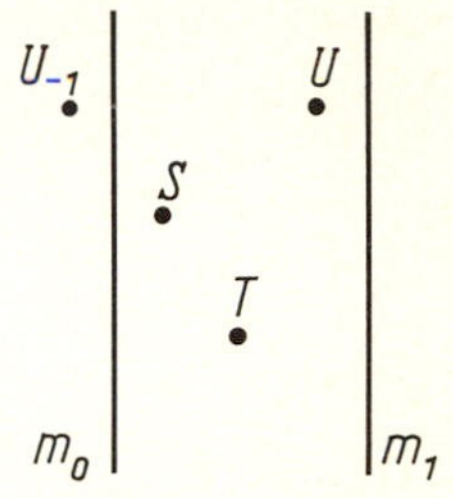

Figure 3.9

projection of S is closer to. This is a very simple method of determining the distance; as an exercise, the reader can easily use it to check that if the points s, t and u of Σ are represented by points S, T and U in the strip as in Figure 3.9, then the distance in Σ from s to t is equal to the distance in the plane from S to T, and the distance from s to u is equal to the distance in the plane from S to the point U_{-1} equivalent to U.

Exercises

1. Prove that in the definition of distance between two points of Σ given in Theorem 2, it is enough to use only tracks f which are either line segments, or break up into two curves f_1 and f_2 each of which is a line segment. [Hint: use Theorem 3.]
2. Let Σ be a cylindrical surface, ℓ' a generator of Σ and c′ a directrix. Write

   r for the distance in Σ from s to t,

   r_1 for the distance between the projections s_1, t_1 of s, t to ℓ', and

   r_2 for the distance between the projections s_2, t_2 of s, t to c′ (that is, the length of the shorter of the two arcs of c′ bounded by s_2, t_2).

   Prove Pythagoras' Theorem on Σ, that is, that $r^2 = r_1^2 + r_2$. [Hint: use Theorem 3.]
3. Note that the representation of Σ on a strip depends on the choice of the generator n′ which is covered by the boundaries m_0, m_1 of the strip; (Σ is cut along n′ to become the strip). Prove that given any two points s, t of Σ the generator n′ can be chosen so that the distance between s, t in Σ is equal to the distance in the strip between the corresponding points S_1, T_1.

3.3. The study of geometry on a cylinder. Now that we have studied the geometry on Σ in such detail, we can proceed with our original aim, and prove that it is locally Euclidean, that is, it is identical with plane geometry in sufficiently small regions. First of all, let us be clear why it is that the process we have described for finding distances leads to different answers for points s and t in Σ on the one hand, and for the corresponding points S and T in the strip m_0m_1 on the other. The answer is now perfectly clear: sometimes the distance is actually equal to that between the corresponding points of the strip, for example, for S and T in Figure 3.9; at other times, for example for S and U in Figure 3.9, it is equal to the distance in the plane not from S to U, but from S to an equivalent point U_{-1}. Therefore the distance from s to t will be equal to that from S to T in the plane provided that S is closer to T than to any of the equivalent points T_i.

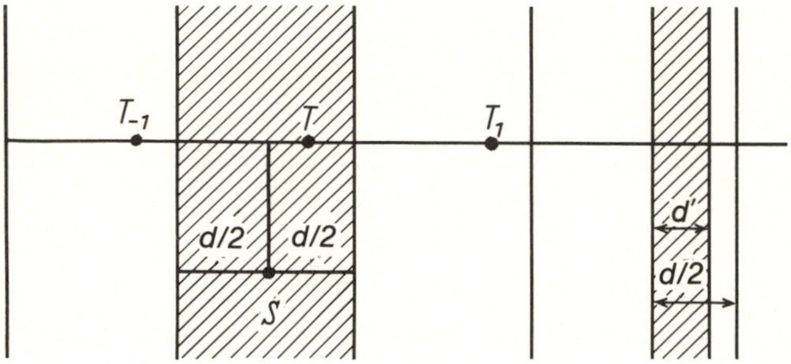

Figure 3.10

Suppose S is a given point in the plane; which points T satisfy this conditions? The answer is obvious is we glance at Figure 3.10: all points T inside a strip bounded by lines parallel to m_0 and at distance $d/2$ from S.

In other words, we need that the distance between the projections of S and T to a line perpendicular to m_0 be less that $d/2$. If we now take any strip parallel to m_0 and of width $d' < d/2$ (see Figure 3.10), then the required property holds for any two points of this strip: the distance between them in the strip equals that between the points of Σ which they cover. Taking the regions covered by these smaller strips, we have therefore found quite large regions in the geometry of Σ within which any measurements will give the same result as on the plane; that is, the geometry of these regions is the same as that of some plane regions. Here we must insert a warning. In our representation of points on the cylinder by points in m_0m_1, not every such region is represented as a strip. The point is that the points of a strip of width $d' < d/2$ must be represented by equivalent points of m_0m_1; in doing this, a strip such as that bounded by p and q in Figure 3.11 is represented in m_0m_1 as two strips. As m_0m_1 covers Σ these glue together to give a single strip on Σ.

It is clear that any disc of radius $r < d/4$ is contained inside a strip of width $d' < d/2$, so that for such a disc the required property holds: the distance in the

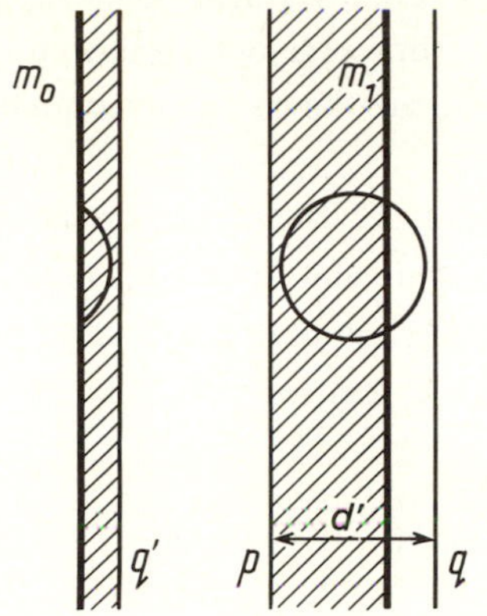

Figure 3.11

plane between its points is equal to that between the corresponding points of Σ. In the strip m_0m_1 such a disc is possibly represented as two segments, which glue together to give a disc in Σ (see Figure 3.11). Obviously a disc of radius $r < d/4$ can be drawn around any point of m_0m_1. This establishes the fundamental property of geometry on Σ.

Theorem 4. The geometry of Σ is locally Euclidean; it is identical with that of the Euclidean plane within, for example, any disc of radius $r < d/4$, where d is the length of the directrix c'.

Now let's see that the geometry of the whole surface Σ is different from that of the plane. For this we deal with a question which is of interest in its own right, the properties of lines in this geometry.

We define a *line* as a curve u' on Σ with the following property: the length of the segment of u' joining any two points a and b is equal to the distance $|ab|$ from a to b, provided that it does not exceed some fixed length (for example $d/4$). Thus just like a line in the plane, a line u' of Σ gives the shortest path joining its points, but not just any of its points, only those which are sufficiently close together along u'.

Now what are the lines of Σ like? As we have seen, any curve f' on Σ can be represented by a plane curve $f^\sim$ (see Figure 3.6); a point $T^\sim$ of $f^\sim$ corresponds to the same point t of Σ as the point T of the strip m_0m_1 equivalent to $T^\sim$. Suppose that a curve u' of Σ is a line, and is represented by a plane curve $u^\sim$. Let $T^\sim$ be a point of $u^\sim$, and draw around $T^\sim$ a disc of radius

$r < d/4$. Then according to Theorem 4, the distance between any two points $A^{\sim}$ and $B^{\sim}$ of this disc equals the distance between the corresponding points a and b of Σ. On the other hand, the length of the arc of the curve u' joining a and b equals the distance from a to b. Hence $u^{\sim}$ has the same property: the length of the arc of $u^{\sim}$ joining $A^{\sim}$ and $B^{\sim}$ equals the distance between $A^{\sim}$ and $B^{\sim}$; it follows from this that the arc is a line segment. Thus any segment of $u^{\sim}$ contained in a disc of radius $r < d/4$ is a line segment, so that $u^{\sim}$ itself is a line. We have thus proved that the lines of Σ are the curves which are represented in the plane by lines.

What do lines look like on Σ? Three cases are possible:

(i) the line $u^{\sim}$ is perpendicular to the m_i; in this case $u^{\sim}$ corresponds in m_0m_1 to a line segment $[Q_0Q_1]$ joining opposite points of m_0 and m_1 and perpendicular to these lines. In Σ this segment links up into a directrix u' (Figure 3.12).

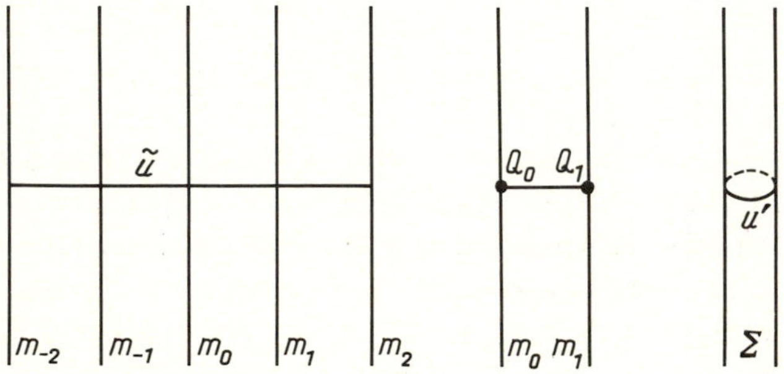

Figure 3.12

(ii) $u^{\sim}$ is parallel to the m_i; in Σ this corresponds to a generator u'.

(iii) $u^{\sim}$ is neither parallel nor perpendicular to the m_i. In the strip, $u^{\sim}$ corresponds to a track made up of an infinite number of parallel line segments. On Σ, this turns into a curve wrapping round Σ infinitely often, spiralling without limit upwards and downwards (Figure 3.13). In the case of a circular cylinder this curve is called a *helix*.

Some of the distinctions between lines in Σ and in the plane are immediately striking. Firstly, lines of type (i) are closed, that is, of finite length. It is interesting to note that through every point of Σ, there is one and only one closed line. Secondly, lines of type (ii) and of type (iii), or two lines of type (iii) which are not parallel will meet infinitely often. These properties show that geometry on Σ is different in many ways from plane geometry, despite the fact that, as we have seen, these distinctions cannot be detected if we do not go outside a disc of radius $r < d/4$.

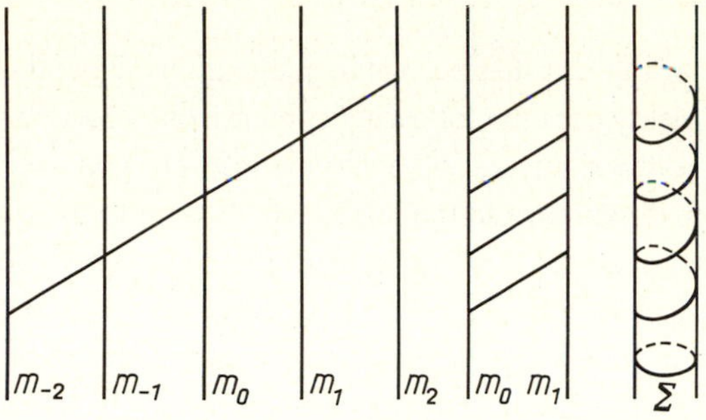

Figure 3.13

Exercises

1. Given two points p and q in Σ, how many different lines join p and q? Consider the various cases for the positions of p and q.

2. Prove that of all line segments joining p and q, there is either one or two shortest segments; when do you get one, and when two?

3. Let ℓ be a shortest segment joining p and q; prove that there exist two different line segments joining p and q and not intersecting ℓ.

4. Prove that any two non-closed lines in Σ either intersect infinitely often or are parallel.

[Hint for all the exercises: use the representation of lines in Σ as lines in the plane.]

§4. A world in which right and left are indistinguishable

Speaking of the fact that our value-judgements depend on the point of view we adopt, Lev Tolstoy uses the following comparison: when you walk along a road, some things are to the right, and others to the left; but you need only turn round to go in the opposite direction, and the things which were to the right turn out to be to the left, and those to the left turn out to be to the right. This comparison appeals to the fact that the words right and left do not define the position of objects relatively to the road: we also need to know the direction in which the observer, from whose point of view the description is made, is looking. If the direction in which the observer is looking is chosen, then right and left-hand sides are uniquely determined. But is this final assertion true? Could it not happen that moving continuously in one direction along a closed road, we return to the point of departure to find that right and left-hand sides have swapped positions? Our experience tells us that this is impossible, and it is in fact impossible in plane geometry. But in this section we construct a geometry which is identical to that of the plane in sufficiently small regions, but for which this paradoxical phenomenon actually happens.

The construction of the new geometry is very similar to that which we used in §3 to describe geometry on a cylinder. We again consider a strip bounded by parallel lines m_0 and m_1, and draw a line ℓ perpendicular to m_0 and m_1; write d for the width of the strip, that is, the distance from m_0 to m_1. Now say that points Q_0 on m_0 and Q_1 on m_1 are *equivalent* if Q_1 is obtained from Q_0 by a translation through a distance d parallel to ℓ followed by a reflection in ℓ; in

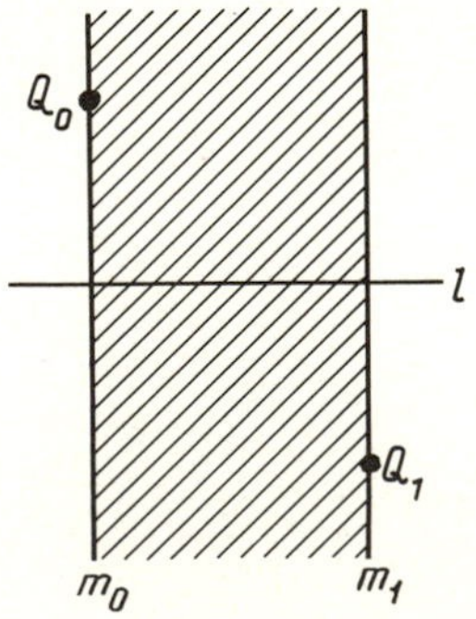

Figure 4.1

other words, Q_0 is a point of m_0 and Q_1 a point of m_1, and they are at same distance from ℓ, but on opposite sides (Figure 4.1). We then view Q_1 as a point equivalent to Q_0 and Q_0 as a point equivalent to Q_1. The same definition can be stated in another way; for this we introduce the important definition: a motion of the plane consisting of a reflection in a line ℓ and a translation in the direction of ℓ is called a *glide reflection* . The line ℓ is called the *axis* of the glide reflection. In particular, if the translation is zero, then we just get ordinary reflection in ℓ.

Using this definition, we can say that points Q_0 and Q_1 are equivalent if Q_0 goes into Q_1 under a glide reflection with axis ℓ and a translation through a distance d.

Now suppose that each pair of equivalent points is joined as before by an instantaneous jet service, and define the distance between points S and T of our strip as the minimum time taken to go from S to T moving within the strip and making free use of the jet service. Under this, equivalent points of m_0 and m_1 should be regarded as one and the same point. This is the definition of our geometry. It differs from the definition of geometry on the cylinder only in the fact that points of m_0 and m_1 are joined by a jet service operating on a different principle.

The precise definition of our geometry is given in the same way as in §3. We take as points of the geometry the points of the strip contained between m_0 and m_1, with points m_0 allowed in, and points of m_1 not allowed. The distance between points S and T is defined as the minimum length of a track f in the strip breaking up as curves $f_1, f_2, ..., f_n$ in such a way that f_1 starts at S and f_n ends at T; and for each $i = 1, ..., n-1$, f_{i+1} starts at the point opposite the endpoint of its predecessor f_i.

Glueing together equivalent points of m_0 and m_1 we can obtain a surface Σ which realises our geometry, in the same way as the geometry in the strip described in §3 can be realised on the cylinder. To visualise this surface Σ we first of all consider the part of the strip bounded by two lines ℓ_0 and ℓ_1 parallel to ℓ and symmetrical with respect to ℓ (Figure 4.2). Glueing it gives the surface illustrated in the diagram; it can easily be constructed from a strip of paper, and is

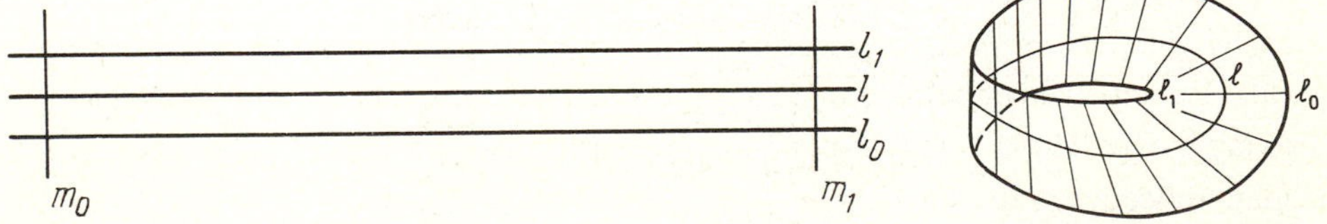

Figure 4.2

called the *Möbius strip*, in honour of the German mathematician A.F. Möbius (1790–1868) who first drew attention to it. Moving the lines ℓ_0 and ℓ_1 arbitrarily far apart we get the required surface, which we can call the *twisted cylinder* .

A convenient description of our geometry can be obtained by developing it onto the plane, as in §3. For this, divide the plane into an infinite number of strips of the same width, bounded by lines $m_0, m_1, m_2, \ldots, m_{-1}, m_{-2}, \ldots$ (Figure 4.3); we write Π_i for the strip bounded by m_i and m_{i+1}. We identify Π_0 with Π_1 by a

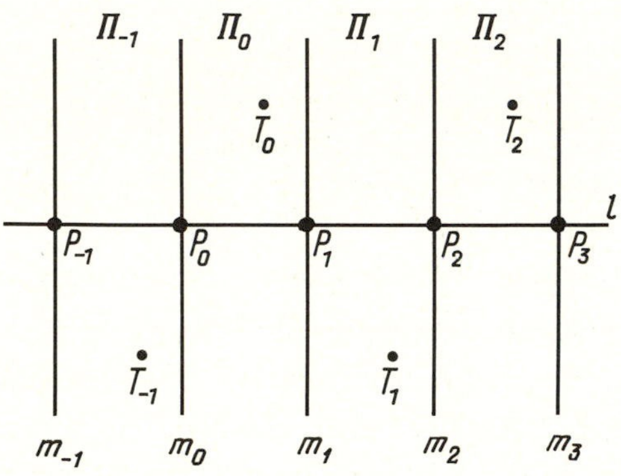

Figure 4.3

glide reflection in the vector $\overrightarrow{P_0P_1}$, that is by a translation in $\overrightarrow{P_0P_1}$ followed by a reflection in ℓ. If a point T_0 of Π_0 goes into a point T_1 of Π_1 under this identification, then we consider T_0 and T_1 to be equivalent. In the same way we identify Π_1 with Π_2; if T_1 goes into a point T_2 of Π_2 then T_2 will be considered to be equivalent both to T_0 and T_1. We continue this process to infinity, and apply it also to the strips $\Pi_{-1}, \Pi_{-2}, \ldots$ to the left of Π_0. We thus get a method of identifying any two strips Π_i and Π_j: if $i < j$ then we must identify Π_i with Π_{i+1} as indicated, then Π_{i+1} with Π_{i+2}, and so on up to Π_j. As one sees easily, the final result can be described as follows: if i and j are the same parity (that is, both even or both odd), then we just make a translation of Π_i parallel to ℓ to lay it over Π_j; if i and j are of opposite parity then we lay Π_i over Π_j by a glide reflection with axis ℓ.

The points of the various strips which correspond to one another under these identifications are said to be equivalent. For example, the points $T_0, T_1, T_2, \ldots,$ $T_{-1}, T_{-2}, \ldots,$ of Figure 4.3 are equivalent. To each point of our geometry corresponds not just a point T_0 of the strip Π_0, but the whole set $T_0, T_1, T_2, \ldots,$ $T_{-1}, T_{-2}, \ldots$ of equivalent points of the plane. Exactly as in §3 it is proved that the distance between points s and t of our geometry is given in the following way: pick any point S in the plane from the set of equivalent points corresponding to s; then the required distance is the minimum of the distances of S to the equivalent points T_i (for $i = 0, 1, 2, ..., -1, -2, ...$) corresponding to t.

It remains to consider the question as to when the distance in the plane between points S and T is equal to that between s and t of Σ. This reduces to the simple geometrical question: when is S closer to T than to any other point equivalent to T? Write T_0 for T, and $T_1, T_2, ..., T_{-1}, T_{-2}, ...$ for the points equivalent to T (see Figure 4.3). First of all, S should be closer to T_0 than to T_2, $T_4, ..., T_{-2}, T_{-4}, ...$ In §3 we saw what conditions S must satisfy for this to happen: S must lie inside the strip bounded by the perpendicular bisectors of the line segments $[T_0T_2]$ and $[T_0T_{-2}]$. In other words, these are the lines perpendicular to ℓ through T_1, T_{-1}. The distance from any point of this strip to any point $T_3, T_5, ..., T_{-3}, T_{-5}, ...$ is greater than to the point T_1 and T_{-1}. Hence we only have to determine when a point S of this strip is closer to T_0 than to T_1 and T_{-1}. Points of the plane equidistant from T_0 and T_1 lie on the perpendicular bisector of the line segment $[T_0T_1]$, that is on the line n_1 of Figure 4.4; points closer to T_0 than to T_{-1} lie above n_1. Similarly, points closer to T_0 than to T_{-1} lie above n_{-1}. Hence points of the plane which are closer to T_0 than to any other point equivalent to T_0 are contained in the shaded area of Figure 4.4.

Write a for the distance from T_0 to T_1; since n_1 is by definition the perpendicular bisector of $[T_0T_1]$, the disc centred at T_0 and with radius r equal

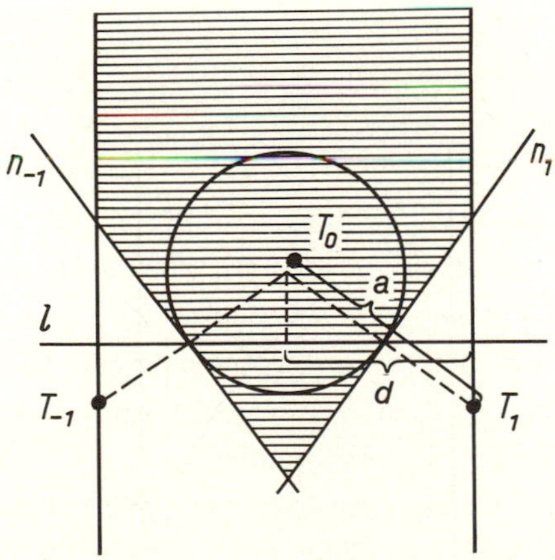

Figure 4.4

to the minimum of $a/2$ and d (the width of the strips $\Pi_0, \Pi_1, ...$) is entirely contained in the shaded area, so that for any point S in this disc, the distance between S and T_0 is equal to the distance in Σ between the corresponding points s and t of our geometry. Now obviously $a \geq d$, so that we can take $r \leq d/2$. If we now consider a disc of radius $d/4$ centred at any point of the plane, then this condition will be satisfied for any two points of the disc: one point will be contained inside a disc of radius $r \leq d/2$ centred at the other, and hence the distance between the two points in the plane will be equal to that between the two corresponding points of Σ. As in §3, we can conclude from this that within any disc of radius $d/4$ our geometry is identical to that of the plane.

In the same way as in §3, it is easy to see that a line of Σ is a curve corresponding to a line in the plane. One particularly interesting line corresponds to the axis ℓ; it is covered by the line segment $[P_0P_1]$; since P_0 and P_1 are equivalent, this is a closed line of length d. Using this line, we now show that in our geometry right and left are indistinguishable (the meaning of this expression was explained at the start of the paragraph). Indeed, imagine a man (that is, an inhabitant of Σ, in the same way as in §2 we considered an inhabitant of the sphere) moving along the line of Σ which corresponds to $[P_0P_1]$. Suppose that he starts moving from P_0, and that he holds his left and right arms outstretched as shown in Figure 4.5. When our man arrives at P_1, he instantly finds himself back at P_0; and simultaneously, his right hand, which has arrived at the point A', finds itself transferred to the point equivalent to A. But this is the point at which his left hand was at the start of the motion. In the same way his left hand finds itself in the position previously occupied by the right hand. Another description of the same phenomenon is provided by a clock moving along the line of Σ corresponding to $[P_0P_1]$: when it gets back to P_0 it will be going backwards!

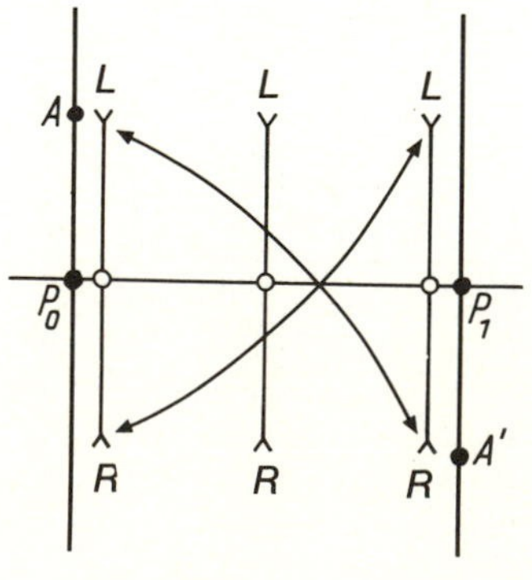

Figure 4.5

One can visualise these unfamiliar features of the geometry of Σ in a completely intuitive way, since they are concerned only with a part of the geometry corresponding to a subset of the strip contained between two lines ℓ_0 and ℓ_1 as in Figure 4.2. This part can be represented as a surface in space, the Möbius strip, as in Figure 4.2.

When considering the Möbius strip in space, it might seem that going around the closed line corresponding to the segment $[P_0P_1]$, we do not return to the same point of the surface, since we find ourselves on the other side. But we must bear in mind that by definition a figure in plane geometry (or in the geometry of a surface) is not lying on one side or other of the surface, but is a subset of the points of the plane (or surface). If you prefer, the figure lies simultaneously on both sides of the plane (or surface).

Exercises

1. Prove that through each point of the twisted cylinder, there is exactly one closed line, exactly one line which does not intersect itself, and an infinite number of lines each of which intersects themselves infinitely often.

2. Prove that on the twisted cylinder there exists exactly one closed line of minimum length, and that all other closed lines are of the same length, twice as long.

3. Does the notion of right and left change for the inhabitant of Σ as he moves around a closed line of length twice that of the minimum?

4. What happens to the twisted cylinder if we cut it along the closed line of minimum length? Does it fall apart into two components, as does the cylinder? What happens if we cut the twisted cylinder along a closed line of length twice that of the minimum? Try out your answers on a paper model of the Möbius strip.

§5. A bounded world

5.1. Description of the geometry. Now we proceed to study a more complicated geometry. This cannot be described as the geometry of some surface in 3-space. To construct it we will make use of the fact that, as we have seen in previous sections, the geometry on a cylinder or on a twisted cylinder can be described without mentioning the surface itself, in terms of their representation in a strip. In our new construction the strip will be replaced by a square.

Consider a square ABDC (Figure 5.1), which we denote by K. We will say that two points on opposite sides on this square are *opposite* if the line segment joining them is parallel to the two other sides. For example, P and Q are opposite in Figure 5.1, as are S and T. The four corners A, B, C and D are all considered to be opposite one another.

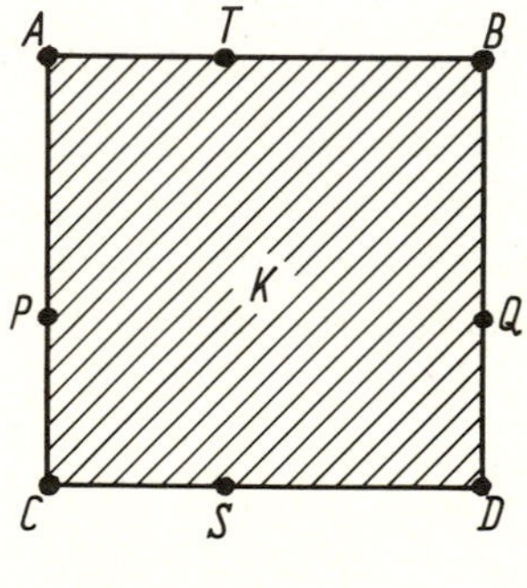

Figure 5.1

Suppose, as we did in §3, that the sides of the square are joined by an instantaneous jet service joining any two opposite points; an inhabitant of the square can move around the square with unit speed, and make free use of the jet service. The minimum time which he needs to travel from one point of the square to another will be called the distance between these points. This is the definition of the geometry which we are going to consider.

According to our definition, the distance between opposite points of the sides of the square is zero, since the jet service operates instantaneously between these points. Hence opposite points should be considered as one and the same in the geometry. Glueing the opposite sides AC and BD as in §3, we get a finite cylinder (Figure 5.2, (a)). Now to glue together the sides ATB and CSD we need

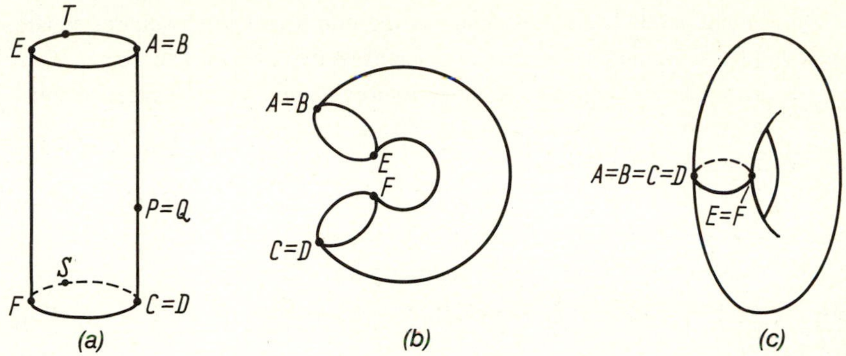

Figure 5.2

to bend our cylinder round into the doughnut-like surface illustrated in Figure 5.2, (c). In mathematics this surface is called a *torus* . But if we carry out this bending, then some segments of generators of the cylinder must stretch, and some shrink. For example, the segment [AC] stretches out to the external meridian of the torus, and [EF] contracts in to the internal meridian. Hence although our geometry can be represented on the torus without tearing, this process will distort distances. The torus can only serve as a distorted map of this geometry, in the same way as a map of a hemisphere of the Earth can only give a distorted representation of the hemisphere.

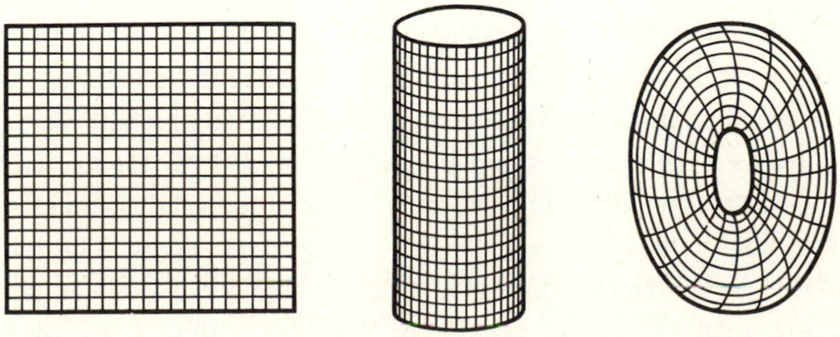

Figure 5.3

Nevertheless, the representation on the torus allows us to imagine our geometry very clearly; for this we need to draw on K a fine-meshed lattice of parallels and meridians, and carry it over to the torus (Figure 5.3). Of course, the mesh squares close to the inner meridian of the torus will be contracted, and those close to the outer meridian will be stretched. Distance in our geometry has to be measured along parallels and meridians by the number of mesh squares separating points, and for any points of the mesh by means of Pythagoras' Theorem.

In view of this (and also to have a short name), we will refer to our new geometry as *geometry on the torus* , and call the square K with the definition of distances indicated above the *torus* .

At first glance our geometry might seem a little artificial. In what follows we explain how we could arrive at it from a more natural starting point: certain physical problems also lead to the geometry of K. Consider for example the planetary system (Figure 5.4) consisting of two planets E and F moving in circular orbits.

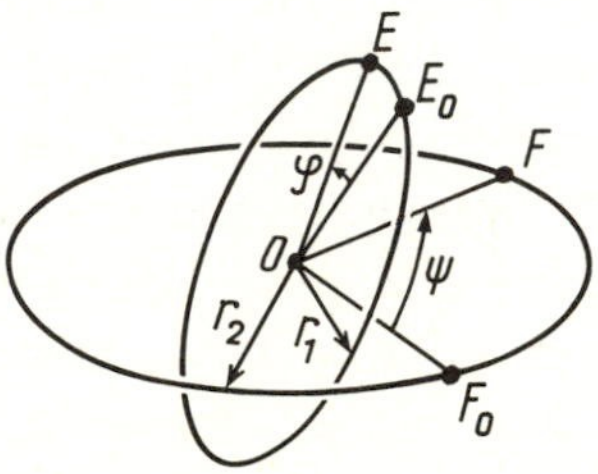

Figure 5.4

To represent the position of this system of planets at some instant, on each of the two orbits we fix the positions E_0 and F_0 which the planets occupied at time $t = 0$, corresponding to the start of observations. The position of E at any instant is determined by the angle φ between the radius in the direction of E_0 and the radius in the direction of the point occupied by E at that instant; in the same way the position of F is determined by an angle ψ. Thus the position of the whole system is determined by the two numbers φ and ψ. Two numbers φ and ψ can be represented as a point P on the plane, taking φ and ψ for its two Cartesian coordinates, so that the point P represents the position of the planetary system. Since angles take values between 0 and 2π, φ and ψ will be within these bounds, so that P will be contained in a square of side 2π. Moreover, the angles 0 and 2π correspond to the same position of the planets, so that the points P with coordinates $(0, \psi)$ and P′ with coordinates $(2\pi, \psi)$ represent one and the same position of the planetary system; the same thing happens for the points with coordinates $(\varphi, 0)$ and $(\varphi, 2\pi)$. These points are the opposite point of the square.

Thus to get a single position of the planetary system represented by a single point, we have to identify together the opposite points of the square. This leads us to geometry on the torus.

We have described the geometry on the torus using the intuitive representation by means of a jet service joining opposite points of the square. We now give a rigorous formal definition; this will be similar to the description of geometry on the cylinder given in §3, Theorem 2. Here is the definition.

We will take points of geometry on the torus to be internal points of the square K and points of its boundary, but only take one of each pair of opposite points. For this we will allow in the square (see Figure 5.1) the sides AB and AC, but not CD and BD. We define the distance between points P and Q to be the length of the shortest track f on the square, breaking up as curves $f_1, f_2, ..., f_n$, in such a way that f_1 starts at P, and f_n ends at Q; and for each $i = 1, 2, ..., n-1$, f_{i+1} starts at a point opposite the endpoint of its predecessor f_i.

Obviously when we glue the square together into the torus, the tracks f glue together to give curves on the torus.

The plan for studying this geometry is roughly the same as that of §§3-4. Many of the arguments will be preserved word-for-word; when this happens we will not repeat ourselves, but only state the final assertions.

First of all, as in §3, we will replace the study of the tracks f appearing in the definition of distance by that of certain curves in the plane. For this purpose we tile the whole of the plane by adjacent squares equal to the original square K from which our geometry was defined (Figure 5.5). Two points P_1 and P_2 belonging to different squares K_1 and K_2 will be called equivalent if the translation which lays K_1 over K_2 takes P_1 into P_2 (the idea is the same as in §§3-4, but the actual notion of equivalence used is different!). In particular, every point of the plane is equivalent to some point of K, that is, it corresponds to some point of the torus, and all the points of the plane corresponding to one point of the torus are equivalent to one another. Using the notions of vector algebra, we can say that Q_1 and Q_2 are equivalent if Q_2 can be obtained from Q_1 by a translation in a vector of the form $k_1\mathbf{e}_1 + k_2\mathbf{e}_2$, where k_1 and k_2 are any whole numbers, and $\mathbf{e}_1 = \overrightarrow{CD}$, $\mathbf{e}_2 = \overrightarrow{CA}$ are the vectors formed by the sides of the square K out of one vertex C.

Let f be a track made up of curves $f_1, f_2, ..., f_n$. Suppose that Q_1 is the endpoint of the curve f_1, so that f_2 starts at the point Q_2 opposite Q_1. We translate K to lay it over the neighbouring square K′, so that Q_2 goes to Q_1; then the curves $f_2, f_3, ..., f_n$ move to curves $f'_2, f'_3, ..., f'_n$ of K′, and now f_1 and f'_2 form one curve. Suppose that Q'_3 is the endpoint of f'_2 in K′, and f'_3 starts at the point Q'_4 opposite Q'_3. Then translate K′ onto its neighbouring square K″ in such a way that Q'_4 goes to Q'_3; the curves $f'_3, f'_4, ..., f'_n$ go to curves $f''_3, f''_4, ..., f''_n$ in K″, and now f_1, f'_2 and f''_3 form one curve. Repeating this process n times we get from the track f a curve $\tilde{f}$ in the plane; $\tilde{f}$ ends at a point $\tilde{Q}$ equivalent to the endpoint Q of f (see Figure 5.5).

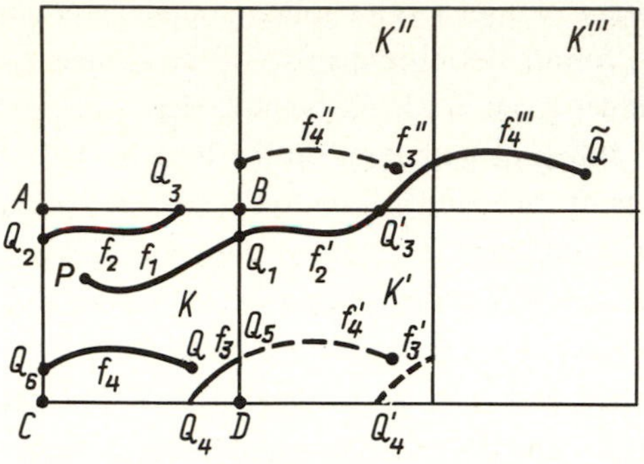

Figure 5.5

Even simpler is the converse process which starts from a curve f˜ in the plane, and constructs the corresponding track f in the square. For this we need only chop f˜ up into pieces contained in individual squares, and then translate the pieces, taking the squares containing them into K.

Since the length of f˜ is equal to that of f, we can use the curves f˜ to define distances (as in §3) in the geometry on the torus. Exactly as in §3 one proves that the distance between points P and Q in this geometry is equal to the length of the shortest plane curve joining P to a point equivalent to Q. From this, again exactly as in §3, it follows that the distance from P to Q in the geometry is equal to the shortest of the distances in the plane from P to points equivalent to Q. Here we can replace P by any point equivalent to P, without affecting the minimum of the distances to points equivalent to Q. Thus we do not need to assume that P lies in the original square K. The proofs of all these assertions are word-for-word the same as those of the similar results in §3, so that we will not give them here.

The subsequent study is rather more closely related to the specific nature of geometry on the torus, and we give it in detail.

Let P and Q be any two points of the plane, and p, q the corresponding points of the torus. When is the distance in the plane from P to Q equal to the distance in the torus from p to q? The first answer is of course as follows: when P is closer to Q than to any other of the points equivalent to Q; for only then will the minimum of the distances from P to points equivalent to Q be simply the distance from P to Q. But how can we find out which points P and Q this holds for?

All points equivalent to Q˜ form the set of vertices of a tiling of the plane by equal adjacent squares with sides parallel to those of K (Figure 5.6). The points of the plane which are closer to Q˜ than to Q_1˜ are to the right of the line (MN); those closer to Q˜ than to Q_3˜ are to the left of (RL); those closer to Q˜ than to Q_2~ are below (NR); and finally, those closer to Q~ than to Q_4~ are above (ML). Thus the points which are closer to Q~ than to any of Q_1~, Q_2~, Q_3~ and Q_4~ fill out the square LMNR. It is easy to see that a point inside this square is closer to Q~ than to any other equivalent point; the reader can carry out the simple elementary verification for himself. Here is the answer to our question: the distance in the plane between P and Q is equal to the distance between the corresponding points p and q of the torus if P is inside a square centred at Q equal and parallel to the square K, that is, a translate of K.

Consider any square K* positioned parallel to K and having side half that of K; then if we take any point P of K* as centre of a square K′ equal and parallel to K, the whole of K* will obviously be contained in K′. Hence the distance on the plane between any two points of K* is equal to the distance in the geometry of the torus between the equivalent points of K. We have therefore found regions in our geometry whose geometry is identical to that of the plane. In place

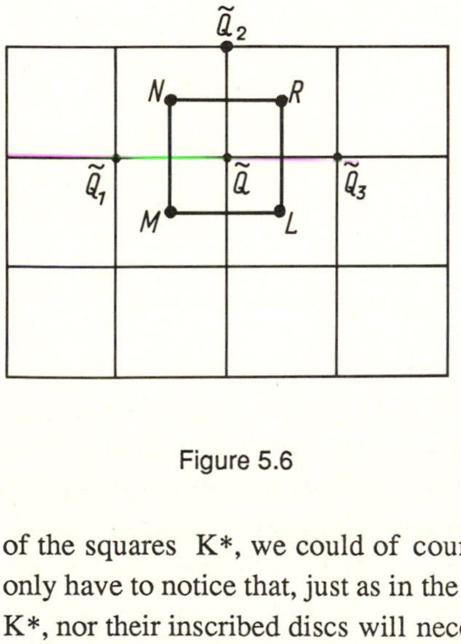

Figure 5.6

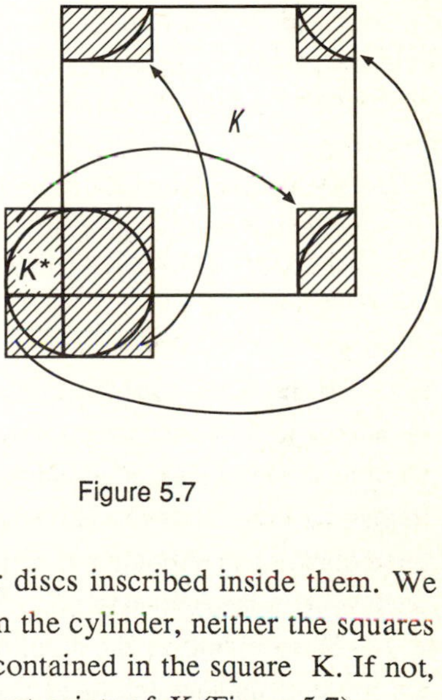

Figure 5.7

of the squares K*, we could of course consider discs inscribed inside them. We only have to notice that, just as in the geometry on the cylinder, neither the squares K*, nor their inscribed discs will necessarily be contained in the square K. If not, then points of K* should be replaced by equivalent points of K (Figure 5.7).

We have thus established that around any point of our geometry we can draw a disc of a certain radius (equal to one quarter of the side of K) within which the geometry is identical to that of the Euclidean plane.

The fact that this geometry is not the same as geometry on the plane is immediately clear. For instance, the distance between any two points of K in our geometry obviously does not exceed the distance between them in the plane, and this distance is bounded by the diagonal of the square. Thus we have constructed a bounded world: the distance between any two points does not exceed some definite magnitude. In this sense, our geometry is closer to spherical geometry than to plane geometry.

Exercises

1. Let P and Q be points lying on a side ℓ of K. Prove that on the torus the side ℓ corresponds to a circle, and that the distance on the torus between the points p and q corresponding to P and Q is equal to the distance between them in this circle (that is, the length of the shorter of the two arcs into which p and q divide it).

2. Let P and Q be points of the square K, P_1 and Q_1 their projections to the side AB, and P_2 and Q_2 their projections to the side AC. Let p, q, p_1, q_1, p_2, q_2 be the corresponding points of the torus, and write

r = distance from p to q,
r_1 = distance from p_1 to q_1, and
r_2 = distance from p_2 to q_2;

prove Pythagoras' theorem: $r^2 = r_1^2 + r_2^2$.

3. Verify that the diagonals of the square K correspond to closed curves of the torus. What happens to the torus if we cut it along the curve corresponding to one diagonal of the square? Will it split into two pieces? What happens if we cut it again along the curve corresponding to the other diagonal of the square?

4. Replace the square in all the constructions of this section by an arbitrary parallelogram ABDC, identifying those points of the opposite sides CD and AB (respectively AC and BD) which are taken into one another by a translation in the vector $\vec{CA}$ (respectively in the vector $\vec{CD}$). Verify that in this way we again arrive at a locally Euclidean geometry. Describe the set of all points P of the plane whose distance from a given point $\tilde{Q}$ is equal to the distance in the geometry between the corresponding points p and q of the geometry (in the case of a square, this is the figure illustrated in Figure 5.6).

5.2. Lines on the torus. In this section we work out the properties of lines in the geometry of the torus: here some new and unusual phenomena appear, which help us to get a better idea of geometry on a torus, and hint at surprising relations of this theory to other questions. The results given here are not required for the understanding of subsequent sections of the book, and the reader may prefer to pass on at once to the following §6.

We will borrow the definition of line from §3; we will not repeat it here, nor the proof of the fact that lines of the torus are the tracks f of K corresponding to lines $\tilde{f}$ in the plane.

Consider then a line $\tilde{f}$ in the plane. The first question which will concern us is this: when is the corresponding line f in the geometry of the torus closed?

First of all we recall that once a system of coordinates has been chosen in the plane, every line not parallel to the y-axis is given by an equation $y = ax + b$. The coefficient a is called the *slope* of the line; to find it, we need to take two points of the line with coordinates (x_1, y_1) and (x_2, y_2). From the system of equations

$$y_1 = ax_1 + b, \qquad y_2 = ax_2 + b,$$

which describes the fact that the two points belong to the line, we get that

$$a = \frac{y_2 - y_1}{x_2 - x_1}.$$

(Figure 5.8). Let us choose as origin the vertex C of K, and as coordinate axes the lines (CD) and (CA) (see Figure 5.1); suppose that in this system of coordinates the line $\tilde{f}$ has equation

$$y = ax + b.$$

Theorem. If the slope a of $\tilde{f}$ is a rational number then f is a closed line on the torus. If the slope is irrational then the line f is infinite, and moreover f passes arbitrarily close to every point of the torus.

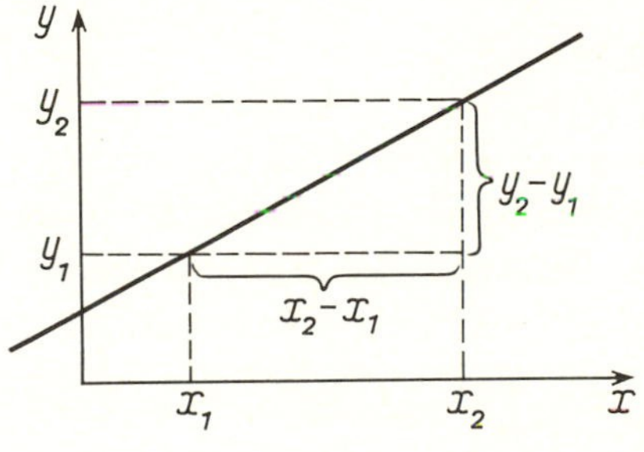

Figure 5.8

We take the side of K to be 1. Then its points have coordinates (x, y) with $0 \le x < 1$, $0 \le y < 1$. Two points of the plane are equivalent if they are obtained from one another by horizontal and vertical translations in an integer number of squares equal to K. This is equivalent to saying that both the x and y-coordinates of the points should differ by integers.

We consider separately the two cases indicated in the statement of the theorem.

I. The slope a is rational. Suppose that $a = p/q$, where p and q are integers. Then points P_1 and P_2 lying on our line and for which the x-coordinates differ by q will be equivalent. Indeed, if they have coordinates (x_1, y_1) and (x_2, y_2) then

$$x_2 - x_1 = q \implies y_2 - y_1 = a(x_2 - x_1) = (p/q)(x_2 - x_1) = (p/q)q = p,$$

so that the y-coordinates of these points differ by p. This means that if we move along the line from P_1 a distance of q squares horizontally, then we move p squares vertically to get to P_2. Therefore P_1 and P_2 are equivalent.

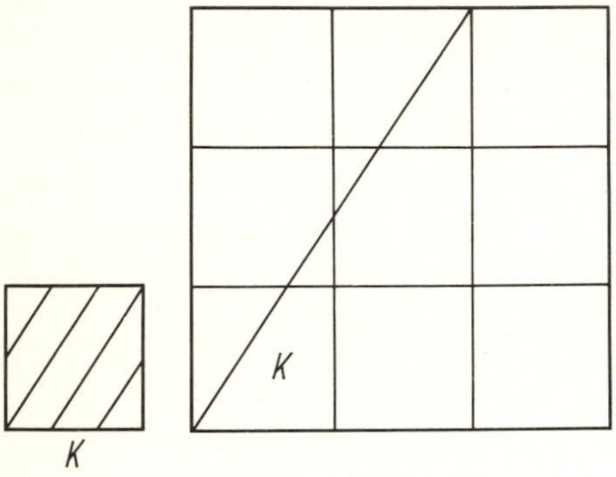

Figure 5.9

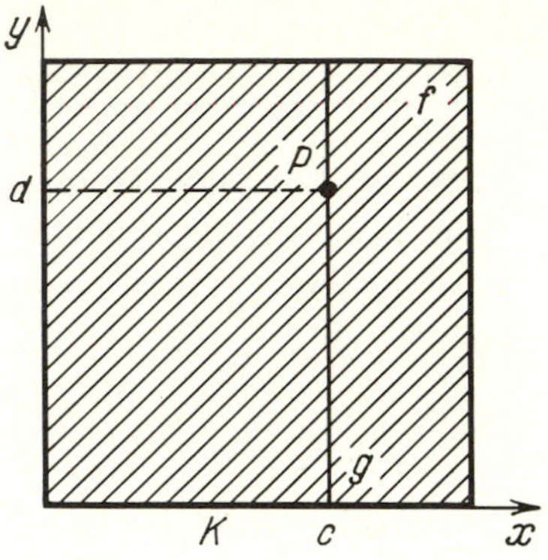

Figure 5.10

Thus if we move along the line $\tilde{f}$, as soon as the x-coordinate increases by q, we return to the starting point on the torus, and from then on we just move around the same curve. This just means that f is a closed curve of the torus. The corresponding track in K may consist of several line segments (see Figure 5.9, where the track is drawn for the case that $\tilde{f}$ is of the form $y = (3/2)x$; recall that the square K has side equal to 1).

II. The slope a is irrational. We are asserting that no matter how small a number we take (for example $1/(1{,}000{,}000)$), then after moving for a sufficiently long time along the line f, we will pass by each point of the torus at a distance smaller than this number. For this it is obviously enough to prove that in the square K the track f which specifies the line passes arbitrarily close to each point of the square K.

Consider a point P of K with coordinates (c, d); through P we draw a line segment g parallel to the vertical sides of the square. On the torus, g becomes a closed line. We are going to show that there even exist points of intersection of f and g arbitrarily close to P (see Figure 5.10).

We determine how to find points of intersection in K of g with the track f, making use of the representation of f as a line $\tilde{f}$ in the plane. For this, recall that the points (x, y) lying on f are the points of K which are equivalent to points (x′, y′) of $\tilde{f}$. We are interested in the points which also lie on g, that is, for which $x = c$. This means that the corresponding point (x′, y′) of $\tilde{f}$ equivalent to (x, y) must have x-coordinate of the form $x' = c + n$, where n is any integer (since the square has side 1). To say that (x′, y′) lies on $\tilde{f}$ means that the y-coordinate is

$$y' = ax' + b = a(c + n) + b = an + ac + b.$$

Thus we have an answer to our question: f and g intersect at the points of K equivalent to the points (x'_n, y'_n) of the plane with

$$x'_n = c + n, \quad y' = an + b', \tag{1}$$

where $b' = ac + b$ and n is any integer.

Thus corresponding to every integer n, there is a point of intersection of f and g in K. Write Q_n for this point, which is determined as the point of K equivalent to the point whose coordinates are given by (1).

Suppose given an arbitrarily large integer N. We will prove that among the points Q_n, there exist some whose distance from P is less that $1/N$.

First of all we note that points Q_n corresponding to different values of n are different. In fact, suppose that $n \neq m$ but $Q_n = Q_m$. Then since these points are equivalent, their coordinates differ by integers. For the x-coordinate this is indeed the case: the first formula of (1) gives

$$x'_n = x'_m + n - m.$$

Suppose that the y-coordinates of these points differ by an integer k, that is

$$an + b' = am + b' + k;$$

from this we get that $a = k/(n - m)$, (division is permissible since $n \neq m$), and this contradicts the irrationality of a. Therefore $Q_n \neq Q_m$.

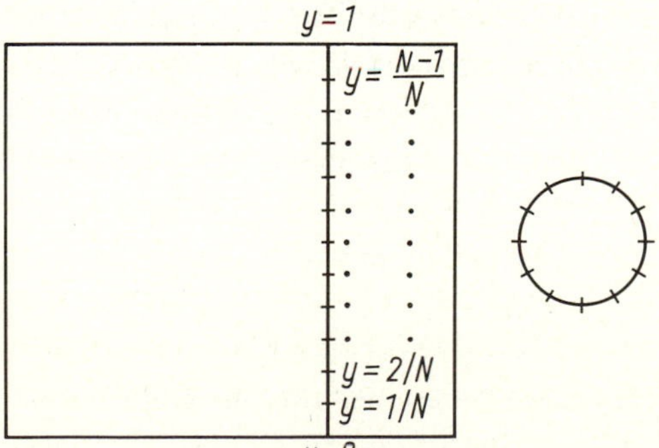

Figure 5.11

We divide the interval g up into N equal parts, which we represent in K as intervals consisting of points with y-coordinates between 0 and 1/N, 1/N and 2/N, ..., (N-1)/N and 1 (Figure 5.11). Since the number of values of n is infinite, and the number of intervals N is finite, some interval contains at least two points with distinct values of n; suppose that these are Q_n and Q_m, with $n \neq m$. This means that for the points whose coordinates are defined by (1) for the two values n and m, there exists an integer k such that y'_n and $y'_m + k$ differ by less than 1/N, that is,

$$an + b' = am + b' + \alpha + k,$$

where $0 < \alpha < 1/N$, and k is an integer. From this we get

$$a(n - m) = \alpha + k.$$

If we set $r = n - m$ then this formula can be written

$$ar = \alpha + k \quad \text{with } 0 < \alpha < 1/N \text{ and } k \text{ an integer.} \quad (*)$$

We now compare the points Q_s and Q_{s+r} for any integer s. From (1) we get

$$x'_{s+r} = x'_s + r, \quad y'_{s+r} = a(s+r) + b' = as + b' + ar = y_s + ar,$$

and from the inequality (*) it follows that

$$y'_{s+r} = y'_s + \alpha + k \quad \text{with } 0 < \alpha < 1/N \text{ and } k \text{ an integer.}$$

As we see, the points Q_s and Q_{s+r} are therefore at a distance $\alpha < 1/N$ apart. Consider, starting from Q_0, the sequence of points $Q_0, Q_r, Q_{2r}, \ldots$ The distance between two successive points equals α, and the length of the interval g is 1. Hence laying down adjacent intervals $[Q_{ir}, Q_{(i+1)r}]$ a sufficient number of times, we can cover the whole segment g (we will obviously need a number $> 1/\alpha$ of intervals). Now the point P belongs to one of these intervals. Therefore P is at distance $\leq \alpha < 1/N$ from one of the points Q_{ir}. This completes the proof of the theorem.

Exercises

1. Prove that any non-closed line in geometry on the torus meets any closed line infinitely often.
2. Prove that a non-closed line on the torus does not intersect itself.
3. Determine the minimum length of closed lines on the torus.

5.3. Some applications. To conclude this section, we give some corollaries and remarks concerning the theorem we have proved.

(1) We return to the example of the planetary system illustrated in Figure 5.4. Suppose that the planets E and F move with constant angular velocity ω_1 and ω_2 on orbits which are circles with the same centre O and radiuses r_1 and r_2, and that $r_1 < r_2$. We will also suppose that the orbits take place in distinct planes. (It should be noted that the planets of the solar system rotate in extremely close planes; however, the moons of Jupiter, for example, are grouped into several planes which form large angles with one another.) Finally, we suppose that the near planet E subtends an angle γ_1 at the centre O of the planetary system, and the far one F an angle γ_2, with $\gamma_1 > \gamma_2$; see Figure 5.12.

We will discuss the question of whether or not the far planet F will be eclipsed by the near planet E, from the point of view of an observer at the centre O. Let's prove that if the ratio ω_1/ω_2 of the angular velocities is irrational, then eclipses will occur infinitely often. For this, note first that an eclipse occurs if at some instant both E and F are close to one of the two rays from O in the line of intersection of the two planes of the orbits (see Figure 5.12). More precisely, if (φ_0, ψ_0) is a point of the torus corresponding to the position of the planets on one of these ray OA_0, then an

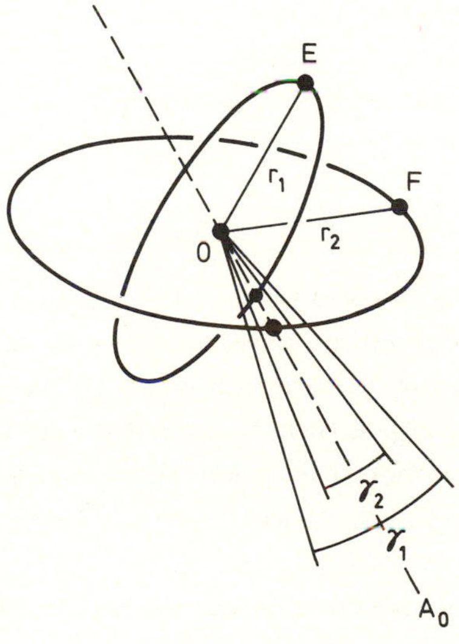

Figure 5.12

eclipse occurs if the point (φ, ψ) of the torus corresponding to the position of the planets at a given instant satisfies the condition

$$|\varphi - \varphi_0| + |\psi - \psi_0| < \gamma_1 - \gamma_2. \tag{2}$$

The points satisfying (2) form a square centred at the point (φ_0, ψ_0) having diagonal $2(\gamma_1 - \gamma_2)$.

If at $t = 0$ the position of the centres of E and F is given by angles α, β, then at time t the corresponding angles will be

$$\varphi = \omega_1 t + \alpha, \quad \psi = \omega_2 t + \beta.$$

From this, we get the relation

$$(\psi - \beta)/(\varphi - \alpha) = \omega_2/\omega_1,$$

that is

$$\psi = (\omega_2/\omega_1)\varphi + \lambda, \quad \text{where } \lambda = -(\omega_2/\omega_1)\alpha + \beta. \tag{3}$$

Now (3) is the equation of a line with slope ω_2/ω_1, and hence the positions of the planets at different times trace out a line on the torus. Since by assumption ω_2/ω_1 is irrational, by what we have proved this line approaches arbitrarily close to any point of the torus, and in particular to the point (φ_0, ψ_0), which proves what we require: there are infinitely many points (φ, ψ) on this line which satisfy the inequality (2).

(2) The central point in the proof of the theorem was the assertion that points of intersection of f and g are arbitrarily close to any point of g. These are the points of the square K equivalent to points whose coordinates are defined by (1). Let us see what the coordinates of the points of intersection themselves are. We restrict ourselves to the case $b' = 0$.

The x-coordinate of the point we are seeking is known; it is c for all these points. To find the y-coordinate, we have to subtract from y'_n an integer, in such a way that the difference is contained between 0 and 1. This difference will be the required coordinate y_n. From (1) we get

$$an = y_n + k \text{ with } 0 \le y_n < 1 \text{ and } k \text{ an integer.}$$

The term k is called the integral part of an, and y_n the fractional part. We can now restate the property of lines on the torus we have proved as follows: for any irrational number a, the fractional part of an approaches arbitrarily close to any number between 0 and 1.

This assertion has many applications to the properties of irrational numbers. As an example, let us determine what are the first digits of the decimal expressions of the powers of 2. Note that $2^1 = 2$, $2^2 = 4$, $2^3 = 8$, $2^4 = 16$, $2^5 = 32$, $2^6 = 64$, $2^7 = 128$, $2^8 = 256$, $2^9 = 512$; hence there are certainly powers of 2 with first digits 2, 4, 8, 1, 3, 6, 5.

Let r be some digit, that is a natural number less than 10. Then r occurs in our series if there exists natural numbers n, m and ℓ such that

$$2^n = r \cdot 10^m + \ell \text{ with } 0 \le \ell < 10^m.$$

This inequality can be rewritten

$$r \le 2^n/10^m < r+1,$$

which is equivalent to

$$\log_{10} r \le n \log_{10} 2 - m < \log_{10} (r+1).$$

In other words, the fractional part of $n \log_{10} 2$ should be contained between $\log_{10} r$ and $\log_{10} (r+1)$. Now since $\log_{10} 2$ is irrational (you should see this for yourself), such an n exists by the property of fractional parts we have proved. In the same way, it can be proved that there exist powers of 2 beginning with any specified string of digits.

(3) An analysis of the proof of the assertion stated and emphasised in Remark (2) above (the proof makes up the final part of the proof of the theorem) allows us to extract a more precise result. The idea of the proof consisted of the fact that among the infinite number of different fractional parts of numbers of the form an, at least two belong to an interval of length $1/N$, and this allows us to construct a number ar having a small fractional part.

Let's try to polish up this argument. Since the number of intervals of length $1/N$ is N, already among the fractional parts of an for $n = 1, \ldots, N+1$, at least two will belong to one interval. Suppose that these are the fractional parts of an and am, then for $r = n-m$ the fractional part of ar is less than $1/N$, that is

$$ar = k + \alpha \quad \text{with } 0 < \alpha < 1/N.$$

In other words $|ar - k| < 1/N$, that is

$$|a - k/r| < 1/Nr. \tag{4}$$

This inequality hold for any N; here k and r are natural numbers, and by construction $r \le N$; it follows from this that a fortiori

$$|a - k/r| < 1/r^2. \tag{5}$$

The inequality (5) shows that the rational number k/r is rather close to the irrational number a. The existence of such approximations for any one value of r does not give anything; for example if $r = 1$ this is trivial. However, from (4) it can be deduced that such approximations exist with r arbitrarily large, that is for $r > M$ for any M.

Indeed, if the set of denominators for which the inequality (5) has a solution were finite, then the number of solutions k/r of (5) would be finite. This follows since all these solutions differ from a by at most 1 (since $r \ge 1$). Then the set of solutions of (4) for all N (with $N > r$) taken together would be finite, since solutions of (4) provide solutions of (5). However, solutions of (4) exist for arbitrarily large N, and differ from a by less than $1/N$. Which would not be possible if the number of them were finite.

Thus for any irrational number, there exists an infinite number of good rational approximations satisfying (5). The existence of such good approximations is not at all obvious. For example we will not find such approximations by using the method which comes at once to mind, truncating the decimal expansion at some definite point. If the infinite decimal expansion of a number a is of the form

$$a = s.r_1 r_2 \ldots, \text{ with } 0 \le r_i \le 9,$$

this means that

$$a = s + r_1/10 + r_2/10^2 + \ldots, \text{ with } 0 \le r_i \le 9.$$

Consider the fraction with denominator 10^n:

$$k/10^n = s + r_1/10 + \ldots + r_n/10^n.$$

Then the error term $|a - k/10^n|$ will be equal to

$$r_{n+1}/10^{n+1} + r_{n+2}/10^{n+2} + \ldots = 1/10^n(r_{n+1}/10 + r_{n+2}/10^2 + \ldots) = q/10^n,$$

where $q = r_{n+1}/10 + r_{n+2}/10^2 + \ldots$ is an arbitrary number less than 1. Thus for this approximation we cannot say anything better than

$$|a - k/10^n| < 1/10^n,$$

whereas according to (5) there exist very much better approximations.

Exercise

Determine a number N depending on k such that among the first N powers $2, 2^2, \ldots, 2^N$ there will necessarily be a number starting with k specified digits $r_1 r_2 r_3 \ldots r_k$. What N works for $k = 2$?

§6. What does it mean to specify a geometry?

6.1. Definition of a geometry. In §§3–5 we considered several examples of locally Euclidean geometries, that is, geometries which are identical to the Euclidean plane in sufficiently small regions. Now we can make use of our accumulated experience and go on to investigate an arbitrary geometry of this kind. In this connection we run into completely new questions: what is a geometry? How do you specify a geometry? When should two geometries be considered the same? In fact up to now, in §§3–5, we have just produced a certain construction and declared this to be a geometry. We are now going to use these examples, and to derive from them a general notion of geometry. As further examples, we will appeal to the ordinary Euclidean geometry of the plane, and to spherical geometry, which was discussed in §2.

Of course by a geometry we do not mean some kind of assembly of assertions (such as the Propositions, Lemmas and Theorems of a Euclidean geometry course), but some definite mathematical object (such as the geometry on the sphere or on the cylinder) which can then be studied, and about which we can prove various assertions.

In all the examples we have seen, a geometry was a set, consisting of elements which we called points. But what are points, how are they specified?

1. For plane geometry we can imagine the plane as being, for example, a blackboard, which we think of as extending unboundedly in all directions. Then a point can be thought of as a mark left after touching the board with chalk; to specify a point, we just have to point it out or mark it with chalk.

1a. If we introduce a coordinate system into the plane, we can specify a point by coordinates, that is by an ordered pair of numbers (x, y). If we wish to describe the points of a plane, freeing ourselves from all geometrical intuition (for example, in order to explain our geometry to an inhabitant of some other world), we can take this as the definition of a point: we say that a point is an ordered pair of numbers.

2. In spherical geometry, a point is a point of the sphere.

2a. Write O for the centre of the sphere Σ and consider all possible rays emanating from O. Every such ray intersects the sphere in a unique point, and through each point of the sphere passes a unique ray through O. Thus to specify a point of spherical geometry, we can give a ray emanating from O.

3. In geometry on the cylinder, a point is a point of the cylinder.

3a. Developing the cylinder (as in §3), we can specify a point on the cylinder as a point of the strip in the plane (excluding the right-hand boundary line of the strip).

3b. Finally in §3 we found another method of specifying a point on the cylinder: each point is given as a set of equivalent points of the plane.

4a. In geometry on the twisted cylinder we defined a point as a point of the strip (excluding the right-hand boundary line).

4b. Here also we indicated a further method of specifying points: each point is given as a set of equivalent points (but the notion of equivalence is different from that in Example 3).

5. In the geometry on the torus described in §5 we had two methods of specifying points, which we could refer to as 5a and 5b, in complete analogy with Examples 3a and 3b or 4a and 4b.

The variety of methods used to specify points leads one to the conclusion that for geometry, it is of no concern what we take points to be, provided that the notion is well-defined.

The second basic notion which we used in all the examples was that of distance between points. We recall how it was defined in all the examples we considered.

1. The distance between points A and B in the plane can be measured, for example, by laying down some unit of length.

1a. If A has coordinates (x, y) and B coordinates (x′, y′), then by Pythagoras' theorem, the distance is given by the formula

$$|AB| = \sqrt{(x - x')^2 + (y - y')^2}\ .$$

2. The distance between points A and B on the sphere is defined as the length of the arc of great circle joining A and B (more precisely, of the shorter of two such arcs).

2a. If we specify a point A by the ray OA emanating from the centre O and passing through A, then the length of the arc joining A and B will be proportional to the angle between the rays OA and OB. If the radius of the sphere is 1 then the distance |AB| is equal to the angle in radians between OA and OB. Thus if we specify points by rays, we can measure distances as angles between rays.

3. On the cylinder, the distance between points a and b is defined as the length of the shortest curve lying on the cylinder and joining a and b.

3a. If points of the cylinder are specified as points of the strip, the distance between points A and B is given as the length of the shortest track consisting of curves $f_1, ..., f_n$, such that f_1 starts at A and f_n has B as its endpoint, and for $i = 1, ..., n-1$, the endpoint of f_i and the starting point of f_{i+1} are opposite points.

3b. If a point of a cylinder is represented as a set of equivalent points on the plane, then the distance from **A** to **B** is defined as the minimum of the distances |AB|, where A is any point of the set of equivalent points defining **A** and B is any point of the set defining **B**.

There are entirely similar methods of defining distances in Examples 4a, 4b, 5a and 5b.

The variety of methods used to define distances leads one to the conclusion that the only essential thing for us is that, somehow, a distance |ab| is defined between any two points a and b of our geometry. However, in all the examples, regardless of the way in which distance was defined, it satisfied some almost obvious general properties, which we now list:

(a) $|ab| \geq 0$, and $|ab| = 0$ if and only if $a = b$;

(b) $|ab| = |ba|$;

(c) $|ac| \leq |ab| + |bc|$.

Property (c) is perhaps less obvious that the others, and we show why it holds in all the examples we have considered.

1. Here the inequality expresses the fact that in a triangle the length of one side is not greater than the sum of the lengths of the other two sides.

1a. In this example we meet a curious algebraic inequality. Suppose that A has coordinates (x, y), B has coordinates (x′, y′), and C has coordinates (x″, y″). Then the inequality (c) becomes

$$\sqrt{(x - x'')^2 + (y - y'')^2} \leq \sqrt{(x - x')^2 + (y - y')^2} + \sqrt{(x' - x'')^2 + (y' - y'')^2},$$

where we always take the positive value of the square roots.

Set

$$x - x' = u, \quad y - y' = v, \quad x' - x'' = u_1, \quad \text{and} \quad y' - y'' = v_1;$$

then

$$x - x'' = u + u_1, \quad \text{and} \quad y - y'' = v + v_1,$$

and our inequality can be rewritten

$$\sqrt{(u + u_1)^2 + (v + v_1)^2} \leq \sqrt{u^2 + v^2} + \sqrt{u_1^2 + v_1^2}.$$

Since this inequality is just an algebraic expression of property (c) of distances of points in the plane (as in Example 1), it must also hold. We have found a geometric proof of an algebraic inequality. This inequality can of course be proved purely algebraically: the reader can easily find his own proof.

2. Here our inequality says that a segment of an arc of great circle gives the shortest distance between two points; this was proved in §2.

2a. The corresponding inequality for angles formed by rays has already been referred to in §2. Properly speaking it is this inequality that is proved directly, and the inequality involved in Example 2 was deduced from it in §2.

3. In this case the inequality is obvious. Let f be a curve joining a and c and of length equal to $|ac|$, f′ a curve joining a and b and of length $|ab|$, and f″ a curve joining b and c and of length $|bc|$. Then the curves f' and f'' together form a curve g joining a and c; its length is the sum of the lengths f' and f'', that is $|ab| + |bc|$. Now since the length of f is the minimum among all such curves, it is at most the length of g, that is

$$|ac| \leq |ab| + |bc|.$$

3a. Since this is just another description of the geometry of Example 3, the inequality also holds here. Of course, it could also be verified directly without appealing to Example 3, just repeating the same arguments as we used in Example 3.

3b. Here again the inequality holds because we have another description of the same geometry. The direct proof of the inequality in this case requires different arguments, and we will go through these in the next section.

Examples 4a, 4b, 5a and 5b are considered in an entirely similar way.

Another fundamentally important property of distances can be noted if we return to an example which we mentioned in passing in §2, namely the sphere with distance defined not as in spherical geomtry, but as in the ambient 3-space; that is, the distance between two points is the length of the chord joining them. For points a and b of the sphere Σ, let us write $|ab|$ for the distance between a and b in the sense of spherical geometry, and $\|ab\|$ for the distance in ambient space. For example, if a and b are the north and south poles of a sphere of radius r, then $|ab| = \pi r$, whereas $\|ab\| = 2r$. Obviously, if $a \neq b$ then $|ab| > \|ab\|$. A certain unnaturalness of the distance $\|ab\|$ for questions of spherical geometry can be felt intuitively: this distance does not come from properties of the geometry of the sphere itself, but is borrowed from properties of the ambient space. Mathematically this is reflected in the fact that the distance $\|ab\|$ is not the length of any curve lying on Σ and joining a and b: indeed, the length of any such curve is at least $|ab|$, and $|ab| > \|ab\|$. Because of this, the distance $\|ab\|$ cannot be imagined in a physical way as the shortest time needed for an inhabitant of the sphere moving with unit speed to get from a to b. To be able to express this property without using the notion of the length of a curve, we introduce a number of notions which will be useful in what follows.

In any geometry, a *chain* joining a and b is a finite sequence of points $p_1, \ldots, p_n$ such that $p_1 = a$ and $p_n = b$; the chains p_ip_{i+1} are the *links* of the chain, and the distance from p_i to p_{i+1} the *length* of the link p_ip_{i+1}; the *length* of the chain $p_1, \ldots, p_n$ is the sum of the lengths of all the links p_ip_{i+1}.

The reader should think of a chain as being analogous to a broken line on the plane, that is a sequence of line segments where each segment starts off at the endpoint of its predecessor; to specify a broken line, we only need to give the endpoints of the segments, which is a chain.

Now the length of a chain is always at least the distance between its endpoints; since $a = p_1$ and $b = p_n$, this assertion can be written as follows:

$$|p_1p_n| \leq |p_1p_2| + \ldots + |p_{n-1}p_n|. \tag{1}$$

For $n = 3$, this inequality is just property (c) of a geometry. In general it can be proved by induction: for by (c), we have

$$|p_1p_n| \leq |p_1p_{n-1}| + |p_{n-1}p_n|,$$

and by the inductive hypothesis, for $n-1$, we have the inequality

$$|p_1p_{n-1}| \leq |p_1p_2| + \ldots . + |p_{n-2}p_{n-1}|.$$

Adding these two inequalities gives (1).

We return to geometry on the sphere. If all the links of a chain are very small (here by distance we mean the distance $\|ab\|$ in the ambient space) then the chain is very close to being a curve on the sphere joining a and b, and the length of the chain is close to the length of this curve. But any such curve has length at least $|ab|$, and $|ab| > \|ab\|$. Therefore for chains with very short links, the length of a chain cannot be very close to $\|ab\|$: any such chain will have length greater than some constant c, where we can take c to be any number strictly less than $|ab|$, but greater than $\|ab\|$.

To underline that this kind of unnatural effect does not occur for distance in the geometries we have considered, we can thus state the following property.

(d) For any two points a and b, and any positive numbers α and β, there exists a chain joining a and b such that the length of the chain differs from the distance from a to b by less than α, and the length of each link is less than β; in other words, there should exist points $p_1, p_2, \ldots, p_n$ such that $p_1 = a$, $p_n = b$, and

$$0 \leq |p_1p_2| + \ldots + |p_{n-1}p_n| - |ab| < \alpha,$$

and

$$|p_ip_{i+1}| < \beta \text{ for } i = 1, \ldots, n-1.$$

In all the examples of geometries we have considered this property is automatically satisfied, and in fact we can even assume that the equality $|p_1p_2| + \ldots + |p_{n-1}p_n| = |ab|$ holds, since any two points can be joined by a line segment of length equal to the distance $|ab|$. In the examples we will consider subsequently, the property (d) will play a similar role: it allows us to use the intuitive representation of the distance betwen two points as the minimal time needed to get from one to the other, moving at constant speed.

After these explanations the definition of geometry which we now give is perhaps a natural one.

A *geometry* is an arbitrary set, whose elements are called points, together with a method of indicating how to find the distance $|ab|$ between any two points a and b, such that the definition of distance satisfies the above conditions (a), (b), (c) and (d).

Thus for us the conditions (a - d) are axioms of a geometry, and in this sense our definition is similar to the axiomatic construction of geometry in a Euclidean geometry course. But in contrast to the axioms of Euclidean geometry, which are many in number, and which are only satisfied by some geometries (the Euclidean plane and 3-space), properties (a - d) are satisfied by very many different kinds of geometry.

Exercises

1. Which of the axioms of Euclidean geometry imply that plane geometry satisfies property (d) of the definition of geometry?

2. Let F be a figure in the plane bounded by a curve c (for simplicity F may be assumed to be a polygon, and c a broken line). Let F be the geometry whose points are the points of F, with the same distance between points as in the plane. Prove that F satisfies the axioms of geometry if and only if it is convex, that is it contains the line segment joining any two of its points. [Hint: consider property (d).]

3. On the line, define the distance between points with coordinates x and y to be $(x - y)^2$. Is this a geometry?

4. On the line, define the distance between points with coordinates x and y to be $\sqrt{(|x - y|)}$. Prove that properties (a), (b) and (c) in the definition of geometry hold. (d) does not hold; verify that (d) cannot be satisfied if we just consider chains all of whose links are the same length.

5. The most degenerate case of a geometry is one having only one point a, and of course $|aa| = 0$. Prove that if a geometry has at least two distinct point, then it already has an infinite number.

6.2. Superposing geometries. We now proceed to the second basic question: when do we consider two geometries to be the same? In Examples 1 and 1a we have of course two different descriptions of one and the same geometry, that is, two geometries which are naturally considered to be the same; similarly for Examples 2 and 2a, Examples 3, 3a and 3b, 4a and 4b, and 5a and 5b. In each of these examples we have two geometries which we think of as the same, despite the fact that the points of which they are composed are entirely different in nature: in Example 1 we have a point on a blackboard, in Example 1a, a pair of numbers. What do these geometries have in common? We emphasised this common nature saying that Geometry 1a gives another description of Geometry 1. By this we meant that points of Geometry 1a correspond to points of Geometry 1, with the distance between any pair of points of Geometry 1 equal to that between the corresponding points of Geometry 1a. The same thing happened in the other examples where we could talk about geometries being the same. This provides the basis for introducing the following natural definition.

Suppose two geometries are given; we will say that these geometries are the same if a rule can be given for assigning to every point of the first geometry some point of the second geometry, in such a way that the following two conditions are satisfied:

(i) every point of the second geometry is assigned to some point of the first;

(ii) the distance between two points of the first geometry is equal to that between the points of the second geometry to which they are assigned.

A rule for assigning points subject to these conditions will be called a *superposition* (or *isometry*) of the first geometry onto the second. Thus a superposition is a map from the set of points of the first geometry onto the set of points of the second which preserves the distance between points.

For example, assigning to each point of the plane its coordinates is a superposition of the geometry of Example 1 onto that of Example 1a; and assigning to any point of the cylinder the set of all points of the plane which cover it is a superposition of geometry of Example 3 onto that of Example 3b.

Example 6. Consider the geometry whose points are the points of a disc in the plane of radius r centred at some point O, and with distance defined as usual in the plane. Choose some system of coordinates centred at O, and assign to each point of our geometry its coordinates, as a point of the geometry of Example 1a. This is not a superposition of the geometry of the disc onto that of Example 1a, since condition (i) of the definition is not satisfied: not just any coordinates (x, y) correspond to some point of the disc (for this they must satisfy the relation $x^2 + y^2 \leq r^2$).

Now at last we can proceed to define the basic notion which will be the main object of our investigation. If a is a point of our geometry and r is any positive number, then the set of all points b of the geometry such that $|ab| \leq r$ is called the *spherical neighbourhood* of radius r of a; it will usually be denoted by $D(a, r)$.

Suppose given a geometry. We say that the geometry is a *2-dimensional locally Euclidean geometry* , or that it is *identical with the plane in sufficiently small regions* if there exists a number $r > 0$ such that for any point a of our geometry the spherical neighbourhood $D(a, r)$ of a of radius r can be superposed on the whole disc of radius r in the plane.

In other words, if we consider $D(a, r)$ as a geometry in its own right (as we did in Example 6), then this geometry will be identical with that of the disc of radius r in the plane. In such a case we will call the spherical neighbourhood of radius r a *disc*.

Obviously, each of Examples 3, 3a, 3b, 4a, 4b, 5a and 5b gives a geometry which comes under this definition: in each of the cases we have proved this. The same is true even more obviously for Examples 1 and 1a. On the other hand, as we have seen, Examples 2 and 2a are not locally Euclidean according to our definition.

Of course, we are not obliged to restrict ourselves to 2-dimensional locally Euclidean geometries, but we can also consider their 3-dimensional analogues, that is, geometries which are identical to Euclidean 3-space in sufficiently small regions. For this we need only amend the preceding definition by replacing the disc of radius r in the plane by the ball of radius r in 3-space. In such geometries we will call a spherical neighbourhood a *ball* .

Having a precise definition, we can now state our main problem:

FIND ALL 2-DIMENSIONAL LOCALLY EUCLIDEAN GEOMETRIES.

To do this, we first of all construct some concrete geometries (as we have done in §§3-5), and then prove that any 2-dimensional locally Euclidean geometry can be superposed onto one of these.

It is a very surprising fact that it is possible, starting from such a general definition, to get a very concrete answer. In fact we already practically know the answer: we will prove below that all the 2-dimensional locally Euclidean geometries are classified by the examples which we have dealt with in §§3-5 (including the generalisation contained in §5.1, Exercise 4), and one further type which will be described in §9 (it is a kind of mixture of geometry on the torus and on the twisted cylinder: this geometry is bounded, and in it right and left are indistinguishable). However, for the proof of this result, the statement of which is now already quite clear, we will need completely new ideas. We proceed to treat these in Chapter II of the book.

Exercises

1. One geometry consists of the interior points of a square, and a second those of a disc, and in both the distance is defined as for points of the plane. Can one geometry be superposed on the other? More generally, given two geometries, each of which is defined as the set of points of a plane region, with the same distance as for points of the plane, when can one be superposed on the other? (Obviously the regions will have to satisfy the condition discussed in §6.1, Exercise 2.)

2. Let us (as a temporary notion) define a *quasi-geometry* to be a set of points with a definition of distance, satisfying all the conditions in the definition of a geometry except for condition (d). The advantage of this notion is that any subset of the points of a geometry, with the definition of distance as in the geometry itself, is a quasi-geometry (but is not necessarily a geometry, see §6.1, Exercise 2). Prove that any quasi-geometry consisting of 3 points can be obtained in this way from the plane, that is, it can be superposed on a quasi-geometry consisting of 3 points in the plane. Prove that this is not necessarily possible for a quasi-geometry consisting of 4 points. [Hint: consider the 4 vertices of a tetrahedron as a subset of space.]

3. Consider the set of points of the surface of a solid angle, defining distance as in §§2-3, that is, as the length of the shortest curve joining two points. Prove that this is a geometry. Prove that it is not locally Euclidean. [Hint: use the result of Exercise 2 above, paying particular attention to the vertex of the angle.]

4. Prove that the geometry of the interior of a disc (considered as a part of the plane) is not locally Euclidean.

5. Let Σ_1 and Σ_2 be geometries, and suppose that $f: \Sigma_1 \to \Sigma_2$ is an invertible map which is a superposition in sufficiently small regions, that is for some $r > 0$, f defines a superposition of the spherical neighbourhood $D(X, r)$ of any point X of Σ_1 onto the spherical neighbourhood $D(f(X), r)$ of its image in Σ_2. Prove that f is then a superposition of geometries (the proof will be given in §10.5, Lemma 5). Here $D(A, r)$ denotes the spherical neighbourhood of A of radius r.

Chapter II
The theory of 2-dimensional locally Euclidean geometries

§7. Locally Euclidean geometries and uniformly discontinuous groups of motions of the plane

7.1. Definition of equivalence by means of motions. In this section we return to the analysis of the examples described in §§3-5, and attempt to include them all within one unified construction. We treat a general method of constructing locally Euclidean geometries in such a concrete way that in §8 we will be able to classify explicitly all the geometries so obtained. On the other hand, this method turns out to be general enough to include any locally Euclidean geometry whatsoever, as will be proved in §10. This will then solve the problem of classifying all possible locally Euclidean geometries.

For each of the three examples of geometries described in §§3-5 we had two descriptions; in the preceding §6, we distinguished these by the letters a and b. The first description defined the geometry as the set of points of some plane figure, and the distance between points as the length of the shortest track of a definite type joining these points. In the second description, a point of the geometry was specified as a set of equivalent points of the plane, where the notion of equivalence was defined differently in each of the three examples; and the distance between points a and b was defined as the minimum of the numbers $|AB|$, where A is any point of the plane belonging to the set of equivalent points defining a, and B is any point of the set defining b. In each of the three cases, we started from the first description, as being the more intuitive, and then proceeded to the second, from which we deduced the main properties of our geometry; this is an indication that it is the second description which is the really effective one, and which is therefore closer to the essence of the problem. This second description will serve as the basis for our more general construction.

The basic idea of the description was that somehow or other (we don't mind how) we have a *notion of equivalence* defined on points of the plane. This means that for any two points A and B of the plane, we know if A is to be considered as equivalent to B or not. This notion of equivalence is only required to satisfy the following conditions:

(α) any point is equivalent to itself;

(β) if B is equivalent to A then A is equivalent to B;

(γ) if A is equivalent to B and B is equivalent to C, then A is equivalent to C.

In view of (β) the property that two points should be equivalent is independent of their order; we are allowed to say simply that A and B are equivalent. Consider the set of points equivalent to some given point A. By (γ), any two points of this set are equivalent to one another, and any point equivalent to some point of this set belongs to it. Hence we will describe such a set for short as a *set of equivalent points*. Each such set defined a point of the new geometry. If a set **A** defined a point a of the geometry, and a set **B** a point b, then the distance $|ab|$ is defined as follows:

$|ab|$ is the minimum of the numbers $|AB|$, where
A is any point of **A**, B is any point of **B**, and (1)
$|AB|$ is the distance in the plane from A to B.

In each of the three examples we proved that our geometry was locally Euclidean by checking that for points A and B of the plane which are sufficiently close (meaning that $|AB|$ should be less than some fixed r), we had $|ab| = |AB|$, where a is the point of the geometry defined by the set **A** of points equivalent to A, and b is the point defined in the same way by B. It follows from this that if A is the centre of a disc of sufficiently small radius r, then no point $A' \neq A$ of the disc is equivalent to A; for otherwise, we would get a contradiction: $0 \neq |AA'| = |aa| = 0$, since equivalent points A, A′ define the same point a of the geometry. Since A is any point of the plane, we could say simply:

(δ) there exists a positive number d such that the distance between any two distinct equivalent points is at least d.

Figure 7.1

Thus a set of equivalent points might for example consist of points on a line, marked out at intervals which are multiples of some given interval (Figure 7.1), but not of the set of points of a line whose distance from a given point is 1, 1/2, 1/3, 1/4, ... (Figure 7.2), and even less of the set of all points of some line. A set of equivalent points satisfying condition (δ) is said to be *discrete*: its points are not 'spread out', but 'isolated'.

There are of course many different ways of defining a notion of equivalence satisfying conditions (α - δ), but by no means just any of these definitions will

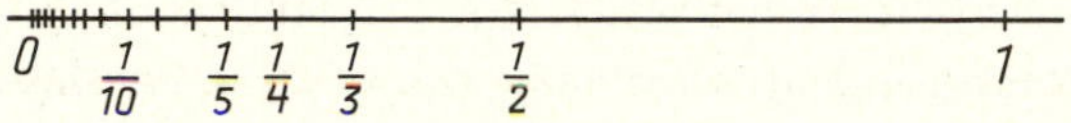

Figure 7.2

give a geometry satisfying conditions (a - d) of §6.1. Which of these conditions could turn out to be false? It is very easy to see that conditions (a) and (b) will always be satisfied; it can also be shown that (d) will hold. Thus the essential point is condition (c). To understand why the inequality (c) holds in our examples, let's try to prove it in Examples 3b, 4b and 5b directly from the construction of these examples (in the preceding section we deduced this from the fact that our geometries were identical with those of Examples 3a, 4a and 5a). Since the arguments in all the cases are entirely similar, we restrict ourselves to Example 3b.

Let a, b and c be points of our geometry corresponding to sets **A, B** and **C** of equivalent points (see Figure 7.3). Suppose that $|ab| = |AB|$, that is, $|AB|$ is the shortest distance from a point A of **A** to any B in **B** (note that $||$ denotes both the distance $|ab|$ in our geometry, and the distance $|AB|$ between points of the plane). Suppose that in exactly the same way, $|bc| = |B'C'|$. Since B and B′ both

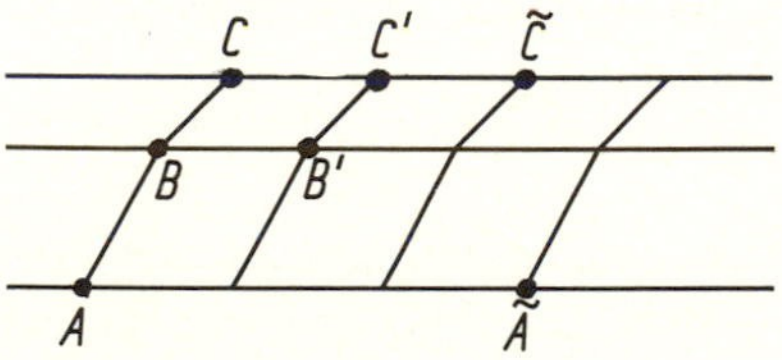

Figure 7.3

belong to the same set **B**, they are equivalent, and by the definition of equivalence in this geometry, there exists a translation in a multiple $k{\cdot}\overrightarrow{P_0P_1}$ of the vector $\overrightarrow{P_0P_1}$ taking B′ into B. Suppose that this translation takes C′ into C. Since translation does not change distances between points, $|BC| = |B'C'|$. In addition, C is equivalent to C′, and hence belongs to **C**. By definition of distance, $|ac|$ is the minimum of the distances $|\tilde{A}\tilde{C}|$ for points $\tilde{A}$ in **A** and $\tilde{C}$ in **C**, and in particular, $|ac| \le |AC|$. By property (c) for distances in the plane, $|AC| \le |AB| + |BC|$. Finally, as we have seen, $|BC| = |B'C'|$. Putting everything together, we get

$$|ac| \le |AC| \le |AB| + |BC| = |AB| + |B'C'| = |ab| + |bc|,$$

which is what we required to proved.

From this proof we can see clearly why it is that (c) holds in our geometry. The reason is contained in the special definition of equivalence: we chose some set of translations (in multiples of the vector $P_0P_1^{\to}$), and said that two points were equivalent if one is taken into the other under one of these translations. Now it was not so essential for us that we were dealing specifically with translations; in the proof, the essential thing was that we had a transformation of the plane taking B′ and C′ into B and C, and such that $|B'C'| = |BC|$, that is, not altering the distance between points. In geometry, such transformations are called motions (or rigid motions). In the definition of geometry on the twisted cyclinder, the role of such motions was played by glide reflections, in addition to translations. Before proceeding further, we need to recall some of the basic facts about motions.

A *motion* is a map from the plane onto the whole plane which preserves distances between points. Using the terminology of §6.2, we can say that a motion is a superposition of the plane onto itself. We recall the examples of motions: rotation, translation, reflection in an axis, and glide reflection.

As with other maps, we can perform one motion followed by another, and speak of the *composite* of motions. Suppose that F is a motion taking A into F(A), and G a motion taking B into G(B). Then the composite of F and G is the map consisting of F followed by G, that is, the map which takes A into G(F(A)); to find this point, we must first of all find the point B = F(A) which F takes A into, then the point G(B) which G takes B into. It follows easily from the definition that this map is also a motion; we write it GF, juxtaposing G and F, with G on the left and F on the right. Writing the composite in this order is convenient, since the definition can be written simply as

$$(GF)(A) = G(F(A)).$$

The composite of two motions will usually depend on the order in which we perform them, that is, which of F or G we carry out first (see Exercise 1). It can even happen that some motion is most conveniently specified as a composite of other motions. For example the glide reflection is given as the result of first reflecting in an axis, then making a translation in a vector parallel to this axis. In this special case, due to the fact that the axis and the vector are parallel, the order in which we perform the motions is immaterial. Glide reflections appeared in our study of the geometry of §4.

A motion of the plane is an invertible map. The inverse map of a motion F is obviously again a motion; it is denoted by F^{-1}. Recall that if F takes X into Y then by definition, F^{-1} takes Y into X. If for example F is a translation in a vector $AB^{\to}$ then F^{-1} is the translation in the vector $BA^{\to}$; if F is a rotation about a centre O through an angle α, then F^{-1} is the rotation about O through $2\pi - \alpha$; if F is a reflection in an axis, then $F^{-1} = F$.

The map which takes every point of the plane into itself is called the *identity*. This is of course a motion, and we denote it by E. So by definition, for every point A we have $E(A) = A$.

The composition of motions satisfies several of the properties of multiplication of ordinary numbers; the identity motion E plays the part of the unit, that is $EF = FE = F$ for every motion F. The inverse motion F^{-1} plays the part of the reciprocal $1/a$ of a number: $F^{-1}F = FF^{-1} = E$.

Furthermore, composition of motions satisfies the associative law

$$F(GH) = (FG)H$$

for any motions F, G, H. For these reasons, composition of motions is also called multiplication. As with multiplication of numbers, a composite of several motions can be simplified using the properties we have listed. We emphasise once more that in doing this, we are not allowed to use the commutative law (that is, $FG = GF$), since it is false in general; because of this, we have to be careful to keep track of the order of factors.

Now we can return to our analysis of possible constructions of geometry. The arguments given after the proof of the inequality (c) lead us to the conclusion that if we start from some definition of equivalence of points of the plane and define distance as in (1), then we will arrive at a geometry, provided that the original definition of equivalence satisfies the following condition: there exists a certain set Γ of motions of the plane such that two points A and B are equivalent if and only if A is taken into B by some motion belonging to Γ.

This is exactly what was going on in the examples which we discussed in §§3-5. In §3, the set Γ consisted of all translations in multiples of some vector $\overrightarrow{P_0P_1}$. In §4, Γ consisted of two types of motions: (a) glide reflections consisting of a translation in any odd multiple of the vector $\overrightarrow{P_0P_1}$ followed by a reflection in the axis (P_0P_1), and (b) translations in any even multiple of $\overrightarrow{P_0P_1}$. In §5, Γ consisted of translations in vectors going from a vertex of some square to the vertexes of squares of a tiling of the plane by adjacent squares of the same size.

We will see that by no means any set Γ of motions of the plane can be taken in this way as the basis of a definition of equivalence of points on the plane. We will first determine some properties of a set Γ which are necessary for this. After that, we will check that any set of motions which satisfy these properties does indeed give rise to a geometry. This will be the road which will lead us in the following sections to the classification of all possible locally Euclidean geometries.

Thus suppose that on the plane we have a notion of equivalence which satisfies conditions (α - δ), defined by any means whatsoever. Suppose in addition that we have a set Γ of motions of the plane, and that the following condition holds: points A and B of the plane are equivalent if and only if $F(A) = B$ for some F in Γ.

Under these conditions, the set Γ of motions has many very special properties; all of these will follow easily from the following assertion.

Theorem 1. Any motion of the plane taking each point into an equivalent point belongs to Γ.

Proof. Let G be a motion of the plane as in the theorem. Consider any point A of the plane, and set B = G(A). By assumption, B and A must be equivalent. Hence, according to the condition on Γ, there must exist some motion F in Γ such that B = F(A). Now consider the motion F^{-1} inverse to F, and the composite $H = F^{-1}G$. We will show a little later that H is the identity motion E; now let's convince ourselves that the theorem will follow from this.

Indeed, if $H = F^{-1}G$ is the identity motion, then $(F^{-1}G)(X) = X$ for every point X. But $(F^{-1}G)(X) = F^{-1}(G(X))$; thus $F^{-1}(G(X)) = X$, so that by definition of the inverse map, G(X) = F(X). This holds for any point X, and therefore G = F is in Γ, which implies the theorem.

Now let's prove that H is the identity map. Let X be any point of the plane such that $|AX| < d/2$, and set Y = H(X). Since H is a motion, $|AX| = |H(A)H(X)|$. But H(A) = A, and H(X) = Y, and therefore $|AY| = |H(A)H(X)| = |AX| < d/2$. It follows from this that $|XY| \leq |XA| + |AY| < d$. Finally, we note that X and Y are equivalent points. Indeed, let G(X) = Z. Then X and Z are equivalent, by assumption on G. By definition, $Y = H(X) = F^{-1}(Z)$, so that Z = F(Y), and since F is an element of Γ, Y and Z are equivalent. Therefore X and Y are equivalent.

We have thus proved that our two equivalent points X and Y satisfy $|XY| < d$. By property (δ) of the notion of equivalence, this is impossible if $X \neq Y$; hence Y = X, and since Y = H(X), we have H(X) = X. This inequality is proved for all points with $|AX| < d/2$, and such points fill out the interior of a disc of radius $d/2$ centred at A.

This proves that the motion H leaves fixed all points of the interior of some disc D(A, R) (in fact with $R = r/2$). Recall that D(A, R) denotes the disc centred at A and of radius R. It follows from this that H fixes all points of the plane, that is,H is the identity motion. Indeed, let X be any point of the plane not lying in

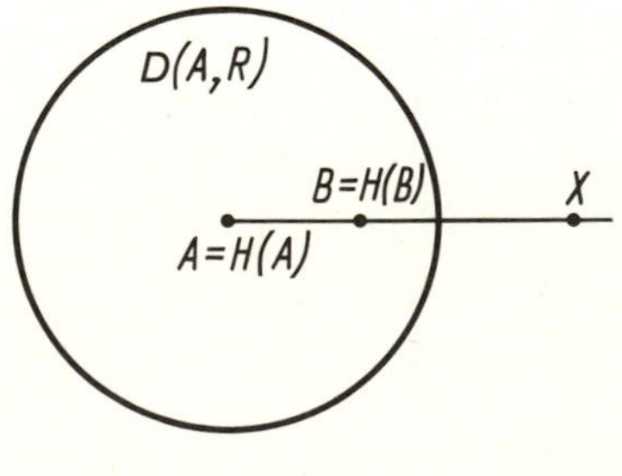

Figure 7.4

D(A, R), and let B be a point of the line segment [AX] distinct from A but lying inside D(A, R) (Figure 7.4).

Note that the distances |AB|, |BX| and |AX| are related by

$$|AB| + |BX| = |AX|.$$

Since H is a motion with H(A) = A and H(B) = B, it follows from this that

$$|AB| + |BH(X)| = |AH(X)|$$

and

$$|AX| = |AH(X)|.$$

The three equalities we have obtained show that X and H(X) both lie on the ray AB, and are at the same distance from A. From this, obviously H(X) = X, which proves that H is the identity motion. (Note that another proof of this fact will follow from §8.1, Lemma 3.) The theorem is proved.

Two important properties of the set of motions Γ follow from the theorem.

Property I. The composite of two motions in Γ is again a motion belonging to Γ.

Let F and G be two motions in Γ and H = GF their composite. By the theorem, to prove that H belongs to Γ, it is enough to prove that H(X) is equivalent to X for any point X; but this is obvious: let Y = F(X), so that H(X) = G(Y). Then X and Y are equivalent since F belongs to Γ, and Y and G(Y) are equivalent since G belongs to Γ. By property (γ) of the notion of equivalence, X and G(Y) are equivalent, and G(Y) = H(X), so that X and H(X) are equivalent.

Property II. For any motion F in Γ, the inverse motion F^{-1} again belongs to Γ.

Again we only need to show that X and $F^{-1}(X)$ are equivalent. Let $Y = F^{-1}(X)$, so that X = F(Y). Since F belongs to Γ, X and Y are equivalent, as required.

Any non-empty set of motions Γ satisfying Properties I and II is called a *group* of motions. This is a very important notion. Here are several examples.

Example 1. The set of all translations is a group.

Example 2. The set of all rotations about a given centre is a group.

Example 3. The set of all translations in vectors parallel to a fixed line is a group.

Example 4. The translations in vectors $\overrightarrow{OA}$ where O is a fixed point, and A is any point on a fixed ray emanating from O do not form a group; although Property I is satisfied, Property II is not.

We can now say that the set of motions Γ involved in our definition of equivalence is a group. In particular, the sets of motions which arose in connection with the examples of geometries considered in §§3–5 form groups. This follows from what we have proved above, but we strongly recommend the reader to check Properties I and II directly in each of these examples.

Now it is time to think about property (δ) of the notion of equivalence. Since two points are equivalent if and only if one can be obtained from the other by some motion in Γ, this property can also be expressed as follows:

Property III. There exists some positive number d such that if F is any motion in Γ, and X any point of the plane such that $X \neq F(X)$, then $|XF(X)| \geq d$.

A group having this property is said to be *uniformly discontinuous*. If we take any point X of the plane, and apply to it all the motions in Γ, then we obtain a discrete set. The word 'uniform' indicates the fact that the number d in the definition can be chosen once and for all, to cover all points X.

The examples of groups of motion which appeared in connection with the geometries of §§3-5 are of course all uniformly discontinuous. On the contrary, none of the groups of the above Examples 1-3 is uniformly discontinuous. Let's give another example: consider the group consisting of just two elements, the identity E and the rotation F through 180° about a centre O (that is, the central reflection in O). It is easy to check that this is a group. Applying all the elements of the group (both of them) to a given point X, we of course get a discrete set of two points, X and F(X). Despite this, the group is not uniformly discontinuous, since the number d in Property III cannot be chosen in a unified way for all X: the closer X gets to the centre O of the rotation, the closer X and F(X) become.

Of course, in our case, not only does the notion of equivalence determine the group of motions Γ, but also conversely, Γ determines the notion of equivalence: points A and B are equivalent if and only if there exists a motion F belonging to Γ such that F(A) = B. We will make this fact the basis of our subsequent arguments: we will start from an arbitary uniformly discontinuous group, and using the fact that it defines a notion of equivalence, construct a locally Euclidean geometry from it. This is the approach that will finally enable us to give a classification of all such geometries.

Exercises

1. a) Prove that the composite $S_O S_{O'}$ of two central reflections centred at O and O′ is a translation in the vector $2\cdot\overrightarrow{O'O}$; when will it happen that $S_O S_{O'} = S_{O'} S_O$?

b) Prove that the composite $S_\ell S_{\ell'}$ of two reflections in parallel axes ℓ and ℓ' is a translation in the vector $2\cdot \mathbf{u}$ where $\mathbf{u} = \mathbf{u}(\ell, \ell')$ is a vector perpendicular to ℓ and ℓ' whose starting-point is on ℓ and whose endpoint is on ℓ'. When will it happen that $S_\ell S_{\ell'} = S_{\ell'} S_\ell$?

c) Prove that the composite $S_{\ell'} S_\ell$ of two reflections in axes ℓ and ℓ' which intersect at a point O is a rotation about O through twice the angle between ℓ and ℓ'. When will it happen that $S_\ell S_{\ell'} = S_{\ell'} S_\ell$?

2. Consider the set of motions of a line ℓ consisting of translations parallel to ℓ and reflections in points of ℓ (it will follow from §8 that these are all the motions of a line); prove that it is a group.

3. Prove that if a motion F belongs to a uniformly discontinuous group Γ and fixes at least

one point of the plane, then it is the identity. [Hint: reread the proof of Theorem 1; this assertion will be proved in §8, Proposition 1.]

4. Let F and G be known motions of the plane; prove that the equation $XF = G$ in the unknown motion X has the unique solution $X = GF^{-1}$. [Hint: multiply through both sides on the right by the motion F^{-1} and use the fact that $FF^{-1} = E$.] Prove that the unique solutions of the equations $FX = G$ and $FXH = G$, where F, G, and H are known motions, are respectively the motions $X = F^{-1}G$ and $X = F^{-1}GH^{-1}$.

7.2. The geometry corresponding to a uniformly discontinuous group. We needed to go through all the arguments of the preceding section mainly in order to motivate a certain construction, the actual treatment of which will take up much less space. We are now going to show how to construct from any uniformly discontinuous group a certain locally Euclidean geometry. This gives us a general construction, which covers in a unified way the examples of §§3-5. In §8, we will classify all possible uniformly discontinuous groups of motion, and thus the geometries which can be constructed by this method. Finally, in §10 we will show that any locally Euclidean geometry can be constructed by this method, which will thus complete our description of all such geometries.

Suppose then that Γ is a given uniformly discontinuous group of motions of the plane. We define a notion of equivalence on points in the plane by saying that B is equivalent to A if there exists a motion F belonging to Γ taking A to B, that is, such that $F(A) = B$. Let's prove that this definition of equivalence satisfies conditions (α - δ). For this, note first that the identity motion E belongs to Γ: for if F is any motion in Γ, then by Property II of the definition of a group, F^{-1} also belongs to Γ, and then by Property I, the composite motion FF^{-1} must also belong to Γ. Since $E = FF^{-1}$, it must belong to Γ.

Since for any point A we have $E(A) = A$, condition (α) follows. If B is equivalent to A, then $F(A) = B$ for some motion F in Γ; but by Property II of the definition of a group, F^{-1} also belongs to Γ. Since $F^{-1}(B) = A$, it follows from this that also A is equivalent to B, so that (β) holds. Finally, suppose that B is equivalent to A, and C is equivalent to B. Then there exists a motion F of Γ such that $F(A) = B$, and a motion G of Γ such that $G(B) = C$. By Property I of the definition of a group, the motion $H = GF$ also belongs to Γ. But $H(A) = G(B) = C$, and it follows from this that C is equivalent to A, so that (γ) holds. Finally since we are assuming that the group Γ is uniformly discontinuous, (δ) will also hold.

We define a point of the new geometry to be a set of equivalent points in the sense of the notion of equivalence we have introduced. Let a and b be two such points, and **A** and **B** the sets defining them. Then we define the distance between a and b to be the shortest of the distances $|AB|$, where A and B are points of the plane with A belonging to **A** and B to **B**.

Remark. The distance can also be defined in a way which involves comparing a smaller set of numbers. Let A_0 be a fixed point of the set **A**. Then $|ab|$ is the minimum of the numbers $|A_0B|$ where B is any point of **B**, that is, we only need compare the numbers $|A_0B|$, with A_0 a fixed point; indeed, if A is any point of **A** and B is a point of **B**, then A is equivalent to A_0, so that there exists a motion F belonging to Γ such that $F(A) = A_0$. Then since F is a motion, $|AB| = |F(A)F(B)| = |A_0F(B)|$. From the definition of equivalence of points it follows that $B' = F(B)$ is a point of **B**. Hence each of the number $|AB|$ can also be represented in the form $|A_0B'|$ with A_0 the fixed point of **A**, and B' a point of **B**.

This notion of distance can be given the same kind of intuitive interpretation which we used in §§3-5. Namely, we can think of the motions of Γ as being an instantaneous jet service, so that an inhabitant of the geometry can move around the plane either at constant speed, or by instantaneously transferring himself from a point A to F(A), where F is any motion of Γ. The shortest time required for him to get from A to B in this way will be the distance from A to B. In this, if F is any element of Γ, then since the distance between two points A and F(A) is equal to 0, the two points should be treated as one.

Now however, we have to determine why the distance $|ab|$ between points a and b given by sets of equivalent points **A** and **B** is well-defined. This is not quite obvious, since as A runs through **A** and B through **B**, the set of distance $|AB|$ will as a rule be an infinite set of positive numbers, and by no means every infinite set of positive numbers will have a minimum element; for example, the set of numbers of the form $1/n$, as n runs through the natural numbers, does not have a minimum.

We fix a point A_0 of **A**. As we have seen above, it is enough to prove that there exists a point B_0 in **B** such that $|A_0B_0| \leq |A_0B|$ for every B in **B**. In this case, $|A_0B_0|$ will give the distance $|ab|$. In Lemma 1 below, we will prove that a disc of any radius R centred at A_0 will contain only finitely many points of **B**. Using this, we take R large enough that this disc contains some point of **B**, and then we choose the point B_0 closest to A_0 from the finite non-empty set of points of **B** in this disc. This B_0 will obviously be the required point. It remains to prove the next result.

Lemma 1. Let **A** be any set of equivalent points in the plane; then any disc in the plane contains at most a finite number of points of **A**.

Proof. Let D be some disc of radius R, and M the part of **A** contained in D. Consider discs D_i of radius $d/2$ centred at the points A_i of M. Note that by Property III of the group Γ the discs D_i do not intersect one another, and on the other hand, they are all contained inside a disc D′ obtained from D by increasing its radius to $R + d/2$. It follows from this that the number of points of M, which is equal to the number of discs D_i, is at most the ratio of the surface area of the disc D′ to that of the D_i, that is, at most

$$\frac{\pi(R+d/2)^2}{\pi(d/2)^2} = \frac{4(R+d/2)^2}{d^2}.$$

This proves the lemma.

We now prove that the distance we have introduced satisfies all the conditions (a - d) required in the definition of a geometry which we gave in §6.1.

(a) The fact that $|ab| \geq 0$ is obvious. We suppose that $a \neq b$, and prove that $|ab| \neq 0$. The sets **A** and **B** specifying a and b do not have points in common, since if some point C is contained in both **A** and **B** then every point of **B** is equivalent to C, and then since C is in **A**, it follows that **B** is contained in **A**; similarly, we get that **A** is contained **B**, so that **A** = **B**, and therefore a = b, which contradicts the assumption. Now let A_0 be any point of **A**; within a disc of radius d/2 centred at A_0 there is at most one point B of **B**: if there were two such points B and B′ then $|A_0B| < d/2$, and $|A_0B'| < d/2$ would imply that $|BB'| < d$, contradicting Property III in the definition of a uniformly discontinuous group. If there is such a point, we set $c = |A_0B|$. Here $c > 0$, since otherwise $A_0 = B$, but we have already observed that **A** and **B** do not have any points in common. If there is no such point B, we set $c = d/2$. Thus there always exists some positive number c such that $|A_0B| \geq c$ for all points B of **B**. According to the remark following the definition of distance, $|ab|$ is the minimum of the numbers $|A_0B|$ where A_0 is the point of **A** we have chosen, and B is any point of **B**, and hence $|ab| \geq c > 0$.

(b) The property $|ab| = |ba|$ follows at once from the definition of a and b, and from the fact that $|AB| = |BA|$ for points of the plane.

(c) In essence, the proof of this property has already been given in §7.1 (and is illustrated in Figure 7.3). The proof basically only used the fact that the notion of equivalence of points is defined by means of some group of motions.

(d) Let a and b be points of our geometry, specified by sets **A** and **B** of equivalent points in the plane. We choose any point A of **A** and let B be the point of **B** closest to A. According to the remark made after the definition of distance, $|ab| = |AB|$.

We now join A and B by a line segment, and divide this segment into parts by points $P_1, \ldots, P_n$ such that $P_1 = A$, $P_n = B$, and the distances $|P_iP_{i+1}|$ are all less than β. Then $|AB| = |P_1P_2| + |P_2P_3| + \ldots + |P_{n-1}P_n|$. We denote by $\mathbf{P}_i$ the set of points of the plane equivalent to P_i, and by p_i the point of our geometry specified by $\mathbf{P}_i$. By definition of distance, we have $|p_ip_{i+1}| \leq |P_iP_{i+1}|$, and hence

$$|ab| \geq |p_1p_2| + \ldots + |p_{n-1}p_n|, \quad \text{with} \quad |p_ip_{i+1}| < \beta.$$

On the other hand, we proved in §6, inequality (1), that

$$|ab| \leq |p_1p_2| + \ldots + |p_{n-1}p_n|,$$

since $a = p_1, b = p_n$. From this we get that

$$|ab| = |p_1p_2| + \ldots + |p_{n-1}p_n|,$$

which is the required assertion, in fact in a stronger form.

Finally, we check that the geometry constructed in this way is locally Euclidean. We will show that the geometry is identical with that of the plane inside any disc of radius $d/4$ centred at any point p; here d is the number involved in the definition of uniformly discontinuous group. We write $\mathbf{P}$ for the set of equivalent points of the plane defining a point p of our geometry, and let P be a point of $\mathbf{P}$. Now let D be the disc of radius $d/4$ in the plane, centred at P. Then for any point A of this disc, let $\mathbf{A}$ be the set of all points equivalent to A, and let a be the point of the geometry defined by $\mathbf{A}$. Then by definition of distance we have $|ap| \le |AP|$, and since $|AP| \le d/4$, the points a belongs to the disc of radius $d/4$ centred at p. We map D to $D(p, d/4)$ by taking A to a.

The assertion which we have to prove is that this gives a superposition of the disc in the plane onto the disc in our geometry. Let A and B be two points of the disc D, $\mathbf{A}$ the set of points equivalent to A, and $\mathbf{B}$ the set of points equivalent to B, so that $\mathbf{A}$ defines a and $\mathbf{B}$ defines b. We have to prove that $|AB| = |ab|$. According to the remark following the definition of distance, $|ab|$ is the least of the numbers $|AB'|$, where A is the chosen point, and B' is any point of $\mathbf{B}$. If $|AB|$ is not the least of these numbers, then for some point B' of $\mathbf{B}$ we get $|AB'| < |AB|$. Since both A and B belong to D, we have $|PA| \le d/4$ and $|PB| \le d/4$, and hence $|AB| \le d/2$. But $|AB'| < |AB|$, so that $|AB'| < d/2$. Hence $|BB'| \le |BA| + |AB'| < d$. But by Property III, the distance between distinct points of $\mathbf{B}$ is at least d. This contradiction proves that $|AB| = |ab|$.

It remains to prove that our map goes onto the whole disc, that is, that any point of our geometry in the disc of radius $d/4$ centred at p comes from some point of the disc D. Let a be such a point, and A the point closest to P of the set $\mathbf{A}$ defining a. We know that $|pa|$ is the least of the numbers $|PA'|$ where A' is a point of $\mathbf{A}$, and hence $|pa| = |PA|$. Since $|pa| \le d/4$, also $|PA| \le d/4$, that is, A is contained in D. The point of our geometry which A maps to will obviously be a.

We have thus proved the following result.

Theorem 2. To every uniformly discontinuous group of motions of the plane there corresponds a 2-dimensional locally Euclidean geometry.

We will denote by Σ_Γ the geometry constructed from a uniformly discontinuous group Γ of motions of the plane, to emphasise the role of Γ.

Exercises

1. Let Γ be the group consisting of rotations about a fixed point of the plane through angles of $0, 2\pi/n, 4\pi/n, \ldots, 2(n-1)\pi/n$. Prove that the construction of the geometry Σ_Γ is still meaningful in this case, although the geometry is not locally Euclidean. Show that the geometry Σ_Γ can be superposed on a cone.

2. We define a motion of an arbitrary geometry to be a map of the set of points of the geometry onto itself which preserves distances. Prove that if Γ is a uniformly discontinuous group of motions of an arbitrary geometry, then we can construct a geometry Σ_Γ from Γ just as in the case of the plane, provided that we assume in addition that every spherical neighbourhood contains at most a finite number of points from any set of equivalent points (a small change is needed for the proof of (d)). Prove that if we started from a locally Euclidean geometry then Σ_Γ will again be locally Euclidean.

3. Prove that translations of a cylinder in multiples of a given vector lying along a generator of the cylinder form a uniformly discontinuous group Γ. What will the geometry Σ_Γ be? The same question for rotations of a circular cylinder about its axis through angles of $0, 2\pi/n, 4\pi/n, \ldots, 2(n-1)\pi/n$.

4. Prove that the map F of the sphere which takes each point into the antipodal point is a motion of spherical geometry, and that F together with the identity E forms a uniformly discontinuous group Γ. The corresponding geometry Σ_Γ is called the Riemann geometry. Prove that it satisfies the incidence axioms of usual geometry: there exists exactly one line through any two distinct points; there exist three non-collinear points; but instead of the parallel axiom, the so-called 'projective plane axiom', that any two lines meet in a point. What happens to this geometry if you cut it along a line? Does it decompose into two components?

5. On the plane, consider two notions (1) and (2) of equivalence of points defined as follows: points A and B are said to be equivalent if

(1) A and B are equidistant from some fixed point P;

(2) A and B lie on one line in some fixed pencil of parallel lines.

Are either of these notions of equivalence given by some group of motions of the plane? What are the resulting sets of equivalent points, and do either of these satisfy condition (δ)? Prove that there are geometries corresponding to either notion of equivalence: (1) the geometry of a ray, and (2) that of the line.

6. Let Φ be a figure in a geometry Σ_Γ. Write $\Phi^\sim$ for the set of points of the plane which specify points of Σ_Γ belonging to Φ. Prove that if $\Phi^\sim$ is a convex figure in the plane then Φ is a geometry.

§8. Classification of all uniformly discontinuous groups of motions of the plane

8.1. Motions of the plane. We start off with a classification of all possible motions of the plane. Recall the examples we have already met: a translation, a rotation, a glide reflection (including the particular case of a reflection in a line). We will prove that these examples exhaust all the motions of the plane. We denote a translation in a vector $\mathbf{x}$ by $T_{\mathbf{x}}$, a rotation about a centre O through an angle φ by $R_O{}^{\varphi}$, a reflection in an axis ℓ by S_ℓ, and a glide reflection with axis ℓ and translation vector $\mathbf{x}$ parallel to ℓ by $S_\ell{}^{\mathbf{x}}$; by definition, $S_\ell{}^{\mathbf{x}} = S_\ell T_{\mathbf{x}}$. First of all we prove two auxilliary propositions.

Lemma 1. If $|AB| = |A'B'|$, then there exists a motion F, which is either a translation or a rotation, such that $F(A) = A'$ and $F(B) = B'$.

Proof. We note an important property of a rotation: on making a rotation through an angle φ, every ray turns through the same angle, equal to φ. Indeed, write AM for the ray, and let OM′ be the ray parallel to AM through the centre of rotation O (see Figure 8.1). For the ray OM′, the property holds by definition of the

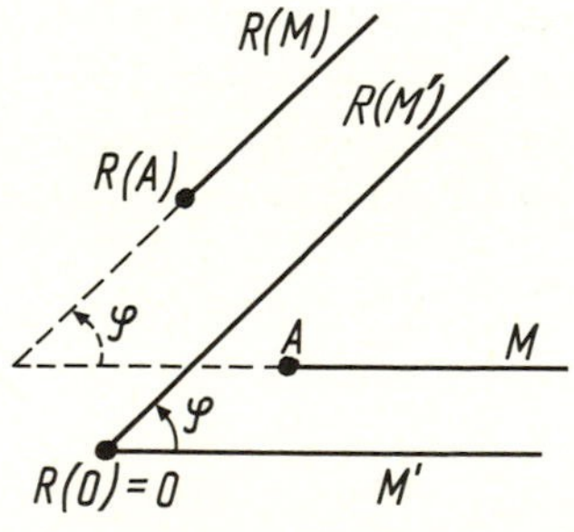

Figure 8.1

rotation; furthermore, the property that rays should be parallel is preserved by any motion. Therefore, the ray R(A)R(M) is parallel to R(O)R(M′), so that the angle between the rays AM and R(A)R(M) is equal to that between OM′ and R(O)R(M′), that is, to φ (see Figure 8.1).

This property suggests how to find the required rotation. Consider first the case that the lines (AB) and (A′B′) are not parallel. Write φ for the angle between the rays AB and A′B′, with φ not equal to 0° or 180°. We use the fact that for two distinct points A and A′, the set of points X such that the angle AXA′^ between the rays XA and XA′ is equal to a given angle φ not equal to 0° or 180° forms an arc of a circle constructed on the chord [AA′] (Figure 8.2).

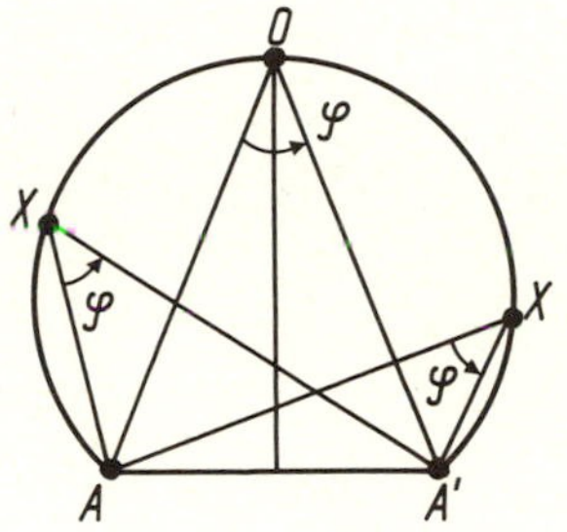

Figure 8.2

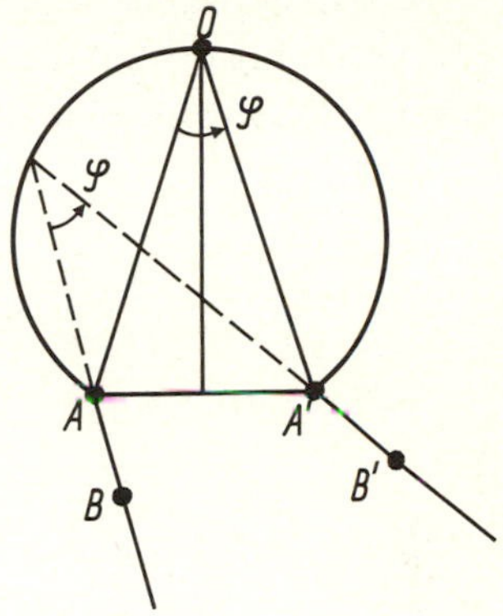

Figure 8.3

Denote by O the point of intersection of this arc with the perpendicular bisector of the line segment [AA′] (see Figure 8.2). Then under the rotation R_O^{φ} with centre O through the angle φ, the point A goes into A′, and by what we have proved, the ray AB turns through φ, and therefore goes into the ray A′B′; then since |AB| = |A′B′|, the point B goes into B′ (see Figure 8.3, which illustrates one case for the position of the line segments [AB] and [A′B′]).

Now suppose that the lines (AB) and (A′B′) are parallel. If the rays AB and A′B′ are opposite, that is if they form an angle of 180°, then the central reflection (or rotation through 180°) around the centre O of the line segment [AA′] takes A into A′ and B into B′ (Figure 8.4).

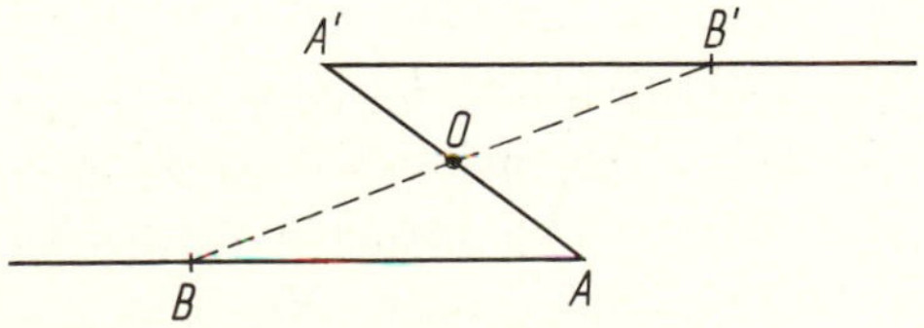

Figure 8.4

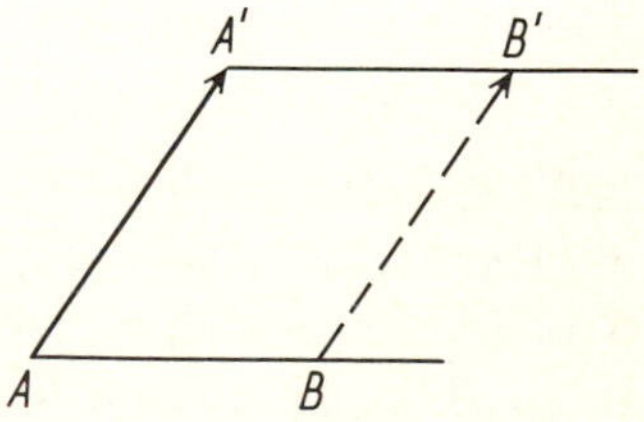

Figure 8.5

Finally, if the rays AB and A′B′ are parallel, then a translation in the vector $AA'^{\rightarrow}$ takes A into A′ and B into B′ (Figure 8.5).

Lemma 2. If $|AB| = |A'B'|$ then there exists a motion S, which is a glide reflection, such that $S(A) = A'$, $S(B) = B'$.

Suppose first of all that the rays AB and A′B′ are not parallel. Write O for the midpoint of the line segment [AA′], and make a translation of the rays AB and A′B′ to O (Figure 8.6). Draw the bisector ℓ of the angle at O between the resulting rays OB_1 (parallel to AB) and OB'_1 (parallel to A′B′). Let C and C′

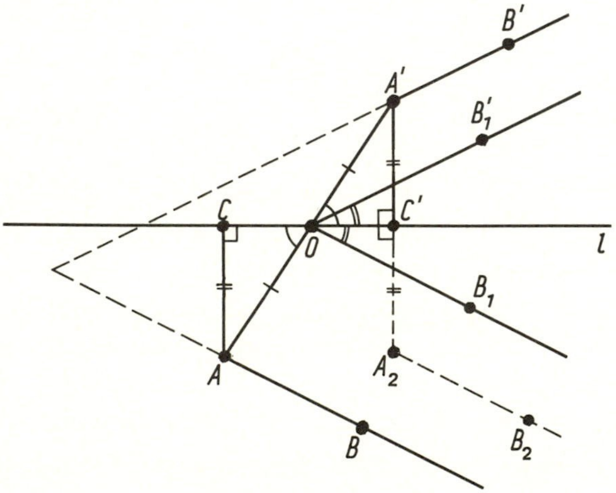

Figure 8.6

be the orthogonal projections to ℓ of A and A′ respectively. Then the right-angle triangles OAC and OA′C′ have equal hypotenuse and angles. Hence $|CA| = |C'A'|$, that is the points A and A′ are the same distance from ℓ.

Under the translation $T_{CC'}$ in the vector $CC'^{\rightarrow}$, the point A goes into a point A_2 such that the line segment $[A_2A']$ perpendicular to ℓ is bisected by C′, and the ray AB goes into a ray A_2B_2 parallel to AB and to OB_1 (see Figure 8.6). Thus on making the reflection S_ℓ in ℓ, the point A_2 goes into A′, and the ray A_2B_2 goes into A′B′, since under this reflection, the ray OB_1 goes into OB'_1, and any motion takes parallel rays into parallel rays. We have thus proved that the composite $S = S_\ell^{CC'} = S_\ell T_{CC'}$ takes the ray AB into the ray A′B′; but by definition, S is the glide reflection with axis ℓ and translation vector $\overrightarrow{CC'}$. Since $|AB| = |A'B'|$, this glide reflection takes B into B′. This proves our assertion for the given position of the rays AB and A′B′.

If the rays AB and A′B′ are parallel, then we take ℓ to be a line parallel to them, again passing through the midpoint of the segment [AA′] (Figure 8.7). Similar arguments prove that the glide reflection $S_\ell^{CC'}$ take A to A′ and B to B′.

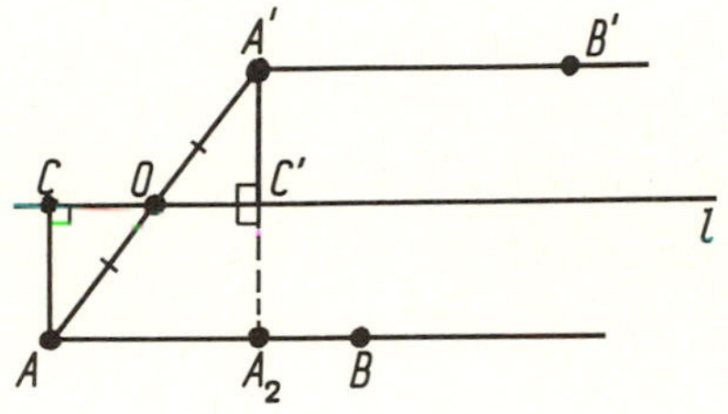

Figure 8.7

Thus any line segment can be moved onto any other line segment of equal length in either of two ways: either (as in Lemma 1) by a motion which is a translation or a rotation, or (as in Lemma 2) by a glide reflection. We have distinguished these two types of motions because they differ in one essential respect.

If F is either a translation or a rotation, and ABC is any triangle whose

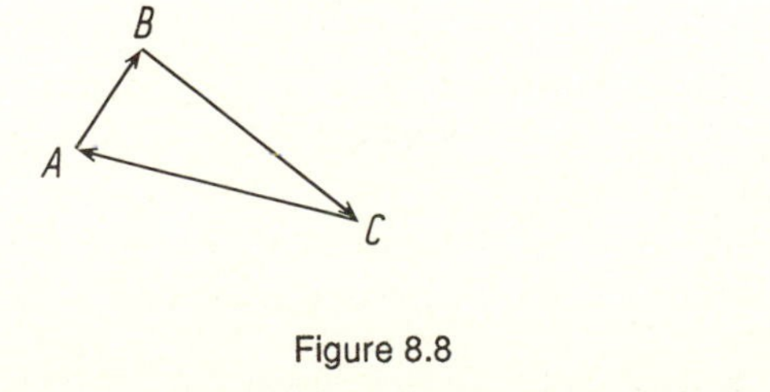

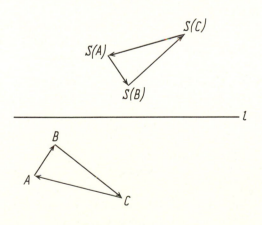

Figure 8.8

angles A, B and C go round in clockwise order, then the angles F(A), F(B) and F(C) of the triangle F(A)F(B)F(C) will go round clockwise. This follows for example from the fact that the motion F can be carried out as a continuous motion of the plane, under which the cyclic order of the angles of a triangle cannot change. On the contrary, if S is a glide reflection, and the vertices A, B and C go round in clockwise order, then the vertices S(A), S(B) and S(C) will go round anticlockwise (Figure 8.8). This is obvious for a reflection in a line, and as we have seen, a translation cannot change the cyclic order. Motions which do not change the cyclic order of the vertices of a triangle (translations and rotations) will be called motions *of the first kind*, and those which reverse the cyclic order of the vertices of a triangle (glide reflections) motions *of the second kind*. From the way that they act on the cyclic order of the vertices of a triangle it follows that if F_1 and F_2 are motions of the first kind, then so is F_1F_2; if F_1 is a motion of the first kind and F_2 is of the second kind, then F_1F_2 and F_2F_1 are of the second kind, and if both F_1 and F_2 are of the second kind then F_1F_2 is of the first kind.

It turns out that Lemma 1 and Lemma 2 give all motions taking one line segment into another.

Lemma 3. Suppose that A, B, A′ and B′ are four points satisfying $|AB| = |A'B'| \neq 0$. Then there exists exactly two motions taking A to A′ and B to B′; one of these is of the first kind and the other of the second, and they are also distinguished by the fact that they take a chosen half-plane bounded by the line (AB) into different half-planes bounded by (A′B′) (Figure 8.9).

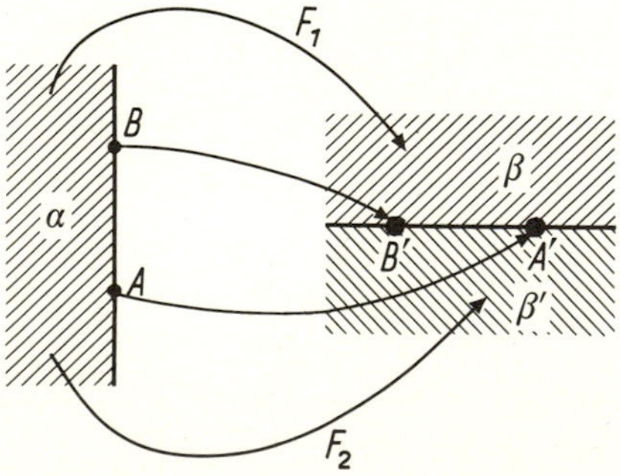

Figure 8.9

Proof. Consider some line ℓ in the plane, and the two half-planes α and α' bounded by ℓ. Under any motion F, where can the line ℓ and the half-planes α and α' go to?

Let's choose a point C in the half-plane α and the point C′ in α'

obtained by reflecting C in ℓ (Figure 8.10). Then clearly, a point X of the half-plane α will satisfy the inequality $|XC| < |XC'|$, a point of the line ℓ the equality $|XC| = |XC'|$, and a point X of the half-plane α' the inequality $|XC| > |XC'|$. Since distances are preserved under motions, the point F(X) will in the respective cases satisfy

$$|F(X)F(C)| < |F(X)F(C')|, \quad |F(X)F(C)| = |F(X)F(C')| \text{ or } |F(X)F(C)| > |F(X)F(C')|.$$

It follows from this that under the motion F, the line ℓ goes into a line, the perpendicular bisector of the line segment [F(C)F(C')], and the half-planes α and α' bounded by ℓ go into the half-planes β and β' bounded by this line and containing respectively the points F(C) and F(C') (see Figure 8.10).

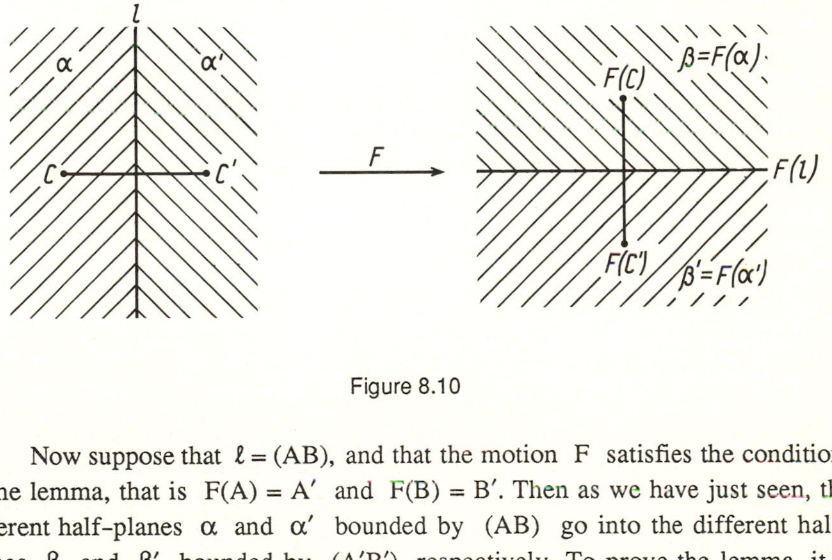

Figure 8.10

Now suppose that $\ell = (AB)$, and that the motion F satisfies the conditions of the lemma, that is $F(A) = A'$ and $F(B) = B'$. Then as we have just seen, the different half-planes α and α' bounded by (AB) go into the different half-planes β and β' bounded by (A'B') respectively. To prove the lemma, it is enough to show that if G is another motion satisfying the same conditions, that is, $G(A) = A'$, $G(B) = B'$, $G(\alpha) = \beta$ and $G(\alpha') = \beta'$ then $G = F$.

Let X be any point of the plane, to be definite, say lying in α. Then since F and G are motions, and $F(A) = G(A) = A'$, $F(B) = G(B) = B'$, we have

$$|F(X)A'| = |G(X)A'| = |XA|, \quad |F(X)B'| = |G(X)B'| = |XB|. \qquad (1)$$

Suppose that $F(X) \neq G(X)$; then the points A' and B' lie on the perpendicular bisector of the line segment [F(X)G(X)], since (1) show that they are equidistant from F(X) and G(X), that is, the line (A'B') coincides with this perpendicular bisector. It follows from this that F(X) and G(X) are in different half-planes β and β' bounded by (A'B'), but this contradicts the fact that X lies in the half-plane α and $F(\alpha) = G(\alpha) = \beta$. Thus $F(X) = G(X)$ for every point X, that is $F = G$. The lemma is proved.

Theorem 1 (M. Chasles (1793-1880)). Every motion of the plane is a translation, rotation or glide reflection.

Proof. Let H be any motion of the plane. Consider two points A and B and their images $A' = H(A)$ and $B' = H(B)$. Since H is a motion, $|A'B'| = |AB| \neq 0$. By Lemmas 1 and 2, there exists both a translation or rotation and a glide reflection taking A to A' and B to B', and these motions are different, since one is of the first kind and the other of the second. By Lemma 3, H must coincide with one of them. The theorem is proved.

Exercises

1. Deduce from the definition of motion that a motion F maps (i) a line segment [AB] onto the whole line segment [F(A)F(B)]; (ii) a line (AB) onto the whole line (F(A)F(B)); (iii) a ray emanating from A onto a whole ray emanating from F(A); (iv) a half-plane bounded by a line (AB) onto a whole half-plane bounded by (F(A)F(B)); (v) a disc D(A, R) onto the whole disc D(F(A), R). [**Hint**: as in the proof of Lemma 3, characterise sets of the indicated type in terms of distances, for example, the segment [AB] consists precisely of the points X for which $|AX| + |XB| = |AB|$, and then use the fact that a motion preserves distance and maps the plane onto the whole plane. The case of the disc will be treated before the statement of §10, Lemma 1.]
2. Using Chasles' theorem, determine which motions of the plane (a) fix the points of some line; (b) fix exactly one point; (c) do not have any fixed points, but take exactly one line to itself; (d) do not have any fixed points, but take more than one line to itself; (e) turn each ray through the same angle; (f) take every line into a line parallel to itself; (g) take every ray into a ray parallel to itself; (h) fix a given point; (i) take a given line into itself.
3. For two given intervals [AB] and [A'B'] of the same length, give another method of finding the motions taking A to A' and B to B', using the facts that the centre of the rotation lies on the perpendicular bisector to the line segment joining a point and its image, and that the axis of a glide reflection passes through its midpoint.
4. Let T_a be a translation in a vector a and S_ℓ the reflection in a line ℓ. Prove that T_aS_ℓ is a glide reflection; what is its axis? The same question for $S_\ell T_a$. For what positions of the vector a will the motions T_aS_ℓ and $S_\ell T_a$ be reflections? When will $T_aS_\ell = S_\ell T_a$? [Hint: represent a as a sum of two vectors, one along ℓ and the other orthogonal to ℓ.]

8.2. Classification: Generalities and groups of Type I and II. We can now proceed to the main problem of this section, namely, that of finding all the uniformly discontinuous groups of motions of the plane. The complete list will be given at the end of §8.3, Theorem 7.

Let's start by excluding those motions which clearly cannot belong to any uniformly discontinuous group of motions. For this, we prove the following general assertion.

Proposition 1. Let F be a motion belonging to a uniformly discontinuous group, and not the identity; then F does not have fixed points, that is, points O such that F(O) = O.

Proof. Suppose that F is a motion belonging to a uniformly discontinuous group, and that F(O) = O for some point O. Let X be any point of the plane for which $|OX| < d/2$, where d is the number of Property III in the definition of a uniformly discontinuous group in §7.1; then

$$|OX| = |F(O)F(X)| = |OF(X)|,$$

and hence also $|OF(X)| < d/2$. From this, by the triangle inequality, we get

$$|XF(X)| \leq |XO| + |OF(X)| < d.$$

But then by definition of a uniformly discontinuous group, it follows that X = F(X). Since X is any point with $|OX| < d/2$, that is, any point within a disc of radius $d/2$ centred at O, the arguments given during the proof of §7.1, Theorem 1 (see Figure 7.4) show that F is the identity motion. The same thing can also be deduced from §8.1, Lemma 3, since there are three non-collinear points A, B and C inside the disc; both F and the identity motion E take A and B into themselves and the half-space containing C bounded by (AB) into itself (since C goes to itself). By Lemma 3, F = E. This proves the proposition.

From this proposition and Chasles' theorem we get.

Corollary. A uniformly discontinuous group of motions of the plane can only contain translations and glide reflections with non-zero translation vectors.

Indeed, any rotation has a fixed point, the centre of the rotation; the rotation through a zero angle is the identity motion, and is the same thing as a translation in a zero vector. A glide reflection with zero vector (that is, a reflection in a line) is not the identity motion, but it has fixed points along its axis. The corollary is proved.

Let Γ be a uniformly discontinuous group of motions of the plane. Write Γ' for the set of all translations in Γ. Obviously Γ' is again a group: if F and G belong to Γ' then FG belongs to Γ (since Γ is a group), and FG is a translation (since both F and G are); hence FG belongs to Γ'. In a similar way one sees easily that if F belongs to Γ' then so does F^{-1}. If Γ' does not exhaust the whole of Γ, then Γ contains at least one further motion, which by Corollary 1 must be a glide reflection. Let S be any glide reflection in Γ. Consider all possible motions of the form TS, where T is any translation in Γ'. Obviously all of these are contained in Γ (by definition of a group), and by Chasles' theorem they must be glide reflections, since they are of the second kind. Let's show that every glide reflection in Γ can be obtained in this way. In fact if S' is a glide reflection in Γ, consider the motion $F = S'S^{-1}$; firstly, it belongs to Γ. Secondly, S^{-1} is also a glide reflection, and hence $S'S^{-1}$ is a motion of the first kind (by the rules for composites of motions of the first and second kinds given at the end of §8.1). But by Corollary 1, the only motion of the first kind which can be contained in a uniformly discontinuous group is a translation. Hence $F = T$ is a translation, and so belongs to Γ'. Let X be any points of the plane and $Y = S(X)$; then $X = S^{-1}(Y)$ and $T(Y) = S'(S^{-1}(Y)) = S'(X)$. Thus $T(S(X)) = S'(X)$, and since this holds for any point X, we have $S' = TS$. This proves the following result.

Corollary 2. Let Γ be a uniformly discontinuous group of motions. Then
either Γ consists only of translations;
or Γ contains a uniformly discontinuous group Γ' consisting of translations, and if S is any glide reflection in Γ then the other motions of Γ not in Γ' are all glide reflections of the form TS with T in Γ'.

In what follows, we divide up all uniformly discontinuous groups into three types according to how many translations they contain. In Type I we put the groups which do not contain any translations other than the identity (the translation in the zero vector). In Type II we put the groups which are not of Type I, but which contain translations in collinear vectors only. Finally, in Type III we put the groups which do not belong to Type I or II. Thus a group of Type III must contain at least two translations in non-collinear vectors. We deal with these three types in succession.

Type I. Let's prove that in this case, the group consists just of the identity. Indeed, as we have seen in Corollary 1, any motion belonging to a uniformly discontinuous group is either a translation, or a glide reflection with non-zero translation vector, and by assumption, the group does not contain any translations other than the identity motion. Thus if the group contains a motion F other than the identity, F must be a glide reflection with translation vector $\mathbf{a} \neq 0$. Since F belongs to our group, the motion FF obtained by performing F twice in succession is also in the group. But as the reader can easily see for himself, FF is a translation in the vector $2\mathbf{a}$, and $2\mathbf{a} \neq 0$, since $\mathbf{a} \neq 0$.

We arrive at a contradiction to the assumption that our group is of Type I. We state the result we have proved.

Theorem 2. A uniformly discontinuous group of motions of the plane of Type I consists only of the identity motion.

Notice that the locally Euclidean geometry corresponding to this group is of course the Euclidean plane itself.

Type II. We start by studying the group Γ' of translations contained in a group Γ

of Type II.

We fix some point O, and lay off from O all the vectors **a** for which the translation $T_{\mathbf{a}}$ belongs to Γ′; by assumption, the set of points $T_{\mathbf{a}}(O)$ (the endpoints of the **a**) is contained in a line ℓ, and since Γ′ is uniformly discontinuous, by §7.2, Lemma 1, there are only finitely many of these contained in some interval of ℓ. Thus among these, there exists one point closest to O, hence among the vectors **a**, there is a shortest non-zero vector; write **c** for this. By definition of a group, Γ′ must contain the translations

$$T_{\mathbf{c}}T_{\mathbf{c}} = T_{2\mathbf{c}},\ T_{2\mathbf{c}}T_{\mathbf{c}} = T_{3\mathbf{c}},\ (T_{\mathbf{c}})^{-1} = T_{-\mathbf{c}},$$

and more generally, all the translations $T_{m\mathbf{c}}$ for any integer m.

Let's prove that these exhaust all the translations contained in Γ′. Suppose that this is not the case, and that Γ′ contains a translation $T_{\mathbf{b}}$ with $\mathbf{b} \neq m\mathbf{c}$ for any integer m. Laying off from O the endpoints of the vectors m**c**, we get a series of points on ℓ, dividing it up into intervals of equal length (see Figure 8.11). The endpoint of **b** falls into one of these intervals, and by our assumption on **b**, into an

Figure 8.11

interior point of an interval. Suppose that it falls between the endpoints of m**c** and (m+1)**c**; then the vector $\mathbf{c}' = \mathbf{b} - m\mathbf{c}$ is shorter than **c**. On the other hand, by definition of a group, $T_{\mathbf{c}'} = T_{\mathbf{b}}T_{m\mathbf{c}}^{-1}$ belongs to Γ′. We have obtained a contradiction to **c** being shortest among all non-zero vectors **a** for which the translation $T_{\mathbf{a}}$ belongs to Γ′.

We have thus shown that Γ′ is of a very simple form: it consists precisely of translations in vectors m**c**, where **c** is a fixed non-zero vector, and m runs through the integers. At this point, the study of groups of Type II ramifies according to two possibilities.

Type II.a. Γ consists only of translations; in this case Γ = Γ′, and we have determined its structure.

Theorem 3. A uniformly discontinuous group of Type II.a is determined by a non-zero vector **c**, and consists of translations in vectors of the form m**c**, for m any integer.

This group is exactly of the type which appeared in the construction of geometry on the cylinder in §3.

Type II.b. Γ contains motions which are not translations. We know by Corollary 1 that such a motion is a glide reflection with non-zero translation vector.

The translations in Γ themselves form a group Γ', and Γ' is of Type II.a; it therefore consists of translations of the form $T_{m\mathbf{c}}$ where $\mathbf{c}$ is some non-zero vector, and m is any integer. Let $S_\ell^{\mathbf{a}}$ be a glide reflection of Γ. As we have already seen, the motion $S_\ell^{\mathbf{a}}S_\ell^{\mathbf{a}}$ is obviously the translation $T_{2\mathbf{a}}$; by definition of a group, it must be in Γ, and hence in Γ', and therefore

$$2\mathbf{a} = m\mathbf{c} \tag{2}$$

for some integer m. It follows in particular that the vectors $\mathbf{a}$ and $\mathbf{c}$ are parallel, and since $\mathbf{a}$ is parallel to the axis of the glide reflection $S_\ell^{\mathbf{a}}$, this axis ℓ is parallel to $\mathbf{c}$.

Let's determine the parity of m in (2). If $m = 2k$, then $\mathbf{a} = k\mathbf{c}$. Consider then the motion

$$F = T_{-k\mathbf{c}}S_\ell^{\mathbf{a}},$$

which must belong to Γ. The reader will easily see (in view of $\mathbf{a} = k\mathbf{c}$) that F is simply the reflection in ℓ. However, by Corollary 1, such a motion cannot belong to a uniformly discontinuous group, and we arrive at a contradiction.

Thus m is odd. Suppose that $m = 2k + 1$, that is $2\mathbf{a} = (2k+1)\mathbf{c}$, so that $\mathbf{a} = k\mathbf{c} + \mathbf{c}/2$; by definition of a group, $S' = T_{-k\mathbf{c}}S_\ell^{\mathbf{a}}$ belongs to Γ. The reader will easily verify that S' is a glide reflection with axis ℓ and translation vector $\mathbf{a} - k\mathbf{c} = \mathbf{c}/2$, that is $S' = S_\ell^{\mathbf{c}/2}$. On the other hand, by Corollary 2, every motion of Γ is either a translation belonging to Γ' (and so of the form $T_{m\mathbf{c}}$), or a glide reflection of the form $TS_\ell^{\mathbf{c}/2}$ for some T in Γ'; then since $T = T_{n\mathbf{c}}$ for some integer n, we have $TS_\ell^{\mathbf{c}/2} = T_{n\mathbf{c}}S_\ell^{\mathbf{c}/2} = S_\ell^{n\mathbf{c}+\mathbf{c}/2}$.

Theorem 4. A uniformly discontinuous group Γ of Type II.b is determined by a line ℓ and a non-zero vector $\mathbf{c}$ parallel to ℓ. It consists of the translations of the form $T_{m\mathbf{c}}$ and the glide reflections of the form $T_{n\mathbf{c}}S_\ell^{\mathbf{c}/2} = S_\ell^{n\mathbf{c}+\mathbf{c}/2}$, where m and n are any integers.

We have thus obtained the group Γ which appeared in the construction of geometry on the twisted cylinder in §4.

Exercise

1. Determine all uniformly discontinuous groups of motions of the line.

8.3. Classification: groups of Type III. As in the case of groups of Type II, we consider first the case III.a, when the group Γ consists of translations only. We will prove that a 2-dimensional version of Theorem 3 then holds.

Theorem 5. A uniformly discontinuous group of Type III.a is determined by two non-collinear vectors **a** and **b**, and consists of all translations in vectors of the form $m\mathbf{a} + n\mathbf{b}$, where m and n are any integers.

Proof. We choose some point O, and draw in the plane the set M of all points equivalent to O under the notion of equivalence corresponding to Γ, as defined in §7.2; M is the set of points of the form $T_\mathbf{x}(O)$ for all translations $T_\mathbf{x}$ in Γ. The set M has an important, although almost obvious, property which we will use throughout what follows. Let P and Q be two points of M; by definition, since P and Q are equivalent, there exists a translation $T_\mathbf{x}$ in Γ such that $T_\mathbf{x}(P) = Q$. Obviously then $\mathbf{x} = \overrightarrow{PQ}$. Suppose that R is a third point of M. Then $T_\mathbf{x}$ is in Γ, so that $T_\mathbf{x}(R)$ is equivalent to R, and is therefore also in M. Thus M goes into itself under a translation in any vector $\mathbf{x}$ whose starting point and endpoint are in M.

Draw any line ℓ through O, passing through at least one further point of M; denote by Γ_1 the set of all translations $T_\mathbf{x}$ in Γ in vectors $\mathbf{x}$ parallel to ℓ. The composite of two such translations $T_\mathbf{x}$ and $T_\mathbf{y}$ is a translation $T_{\mathbf{x}+\mathbf{y}}$ with the same property. In exactly the same way, for any translation $T_\mathbf{x}$ in Γ_1, the translation $T_\mathbf{x}^{-1} = T_{-\mathbf{x}}$ belongs to Γ_1. Thus Γ_1 is a group, and since it is contained in the uniformly discontinuous group Γ, it is itself uniformly discontinuous. Γ_1 is obviously of Type II.a, as considered in Theorem 3. Hence by Theorem 3, Γ_1 consists of the translations $T_{n\mathbf{a}}$ for any integer n, where **a** is the shortest of the non-zero vectors $\mathbf{x}$ for which $T_\mathbf{x}$ belongs to Γ_1. Therefore, the points of M lying on ℓ form an equally spaced series of points, the endpoints of the vectors $n\mathbf{a}$ laid off from O.

As we will prove later, the vector **a** is one of the two vectors of the conclusion of Theorem 5 (that is, **a** or **b**). To find the other vector **b**, draw a line ℓ' through any point P of M parallel to ℓ. By the basic property of M stated at the beginning of the proof, M goes into itself under the translation in $\overrightarrow{OP}$; under it, the series of points on ℓ is taken into a series on ℓ'. Since conversely, M goes into itself under the translation in $\overrightarrow{PO}$, which takes ℓ' back into ℓ, there are no points of M on ℓ' other than the points of this series.

Thus for any line ℓ' parallel to ℓ and passing through a point of M, the points of M on ℓ' form a series which is a translation of that on ℓ (Figure 8.12). Let $d = |\mathbf{a}|$, and consider a strip Φ of width d, bounded by two parallel lines through the ends O and A of the vector **a**, and perpendicular to ℓ. Since ℓ' contains a series of equally spaced points at a distance d, at least one of these points will be contained in or on the boundary of Φ. This has the important corollary: among all the lines ℓ' passing through a point of M, and parallel to ℓ but distinct from it, there exists one which is closest to ℓ.

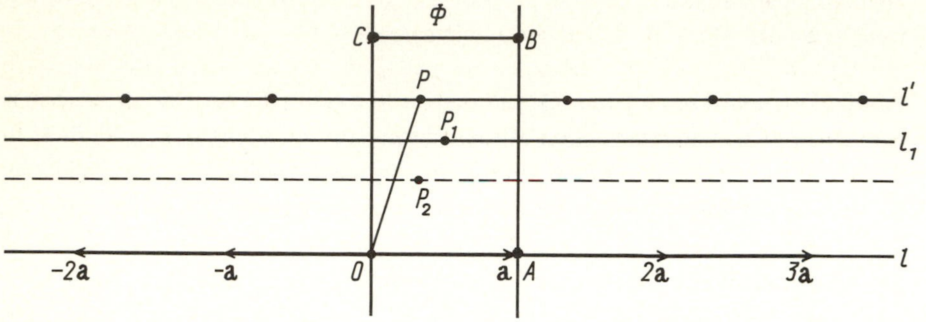

Figure 8.12

To prove this, cut out of the strip Φ a rectangle OABC which is big enough so that it meets at least one of the lines ℓ'. Since a rectangle is a bounded set, by §7.2, Lemma 1, it contains only a finite number of points of M. In particular, there is some point P_1 in OABC closest to ℓ. The line ℓ_1 through P_1 will be the line we are looking for. Indeed, if there existed a line ℓ_2 containing at least one point of M, and parallel to ℓ, but closer to ℓ than ℓ_1, then by what we have just said, ℓ_2 would contain a point P_2 belonging to the rectangle OABC, which would then be closer to ℓ than P_1, and this would contradict the choice of P_1.

Let P_1 be a point of the line ℓ_1 we have just constructed. Let's write **b** for the vector $\vec{OP_1}$, and prove that together with the previously constructed vector **a**, **b** satisfies the requirements of Theorem 5. Let $T_{\mathbf{x}}$ be any translation belonging to Γ. Laying off **x** from O, we get a point Q, such that $\mathbf{x} = \vec{OQ}$; we draw the line ℓ' through Q parallel to ℓ. If on the other hand, we lay off from O the vectors **b**, 2**b**, 3**b**, ... , we get points $B_1, B_2, B_3, \ldots$; through each of these points, we also draw a line ℓ_m parallel to ℓ (Figure 8.13).

Let's prove that ℓ' is one of the ℓ_m. By contradiction, assume that this is false, and that ℓ' intersects the line (OB_1) in a point contained between B_{m-1} and B_m. Translate M by the vector (1-m)**b**; then M goes into itself, and ℓ' goes into a line ℓ'' which is closer to ℓ than ℓ_1, which contradicts the construction of ℓ_1.

Thus ℓ' must be one of the lines ℓ_m. We translate M by m**b**. This takes ℓ' $= \ell_m$ into ℓ, and therefore takes Q into some point P of M lying on ℓ. As we have seen, P is then the endpoint of some vector n**a**. In vector notation, the relation between the vectors **x** (with endpoint Q), m**b** (with endpoint B_m) and n**a** (with endpoint P) is of the form

$$\mathbf{x} - m\mathbf{b} = n\mathbf{a};$$

that is $\mathbf{x} = n\mathbf{a} + m\mathbf{b}$. This proves Theorem 5.

Remark. Theorem 5 proves that the set M, obtained by laying off from some point O all the vectors $\mathbf{x}$ such that $T_{\mathbf{x}}$ belongs to Γ, forms the lattice of vertices of equal parallelograms, one of which is constructed on the vectors **a** and **b** (Figure 8.14). This is a 2-dimensional generalisation of the picture on a line which we had in Theorem 3 (Figure 8.11). But passing from the 1-dimensional case to the 2-dimensional case, a new phenomenon also appears. In the case of the series of equally spaced points on the line, obtained by laying off the vectors $n\mathbf{c}$ from O,

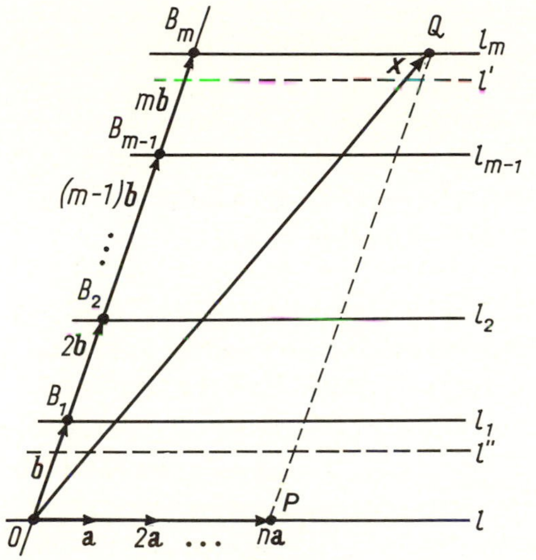

Figure 8.13

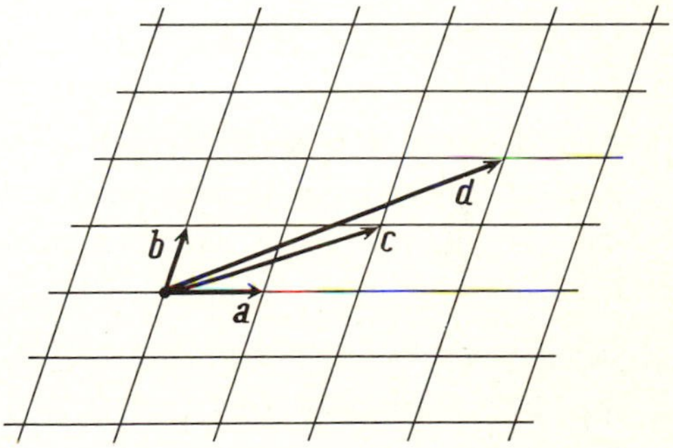

Figure 8.14

the generator **c** is determined almost uniquely: the only ambiguity is that we could take $-\mathbf{c}$ instead of **c**; indeed, **c** was characterised as the shortest of all (non-zero) vectors $\mathbf{x}$ for which $T_{\mathbf{x}}$ belongs to Γ, that is, its endpoint is the nearest point of the series to O. However, in the plane, the same lattice of points (that is, the same translation group Γ) can be obtained from different pairs of vectors; for example, in the case illustrated in Figure 8.14, the pairs (**a**, **b**), (**a**, **c**) and (**c**, **d**) all generate the same lattice.

The proof of Theorem 5 not only establishes the existence of a pair of vectors (**a**, **b**) as required, but it also determines which pairs of vectors are generators. That is, **a** must be a shortest vector among all non-zero vectors $\mathbf{x}$

collinear to $\mathbf{a}$ for which the translation $T_{\mathbf{x}}$ is in Γ. In other words, if $\mathbf{a} = \overrightarrow{OP}$, then the line segment [OP] must not contain any points of M other than O and P; in particular, we could take P to be any point of M closest to M. If $\mathbf{a}$ is already chosen, then $\mathbf{b}$ must be taken to be a vector non-collinear to $\mathbf{a}$ such that $T_{\mathbf{b}}$ is in Γ, and such that the translation $\ell' = T_{\mathbf{b}}(\ell)$ of the line $\ell = (OP)$ is a line ℓ' closest to ℓ among all lines of the form $T_{\mathbf{x}}(\ell)$, where $T_{\mathbf{x}}$ is a translation in Γ. In other words, if $\mathbf{b} = \overrightarrow{OQ}$ and ℓ' is the line parallel to ℓ through Q, then ℓ' must be a nearest line to ℓ of all lines parallel to ℓ and passing through a point of M. In particular, we can take Q to be the point of ℓ' in the strip Φ of Figure 8.13 (or on its left-hand boundary line), that is, it can be chosen so that the orthogonal projection of the vector $\mathbf{b}$ onto the direction of $\mathbf{a}$ should be shorter than $\mathbf{a}$.

Now we proceed to the final case of groups of Type III.b. Here the result is very similar to that of Theorem 4.

Theorem 6. A group Γ of Type III.b is determined by a line ℓ, a vector $\mathbf{a}$ parallel to ℓ, and a vector $\mathbf{b}$ perpendicular to ℓ; the group consists of the translations $T_{\mathbf{x}}$, where $\mathbf{x}$ is a vector of the form $m\mathbf{a} + n\mathbf{b}$, for m and n any integers, and of the glide reflections $T_{\mathbf{x}}S_\ell^{\mathbf{a}/2}$, where $\mathbf{x}$ is a vector of the same form.

Proof. Consider the group Γ' of translations in Γ. By assumption this does not exhaust the whole group, so that Γ must contain some glide reflection S in some line ℓ. As we saw, SS is a translation in a vector $\mathbf{f}$, that is $SS = T_{\mathbf{f}}$, with $T_{\mathbf{f}}$ in Γ'. As in the proof of Theorem 5, let Γ_1 be the group of translations $T_{\mathbf{x}}$ contained in Γ' and such that $\mathbf{x}$ is collinear to $\mathbf{f}$. As we saw, all $\mathbf{x}$ in Γ_1 are given by $\mathbf{x} = m\mathbf{a}$, where $\mathbf{a}$ is the shortest non-zero vector with this property; in particular, $\mathbf{f} = m\mathbf{a}$. Now we argue as in the proof of Theorem 4. If m were even, say $m = 2k$, then the motion $T_{-k\mathbf{a}}S$ would be the reflection in ℓ, and no such motion is allowed in our group; therefore m is odd. Set $m = 2k + 1$, and $S_1 = T_{-k\mathbf{a}}S$; then as in the proof of Theorem 4, $S_1 = S_\ell^{\mathbf{a}/2}$. On the other hand, we have chosen $\mathbf{a}$ as indicated in the remark after Theorem 5: it is the shortest vector collinear to $\mathbf{a}$ (or equivalently, to $\mathbf{f}$) for which the corresponding translation is in Γ'. Hence we can adjoin a vector $\mathbf{b}$ to $\mathbf{a}$ such that all the translations $T_{\mathbf{x}}$ belonging to Γ' have vectors $\mathbf{x} = m\mathbf{a} + n\mathbf{b}$, where m and n are any integers.

We now recall that by Corollary 2, any element of Γ must be either a translation, and hence in Γ', or a glide reflection, and hence can be represented in the form $T_{\mathbf{x}}S_\ell^{\mathbf{a}/2}$, with $T_{\mathbf{x}}$ an element of Γ'. In other words, Γ is a union of two sets: the translations $T_{\mathbf{x}}$ and the glide reflections $T_{\mathbf{y}}S_\ell^{\mathbf{a}/2}$, where $T_{\mathbf{y}}$ and $T_{\mathbf{x}}$ are in Γ'. However, if we just choose a translation group Γ' and a glide reflection $S_\ell^{\mathbf{a}/2}$ completely independently, then this union of two sets need not be a group. Indeed, the composite $S_\ell^{\mathbf{a}/2}T_{\mathbf{x}}$ of $S_\ell^{\mathbf{a}/2}$ and $T_{\mathbf{x}}$ in the other order must also be contained in Γ for any $T_{\mathbf{x}}$ in Γ'; we will see below that this condition is also sufficient to get a group. For any glide reflection $S_\ell^{\mathbf{c}}$ and any translation $T_{\mathbf{x}}$, the reader will easily check the equality

$$S_\ell^c T_{\mathbf{x}} = T_{\mathbf{x}'} S_\ell^c, \qquad (3)$$

where $\mathbf{x}'$ is the vector obtained by reflecting $\mathbf{x}$ in the line ℓ; for instance, it is easy to see that for any X, the points $S_\ell^c(T_{\mathbf{x}}(X)) = T_{\mathbf{x}'}(S_\ell^c(X))$ have the same projections both to ℓ and to a line perpendicular to ℓ.

The motion $T_{\mathbf{x}'} S_\ell^c$ must be an element of Γ, since the right-hand side of (3) is. Hence by Corollary 2, it can be written in the form $T_{\mathbf{y}} S_\ell^c$, where $T_{\mathbf{y}}$ is a translation in Γ'. It follows that $T_{\mathbf{x}'} = T_{\mathbf{y}}$ is in Γ'. We thus get the condition that if a translation $T_{\mathbf{x}}$ belongs to Γ', then so does $T_{\mathbf{x}'}$. Laying off the vectors $\mathbf{x}$ with $T_{\mathbf{x}}$ in Γ' from a point O chosen to lie on the axis ℓ, we get a lattice M, as in the proof of Theorem 5. Our condition shows that M has reflectional symmetry about some line ℓ through O and parallel to $\mathbf{a}$.

Which lattices have a symmetry of this kind? One answer springs directly to mind: the lattices for which the generating vectors $\mathbf{a}$ and $\mathbf{b}$ can be chosen to be perpendicular (Figure 8.15). But there is another, not so obvious solution: we could put in a further point at the centre of each rectangle of the lattice of Figure 8.15 to get a new lattice (Figure 8.16).

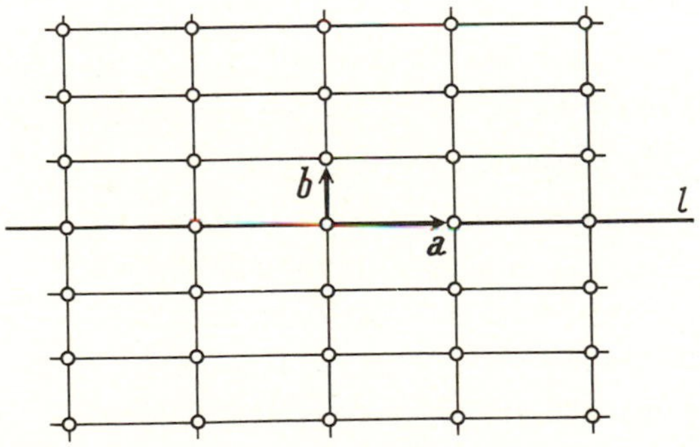

Figure 8.15

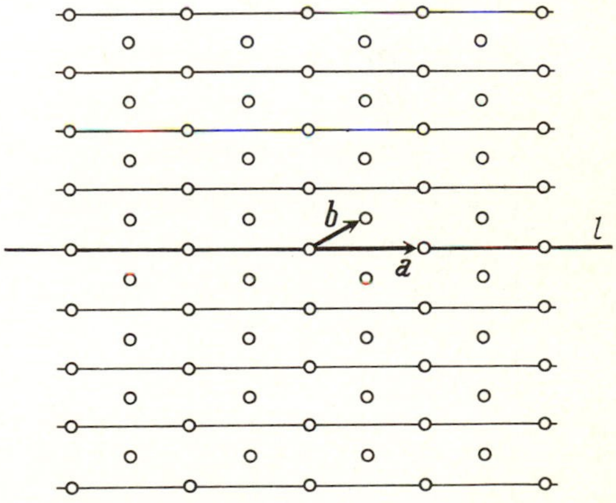

Figure 8.16,

We now prove that only these two types of lattices have the kind of symmetry we are interested in. After this, we will see that the second kind cannot arise in connection with uniformly discontinuous groups. It will follow from this that in our case the first kind occurs, so that $\mathbf{a}$ and $\mathbf{b}$ can be chosen to be perpendicular, and this is the assertion of Theorem 6.

Suppose then that the lattice made up of the endpoints of the vectors $m\mathbf{a} + n\mathbf{b}$ has reflectional symmetry in the line parallel to $\mathbf{a}$. By the remark after Theorem 5, we can take $\mathbf{b}$ so that its projection in the direction of $\mathbf{a}$ is shorter than $\mathbf{a}$. Since the reflected vector $\mathbf{b}'$ has the same length as $\mathbf{b}$, the projection of $\mathbf{b} + \mathbf{b}'$ is shorter than $2\mathbf{a}$. But $\mathbf{b} + \mathbf{b}'$ is collinear to $\mathbf{a}$, and by choice of $\mathbf{a}$, it must be a multiple of $\mathbf{a}$ (Figure 8.17). But since it is shorter than $2\mathbf{a}$,there are just two possibilities: either $\mathbf{b} + \mathbf{b}' = \mathbf{0}$, or $\mathbf{b} + \mathbf{b}' = \mathbf{a}$. In the first case, the projection of $\mathbf{b}$ to the direction of $\mathbf{a}$

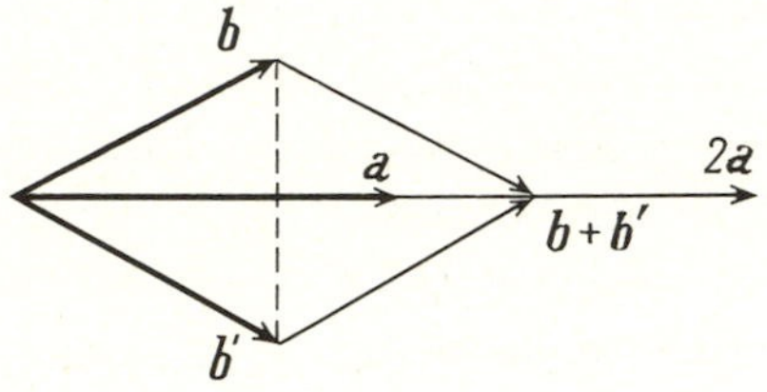

Figure 8.17

is $\mathbf{0}$, and so $\mathbf{a}$ and $\mathbf{b}$ are perpendicular. In the second, the projection is $\mathbf{a}/2$, as in Figure 8.16.

We return now to our uniformly discontinuous group Γ. We have established that Γ' corresponds to one of the two lattices illustrated in Figure 8.15 and Figure 8.16. Suppose that the second case occurs, that is $\mathbf{b} + \mathbf{b}' = \mathbf{a}$. Consider the motion $T_{-\mathbf{b}}S_\ell^{\mathbf{a}/2}$. The reader can see that this is a reflection (in what axis?). But we can also give a different argument. Since $T_{-\mathbf{b}}S_\ell^{\mathbf{a}/2}$ is a motion of the second kind, it must be a glide reflection $S_{\ell'}^{\mathbf{c}}$, and then composing it with itself would give the translation $T_{2\mathbf{c}}$. On the other hand, by (3) we can write $(T_{-\mathbf{b}}S_\ell^{\mathbf{a}/2})(T_{-\mathbf{b}}S_\ell^{\mathbf{a}/2})$ in the form $T_{-\mathbf{b}-\mathbf{b}'}S_\ell^{\mathbf{a}/2}S_\ell^{\mathbf{a}/2} = T_{-\mathbf{a}}T_{\mathbf{a}} = E$. Hence $T_{2\mathbf{c}} = E$, and from this we get $\mathbf{c} = \mathbf{0}$, that is, our glide reflection is just a reflection. However, our group is not allowed to contain a reflection, and therefore the second case is impossible. This proves the harder part of Theorem 6, that every uniformly discontinuous group of Type III.b is given by the method indicated in Theorem 6.

It remains to prove the converse, that the set of motions described in Theorem 6 actually forms a uniformly discontinuous group. First we show that they form a group. For this, recall that every motion of Γ is of the form $T_{\mathbf{x}}$ or $T_{\mathbf{y}}S_\ell^{\mathbf{a}/2}$, where $T_{\mathbf{x}}$ and $T_{\mathbf{y}}$ are translations in Γ'. We need to check that the composite of two such motions is again of the same form. For this we must consider the four motions

$$T_{\mathbf{x}}T_{\mathbf{y}},\ T_{\mathbf{x}}T_{\mathbf{y}}S_\ell^{\mathbf{a}/2},\ T_{\mathbf{y}}S_\ell^{\mathbf{a}/2}T_{\mathbf{x}},\ \text{and}\ T_{\mathbf{x}}S_\ell^{\mathbf{a}/2}T_{\mathbf{y}}S_\ell^{\mathbf{a}/2}.$$

The first two motions are of the form $T_{\mathbf{x}+\mathbf{y}}$ and $T_{\mathbf{x}+\mathbf{y}}S_\ell^{\mathbf{a}/2}$, and are therefore obviously in Γ. To consider the two last motions, we use (3), and the fact that if Γ' contains a translation $T_{\mathbf{x}}$, it also contains the translation $T_{\mathbf{x}'}$ in the reflected vector $\mathbf{x}'$; this follows from the fact that the vectors $\mathbf{a}$ and $\mathbf{b}$ of Theorem 6 are perpendicular, so that if $\mathbf{x} = m\mathbf{a} + n\mathbf{b}$ then $\mathbf{x}' = m\mathbf{a} - n\mathbf{b}$. The result we obtain is

$$T_{\mathbf{y}}S_{\ell}^{\mathbf{a}/2}T_{\mathbf{x}} = T_{\mathbf{y}+\mathbf{x}'}S_{\ell}^{\mathbf{a}/2}$$

and

$$T_{\mathbf{x}}S_{\ell}^{\mathbf{a}/2}T_{\mathbf{y}}S_{\ell}^{\mathbf{a}/2} = T_{\mathbf{x}+\mathbf{y}'}S_{\ell}^{\mathbf{a}/2}S_{\ell}^{\mathbf{a}/2} = T_{\mathbf{x}+\mathbf{y}'+\mathbf{a}},$$

where $\mathbf{y}'$ is the reflection of $\mathbf{y}$; these motions obviously belong to Γ.

In the same way we have to consider the inverse motions of $T_{\mathbf{x}}$ and $T_{\mathbf{y}}S_{\ell}^{\mathbf{a}/2}$. Obviously $T_{\mathbf{x}}^{-1} = T_{-\mathbf{x}}$, and the reader can check in exactly the same way that $(T_{\mathbf{y}}S_{\ell}^{\mathbf{a}/2})^{-1} = T_{-\mathbf{a}-\mathbf{y}'}S_{\ell}^{\mathbf{a}/2}$, since $T_{-\mathbf{a}-\mathbf{y}'}S_{\ell}^{\mathbf{a}/2}T_{\mathbf{y}}S_{\ell}^{\mathbf{a}/2} = E$; these inverses obviously belong to Γ.

Now let's check that Γ is a uniformly discontinuous group. Let X be any point of the plane, and F any motion in Γ not equal to the identity; we consider the distance $|XF(X)|$. If F is a translation then $F = T_{m\mathbf{a}+n\mathbf{b}}$, and

$$|XF(X)| = |m\mathbf{a} + n\mathbf{b}| = \sqrt{m^2|\mathbf{a}|^2 + n^2|\mathbf{b}|^2} \geq r,$$

where r is the smaller of $|\mathbf{a}|$ and $|\mathbf{b}|$, since the vectors $\mathbf{a}$ and $\mathbf{b}$ are perpendicular, and at least one of the integers m and n is non-zero. If F is a glide reflection then $F = T_{m\mathbf{a}+n\mathbf{b}}S_{\ell}^{\mathbf{a}/2}$ with integers m and n, and

$$|XF(X)| \geq |X^{\sim}F(X)^{\sim}| = |m\mathbf{a} + \mathbf{a}/2| = |m + 1/2|\cdot|\mathbf{a}| \geq |\mathbf{a}|/2,$$

where $X^{\sim}$ and $F(X)^{\sim}$ are the projections of X and $F(X)$ to the line ℓ, since $\mathbf{a}$ is parallel to ℓ, $\mathbf{b}$ is perpendicular to ℓ and m is an integer. Therefore for every non-identical motion F in Γ and any point X in the plane, we have the inequality

$$|XF(X)| \geq d > 0,$$

where d is the smaller of the two numbers $|\mathbf{a}|/2$ and $|\mathbf{b}|$. Since d is independent of the choice of X and F, the group Γ is uniformly discontinuous.

This completes the proof of Theorem 6.

The result of Theorem 6 might leave the reader with a certain feeling of incompleteness, on account of the fact that the glide reflections of a group Γ of Type III.b are not described in Theorem 6 in the usual way, by specifying their axes and vectors, but as a composite $T_{\mathbf{x}}S_{\ell}^{\mathbf{a}/2}$ of two motions, where $\mathbf{x}$ and $\mathbf{a}$ are vectors of the given form. However, they can be described without difficulty in the usual way as follows, by finding the axes and vectors of these composites.

Proposition 2. In the notation of Theorem 6, for any integers m and n we have

$$T_{\mathbf{x}} S_{\ell}^{\mathbf{a}/2} = S_{\ell(n)}^{m\mathbf{a}+\mathbf{a}/2},$$

where $\mathbf{x} = m\mathbf{a} + n\mathbf{b}$ and $\ell(n) = T_{n\mathbf{b}/2}(\ell)$.

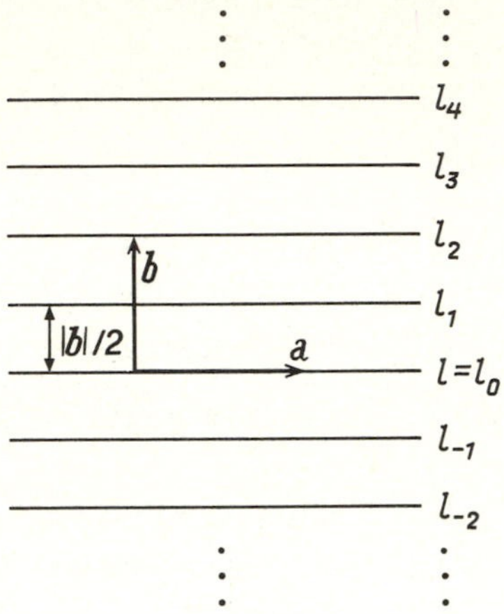

Figure 8.18

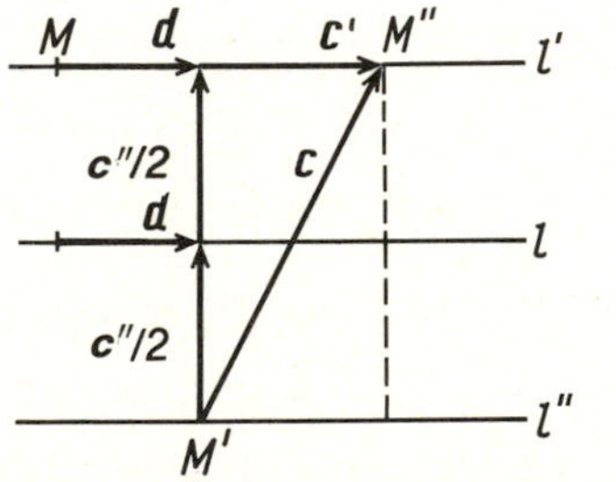

Figure 8.19

Thus the glide reflections of Γ have as their axes parallel lines $\ell(n)$ dividing the plane into strips of equal width $|\mathbf{b}|/2$, and as translation vectors $m\mathbf{a} + \mathbf{a}/2$, where m and n run independently through the integers (Figure 8.18). The proof of the proposition follows at once from the following lemma, since $\mathbf{a}$ is parallel to ℓ and $\mathbf{b}$ orthogonal to ℓ.

Lemma 4. The composite of a glide reflection and a translation is given by

$$T_{\mathbf{c}} S_{\ell}^{\mathbf{d}} = S_{\ell'}^{\mathbf{d}+\mathbf{c}'}, \text{ where } \ell' = T_{\mathbf{c}''/2}(\ell);$$

here $\mathbf{c}'$ and $\mathbf{c}''$ are respectively the projections of $\mathbf{c}$ to the axis ℓ and to a line orthogonal to ℓ (Figure 8.19).

Proof. Note first that $T_{\mathbf{c}} S_{\ell}^{\mathbf{d}}$ is a motion of the second kind, since $T_{\mathbf{c}}$ is of the first kind and $S_\ell^{\mathbf{d}}$ is of the second kind; thus by Chasles' theorem, $T_{\mathbf{c}}S_\ell^{\mathbf{d}}$ is a glide reflection, and it only remains to determine its axis and translation vector. The reader will easily check that the axis of a glide reflection with non-zero translation vector is the unique line of the plane which goes to itself under the motion, and that all points of this line are translated by the same vector, which will then obviously be the translation vector of the glide reflection. The line $\ell' = T_{\mathbf{c}''/2}(\ell)$ obviously goes to itself under $T_{\mathbf{c}}S_\ell^{\mathbf{d}}$, since the glide reflection $S_\ell^{\mathbf{d}}$ takes ℓ' into the line ℓ'', and the translation $T_{\mathbf{c}}$ takes ℓ'' into ℓ' (see Figure 8.19). A point M on ℓ' is translated by the vector $\mathbf{d} + \mathbf{c}'$, (see Figure 8.19, where $S_\ell^{\mathbf{d}}$ takes M into M′, and $T_{\mathbf{c}}$ takes M′ into M″). This proves that $\ell' = T_{\mathbf{c}''/2}(\ell)$ is the axis of $T_{\mathbf{c}}S_\ell^{\mathbf{d}}$ and $\mathbf{d} + \mathbf{c}'$ its translation vector. The lemma is proved.

This completes the classification of all uniformly discontinuous groups of motion of the plane: they are precisely the groups of Types I, II.a, II.b, III.a and III.b described in Theorems 2-6.

If we attempt to describe all of these groups in a single theorem by just writing out Theorems 2–6 one after another, then the length of the resulting theorem should convince us that this is not a reasonable approach. Despite this, this can all be done using a different description of the groups. The point is that in Theorems 2-6, we specified the groups by simply listing all of their elements, and there is no particular need to do this. The reader can convince himself that any motion of a group of Type II.a can be obtained as a composite of the translation $T_{\mathbf{c}}$ and its inverse any number of times, so that for example

$$T_{3\mathbf{c}} = T_{\mathbf{c}}T_{\mathbf{c}}T_{\mathbf{c}}, \text{ and } T_{-3\mathbf{c}} = (T_{\mathbf{c}})^{-1}(T_{\mathbf{c}})^{-1}(T_{\mathbf{c}})^{-1}.$$

In exactly the same way, the motions of a group of Type II.b can all be obtained as composites of the glide reflection $S_\ell^{\mathbf{c}/2}$ and its inverse any number of times; those of a group of Type III.a as composites of any number of translations $T_{\mathbf{a}}$ and $T_{\mathbf{b}}$ and their inverses, for example

$$T_{2\mathbf{a}-3\mathbf{b}} = T_{\mathbf{a}}T_{\mathbf{a}}(T_{\mathbf{b}})^{-1}(T_{\mathbf{b}})^{-1}(T_{\mathbf{b}})^{-1};$$

and those of a group of Type III.b as composites of the glide reflection $S_\ell^{\mathbf{a}/2}$ and the translation $T_{\mathbf{b}}$ and their inverses.

In the general case, to specify a group of motions Γ, it is enough to specify some elements $F_1, \dots, F_k$ of Γ such that all the remaining motions of Γ can be obtained as a composite of any number of $F_1, \dots, F_k$ and their inverses. In this situation we say that Γ is *generated* by the motions $F_1, \dots, F_k$, and write $\Gamma = \langle F_1, \dots, F_k \rangle$; the motions $F_1, \dots, F_k$ are called *generators* of Γ. Thus to determine a group Γ it is sufficient to specify its generators.

As an example, let's prove that as mentioned above, a group of Type III.b is generated by the motions $S_\ell^{\mathbf{a}/2}$ and $T_{\mathbf{b}}$. Indeed,

$$S_\ell^{\mathbf{a}/2} S_\ell^{\mathbf{a}/2} = T_{\mathbf{a}}, \quad (T_{\mathbf{b}})^{-1} = T_{-\mathbf{b}}, \quad \text{and} \quad (T_{\mathbf{a}})^{-1} = T_{-\mathbf{a}}.$$

Considering the different composites of $T_{\mathbf{a}}, T_{-\mathbf{a}}, T_{\mathbf{b}}$ and $T_{-\mathbf{b}}$, we get all the translations of the form $T_{\mathbf{x}}$, where $\mathbf{x} = m\mathbf{a} + n\mathbf{b}$, and the composite of these with $S_\ell^{\mathbf{a}/2}$ give us the motions $T_{\mathbf{x}} S_\ell^{\mathbf{a}/2}$. But by Theorem 6, these motions give all the elements of a group of Type III.b. In the same way it can be proved that the groups of the remaining types are generated by the motions indicated above.

Moreover, since a translation of the plane is determined by a vector, and a glide reflection by an axis together with a vector on it, representing the generators of a group in the plane by means of these, we can even represent our groups by simple geometrical diagrams in the plane.

Theorem 7. There are five different types of uniformly discontinuous groups of motions of the plane, Types I, II.a, II.b, III.a and III.b. These can be generated as follows:

I. $\Gamma = \langle E \rangle$ (Figure 8.20);

II.a. $\Gamma = \langle T_{\mathbf{a}} \rangle$, where $\mathbf{a} \neq \mathbf{0}$ (Figure 8.21);

II.b. $\Gamma = \langle S_\ell^{\mathbf{a}/2} \rangle$, where $\mathbf{a} \neq \mathbf{0}$ (Figure 8.22);

Figure 8.20

Figure 8.21

Figure 8.22

III.a. $\Gamma = \langle T_{\mathbf{a}}, T_{\mathbf{b}} \rangle$, where $\mathbf{a}$ and $\mathbf{b}$ are non-collinear vectors (Figure 8.23).

III.b. $\Gamma = \langle S_\ell^{\mathbf{a}/2}, T_{\mathbf{b}} \rangle$, where $\mathbf{a}$ and $\mathbf{b}$ are non-zero orthogonal vectors with $\mathbf{a}$ parallel to ℓ (Figure 8.24).

All the motions of groups of each type are listed in Theorems 2–6.

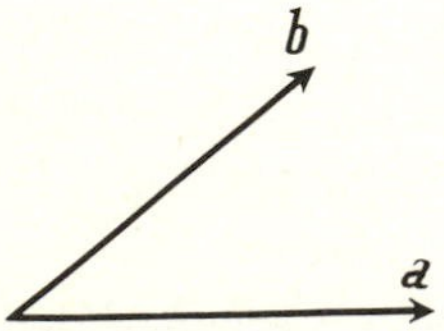

Figure 8.23

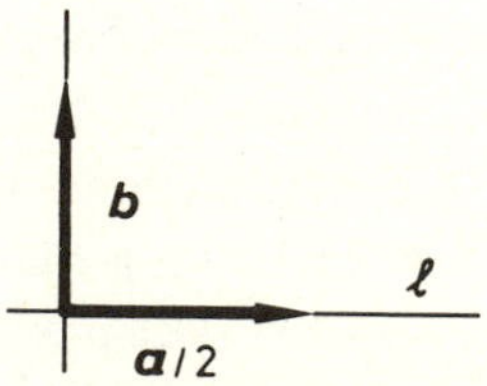

Figure 8.24

Exercises

1. Let Γ be a uniformly discontinuous group of Type III.a, and M the corresponding lattice in the plane, with O, A and B three points of M; set $\mathbf{a} = \overrightarrow{OA}$ and $\mathbf{b} = \overrightarrow{OB}$. Then prove that Γ is generated by $T_\mathbf{a}$ and $T_\mathbf{b}$ (that is, that $\mathbf{a}$ and $\mathbf{b}$ satisfy the requirements of Theorem 5) if and only if the parallelogram with sides OA and OB does not contain any points of M other than its vertices. The same thing if this parallelogram has smallest area among all parallelograms with vertices at points of M; if Δ is this smallest area then any other parallelogram with vertices at points of M has area $N\Delta$ for some integer N.

2. Prove that any motion of a group of Type III.b can be represented, in fact in a unique way, in the form

$$(T_\mathbf{b})^k(S_\ell^{\mathbf{a}/2})^m,$$

where k and m are integers.

3. Prove that the set of motions which can be written as a composite of any number of the motions $F_1, ..., F_k$ and their inverses $F_1^{-1}, ..., F_k^{-1}$, taken in any order, forms a group; this will obviously be the group $< F_1, ..., F_k >$ generated by $F_1, ..., F_k$. [Hint: note that $(G_1G_2 ... G_m)^{-1} = G_m^{-1} ... G_2^{-1}G_1^{-1}$.]

4. Consider a composite of two rotations $F = (R_{O''}^{\beta})(R_{O'}^{\alpha})$; prove that

(i) if $\alpha + \beta = 2k\pi$ for some integer k then F will be a translation $T_\mathbf{a}$ in the vector $\mathbf{a} = \overrightarrow{(O'R_{O''}^{\beta}(O'))}$,

(ii) if $\alpha + \beta \neq 2k\pi$ then F will be a rotation $R_O^{\alpha+\beta}$ through an angle $\alpha + \beta$, whose centre is determined as in Figure 8.25.

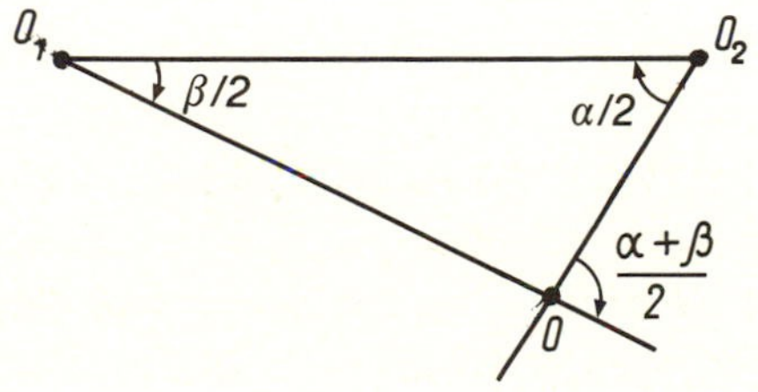

Figure 8.25

§9. A new geometry

The examples of locally Euclidean plane geometries given in §§3-5 led us in §§6-7 to uniformly discontinuous groups of motion of the plane, which provided a unified method of constructing such geometries, including all the examples previously encountered. In §8, we succeeded in giving a general classification of all such groups. At this point, before developing the general theory any further, it is interesting to look back.

A group of Type I consists only of the identity motion, and the geometry which it defines is of course the Euclidean plane. Groups of Type II.a, II.b and III.a have already appeared in §§3-5 (see also §5.1, Exercise 4), and lead to the geometries considered in these sections. There remain the groups of Type III.b, which provide some new kind of uniformly discontinuous group, and hence should give some new type of 2-dimensional locally Euclidean geometries, which we will find interesting to describe and study. When we speak of describing a geometry, recall that for each of the geometries we have met up to now, the set of points of the geometry had an entirely transparent description, in addition to their specification in terms of uniformly discontinuous groups: the cylinder in §3, the twisted cylinder in §4, and the torus in §5. We would like to have an equally transparent description of the set of points of the new geometry, which, in contrast to the cases we have seen up to now, is so far only specified by a uniformly discontinuous group. And of course we must not forget about the distances between points.

To do this, recall that the cylinder of §3 could be obtained from a strip in the plane, the twisted cylinder of §4 also from a strip, and the torus of §5 from a square in the plane by glueing together certain points of the boundaries of these plane regions in a manner specified separately in each section. We might hope that the set of points of the new geometry is also a certain surface, obtained from some similar kind of region in the plane by identifying or glueing together certain points of the boundary of the region. We will see that this is indeed the case.

Recall that a point a of the geometry defined by a uniformly discontinuous group Γ is specified by a set **A** of equivalent points of the plane, where two points are equivalent if one is taken into the other by a motion of Γ. However, in order to specify a, there's absolutely no need to know all points of **A**; we need only know one point A of **A**, and then all the others are obtained from A by applying motions in the given group Γ. Therefore in order to determine the set of all points of the geometry we need only specify some region of the plane, for example a polygon, satisfying the following property:

(1) The region contains one point from every set of equivalent points of the plane.

At the same time, we would like distinct points of the region to be inequivalent, so that different points of the region determine different points of the geometry. A requirement which on the one hand is not too far from this, and on the other hand is more convenient and agrees with our plans is the following:

(2) No interior point of the region is equivalent to any other point of the region; that is, equivalent points of the region can only lie on the boundary.

Given a region satisfying (1) and (2), we see that the set of points of the geometry is given as a surface obtained from the region by identifying or glueing together equivalent points of its boundary. This is exactly what we need: the strip of §§3-4 and the square of §5 obviously satisfy conditions (1) and (2) for the uniformly discontinuous groups constructed in those sections, and the opposite points of the boundaries of the regions which we identified there are exactly the equivalent points. A region in the plane satisfying (1) and (2) is called a *fundamental domain*, in view of the importance of this notion.

After these preparations, suppose that Γ is a uniformly discontinuous group of Type III.b; we use the description of the motions of Γ given in §8.3, Theorem 6 and Proposition 2. Let's prove that the rectangle PQRS of Figure 9.1 is a fundamental domain for Γ; in this figure $\ell_{-1}, \ell_0 = \ell$ and ℓ_1 are the consecutive

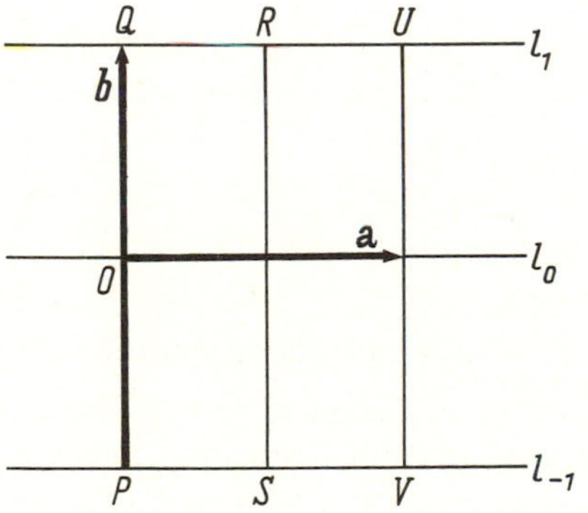

Figure 9.1

axes of glide reflections belonging to Γ, and $\overrightarrow{PS} = \overrightarrow{QR} = \mathbf{a}/2$, $\overrightarrow{OQ} = \overrightarrow{PO} = \mathbf{b}/2$.

To check property (1), note that we can use a translation of Γ in a vector of the form $n\mathbf{b}$ (with n an integer) to take any point X of the plane to a point X' of the strip bounded by ℓ_{-1} and ℓ_1; then we can use a translation of Γ in a vector of the form $m\mathbf{a}$ (with m an integer) to take X' into a point X'' in the rectangle PQUV (see Figure 9.1). If X'' is not in the rectangle PQRS then it is in the rectangle SRUV, so that we can use the glide reflection $S_\ell^{-\mathbf{a}/2} = T_{-\mathbf{a}} S_\ell^{\mathbf{a}/2}$ of Γ to take X″ into a point X‴ in PQRS. Thus any point X of the plane is equivalent to one of the points X″ or X‴ of PQRS, and this proves (1).

To check (2) let's determine when distinct points X and Y of PQRS can be equivalent. We distinguish two cases. If $Y = F(X)$, where F is a translation of

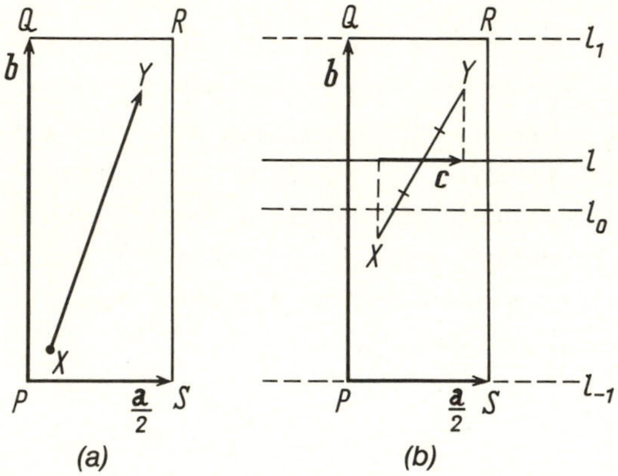

Figure 9.2

Γ then $\vec{XY} = m\mathbf{a} + n\mathbf{b}$, where m and n are integers. From Figure 9.2, (a), it is immediate that this is only possible (up to interchanging X and Y) in the case

$$X \in [PS],\ Y \in [QR] \text{ and } F = T_{\mathbf{b}} \tag{1}$$

(see Figure 9.3, (a)).

If $Y = F(X)$ where F is a glide reflection, then the axis of F must bisect [XY], and its translation vector $\mathbf{c}$ is equal to the projection of $\vec{XY}$ onto this axis. Hence from the description of the glide reflections of Γ and from Figure 9.2, (b) one sees at once that this is only possible in the following cases (up to interchanging X and Y): either

$$X \in [PQ] \text{ and } Y = [RS] \text{ and } F = S_\ell^{\mathbf{a}/2}, \tag{2}$$

or

$$X = P,\ Y = S \text{ and } F = S_{\ell_{-1}}^{\mathbf{a}/2} = T_{-\mathbf{b}} S_\ell^{\mathbf{a}/2}, \tag{3}$$

or

$$X = Q,\ Y = Y \text{ and } F = S_{\ell_{-1}}^{\mathbf{a}/2} = T_{\mathbf{b}} S_\ell^{\mathbf{a}/2}, \tag{4}$$

(see Figure 9.3, (b)).

Since the boundary of a polygon is formed by its sides, this proves (2) and completes the proof that the rectangle PQRS is a fundamental domain for Γ.

Moreover, in (1), (2), (3) and (4) we have found all pairs of equivalent points on the sides of the rectangle PQRS (see Figure 9.3). We can now say that the points of our new geometry form the surface which is obtained from the

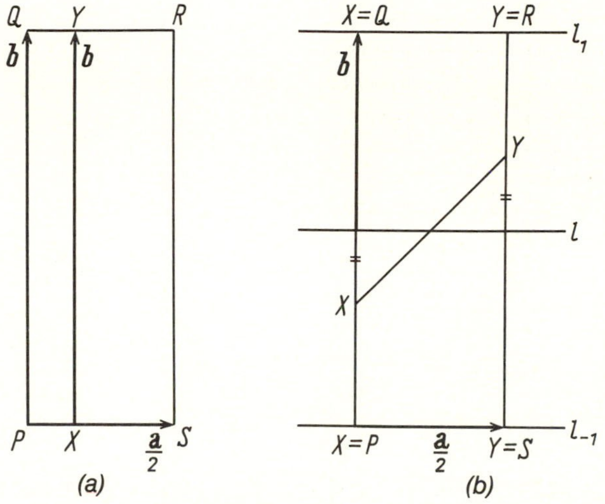

Figure 9.3

rectangle PQRS by glueing its sides [PS] and [QR] by means of a translation in $\vec{PQ}$, and glueing its sides [PQ] and [SR] by means of a glide reflection whose axis ℓ bisects the sides [PQ] and [SR]. This surface is called the *Klein bottle* in honour of the German mathematician Felix Klein (1849-1925) who was the first to draw attention to it.

In Figure 9.4 we carry out the process of glueing the Klein bottle: starting from the rectangle PQRS (Figure 9.4, (a)), we glue the sides [PS] and [QR] by a translation to get the finite cylinder with ends (Figure 9.4, (b)), the ends being circles obtained by glueing together the endpoints of the sides [QP] and [RS]; it now remains to glue these boundary circles in such a way that the corresponding points on the sides [QP] and [RS] are glued by a glide reflection in [MN] (see Figure 9.4, (a)). This can be done if we take one end of the cylinder, and bending it, push this edge to penetrate the cylinder from outside and then glue it to the other edge from within (Figure 9.4, (c)). The resulting representation in 3-space of the Klein bottle does indeed resemble a bottle.

Thus the Klein bottle, like the torus, is obtained from a finite cylinder by glueing its ends. However, these glueings are different: if we mark directions round the boundary circles of the cylinder as shown in Figure 9.4, (b) then along the circle along which the glueing is carried out, these two directions agree in the case of the torus, and are opposite in the case of the Klein bottle. It can moreover be proved that the Klein bottle, in contrast to the torus, cannot be embedded in 3-space without self-intersections; in Figure 9.4, (c), the points of self-intersection of the

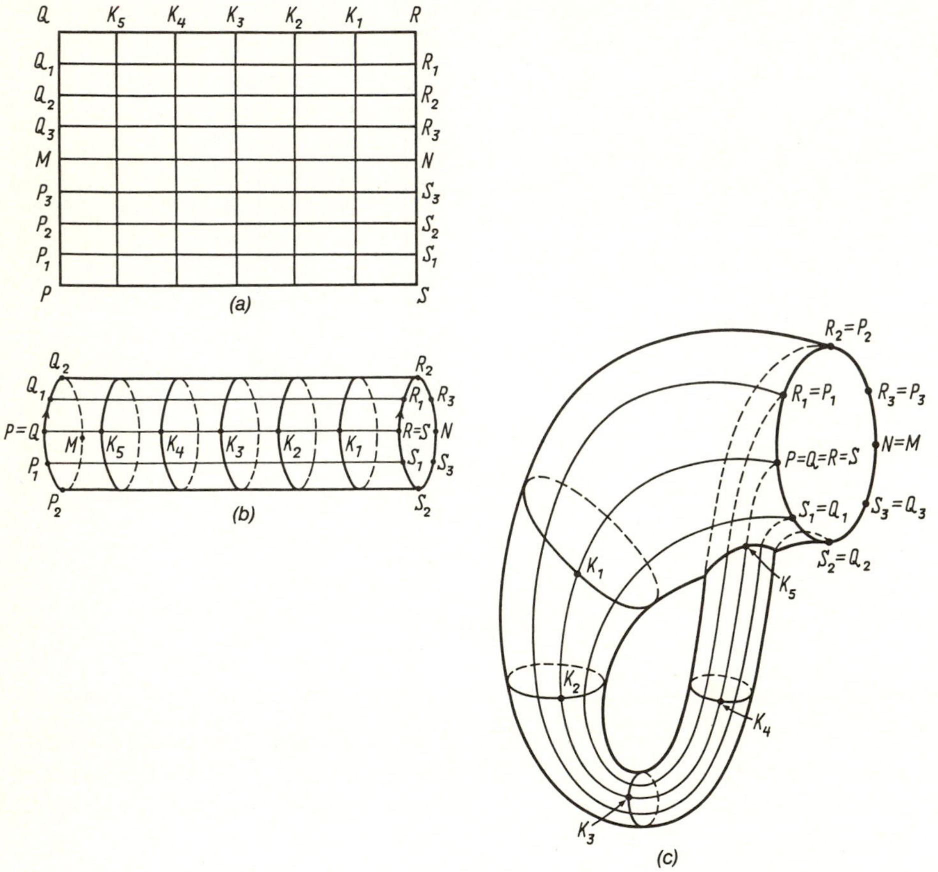

Figure 9.4

Klein bottle form a circle c, and each point of c is to be thought of as two points of the Klein bottle, one of them lying on each of the local sheets.

Now let's consider the distance between points of our geometry. For this, as in the case of a torus, we can draw on the rectangle PQRS a sufficiently fine-meshed net of line segments parallel to the sides (see Figure 9.4, (a)), and carry it over to the Klein bottle (see Figure 9.4, (b) and (c)). Under this, a line segment parallel to [PQ] and [RS] goes into a closed curve of the Klein bottle which is obviously a closed line; we will call this a *parallel*. Also, each pair of line segments parallel to the sides [QR] and [SP] and symmetric with respect to the line ℓ_0 joining the midpoints of [PQ] and [RS] goes over into a closed curve on the Klein bottle which is again obviously a closed line of the geometry, and which we will call a *meridian* . Clearly different parallels do not intersect one another, and similarly for different meridians; and at each point of intersection of a parallel and a meridian the two curves are perpendicular in the sense of our geometry. As a result we get a net of closed curves, the parallels and the meridians, which cover the whole Klein bottle; in our geometry these curves are all closed lines, and each parallel is perpendicular to each meridian at their points of intersection. Hence by Pythagoras' theorem, and more particularly from the fact that our geometry is identical in sufficiently small regions with the plane containing the rectangle PQRS, we get the following intuitive representation of distance of our geometry: a curve in the geometry is a curve in the Klein bottle; its length is determined by the number of parallels and meridians which one crosses when moving along the curve; and the distance between two points is equal to the length of the shortest curve joining them.

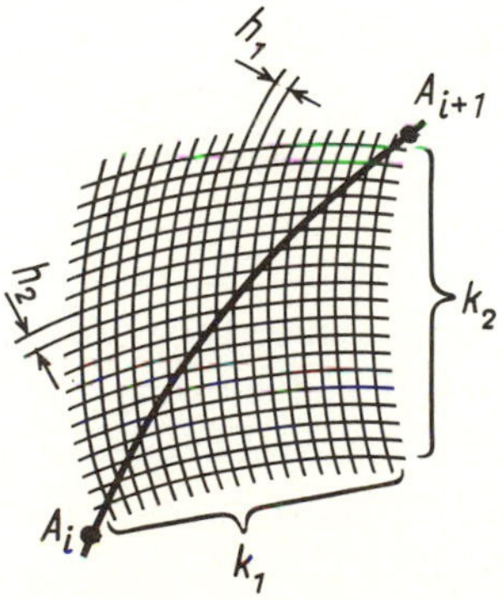

Figure 9.5

One could for example subdivide the curve by points A_i into sufficiently short pieces A_iA_{i+1}, which we can think of as being approximately line segments; then the length of the curve is the sum of the lengths of these pieces, and the length of each piece, by Pythagoras' theorem, is roughly equal to

$$\sqrt{h_1^2k_1^2 + h_2^2k_2^2},$$

where k_1 and k_2 are the numbers of parallels and meridians which A_iA_{i+1} meets, and h_1, h_2 are the meshes of the net in the parallel and meridian directions (see Figure 9.5). If we divide the curve up into smaller and smaller pieces, and at the same time choose smaller and smaller meshes, then we will get a more and more exact value for the length of a curve.

We can now summarise the result of our study.

Theorem 1. A uniformly discontinuous group of motions of the plane of Type III.b defines a locally Euclidean geometry on the Klein bottle.

To conclude our description, let us spend some time on the similarities and differences between plane geometry and the geometries on the torus and on the Klein bottle, in terms of parallels and meridian; of course, in the plane, by parallels and meridians we mean the lines of two perpendicular pencils of parallel lines. In

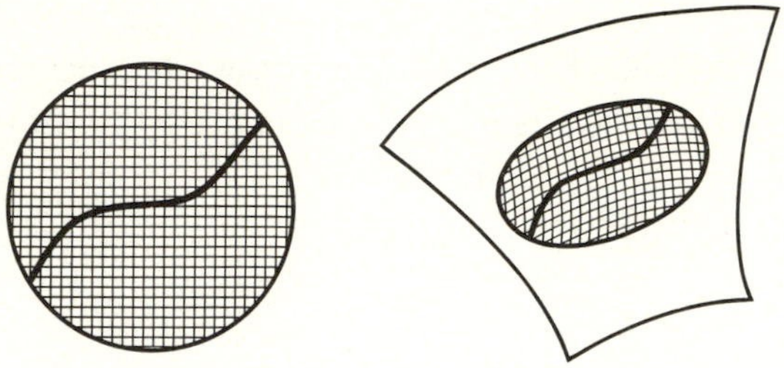

Figure 9.6

each of these geometries, the parallels and meridians are lines of the geometry, forming a net which in a small neighbourhood of each point has the same structure as that of the plane (Figure 9.6), and allowing one as in the plane to define the length of a curve in the geometry in terms of the number of parallels and meridians which the curve crosses. But the global structure of the net in these geometries is different. On the plane, parallels and meridians are open curves having infinite length, whereas on the torus and on the Klein bottle they are closed curves of finite length. Moreover, in contrast to the plane and the torus, where each parallel intersects each meridian in one point, on the Klein bottle all the meridians (with the exception of the two central meridians obtained by glueing together the line segments [MN] and [QR] or [PS] in Figure 9.4, (a)) intersect every parallel in two points, as the reader can easily verify.

We now turn to other properties of geometry on the Klein bottle. In §§4-5 we met two unusual properties of geometries: in the geometry on the twisted cylinder of §4, right and left were indistinguishable, which means that there exists a closed curve in the geometry such that on going round the curve the right-hand and left-hand sides change positions; and the geometry on the torus of §5 is bounded, that is, the distance between any two of its points never exceeds a certain constant.

It turns out that geometry on the Klein bottle has both of these properties at the same time!

Theorem 2. Geometry on the Klein bottle is bounded, and right and left are indistinguishable in it.

The first property is obvious: by definition of distance in a geometry defined by a uniformly discontinuous group, the distance in the geometry on the Klein bottle between two points A and B of the rectangle PQRS is not greater than the distance |AB| in the plane between these points. Since the points of PQRS give all the points of geometry on the Klein bottle, the distance between any two points of this geometry is not greater than the diagonal of PQRS. To prove the second property, consider the closed line (a central meridian) of geometry on the Klein

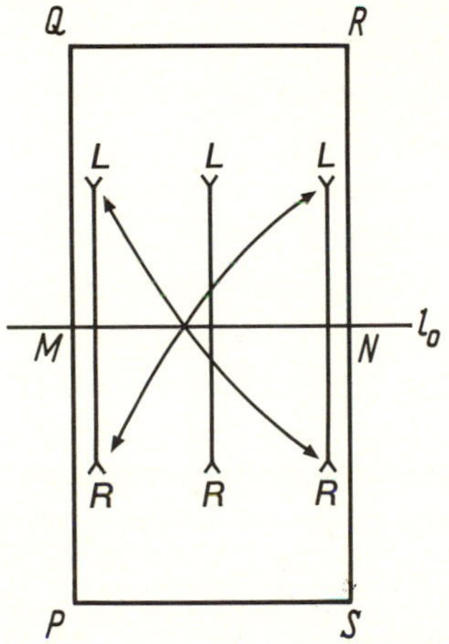

Figure 9.7

bottle which is obtained by glueing the endpoints of the line segment [MN] joining the midpoints of the sides [PQ] and [RS] of PQRS. Then exactly as in §4, we can see that on going round this curve right and left change sides (see Figure 9.7). The reader can verify for himself that the other central meridian obtained from [PS] or [QR] has the same property; see Exercise 3.

Exercises

1. Let Φ be a fundamental domain for a uniformly discontinuous group Γ. Prove that Φ defines a subdivision of the plane into regions $\gamma(\Phi)$ as γ runs through Γ, such that (a) the regions $\gamma(\Phi)$ fill the whole of the plane; (b) they only intersect along their boundaries; (c) for any two such regions there is exactly one motion of Γ which takes one into the other. What is the subdivision of the plane defined by the fundamental domain of Figure 9.1 for the group of Type III.b?

Prove that conversely, properties (a), (b) and (c) imply that Φ is a fundamental domain.

2. Let Γ be a uniformly discontinuous group, A a fixed point of the plane, and $\Delta(A)$ the region consisting of points which are closer to A than to any other point equivalent to A. Prove that $\Delta(A)$ is a fundamental domain for Γ, and is moreover a convex polygon. Prove that the fundamental domain of Figure 9.1 for a group of Type III.b can be obtained in this way from some point A; where is A? Which fundamental domain $\Delta(A)$ will correspond to a group of Type III.a? What subdivision of the plane?

3. Verify that on going round the central meridian of the Klein bottle obtained from the line segment [PS] or [QR], right and left also change sides.

4. Show that the lengths of the central meridians of a Klein bottle are half those of the other meridians.

5. Parallels and meridians in geometries on the cylinder and on the twisted cylinders are defined as the lines obtained respectively from lines in the plane parallel to the vector **a** and perpendicular to **a** (see §8.3, Theorem 7, groups of Type II.a and II.b for the notation). What are the differences, as discussed after Theorem 1 above, between the structure of the net of parallels and meridians in these geometries and those of geometries on the plane, the torus and the Klein bottle?

6. Consider the uniformly discontinuous group of motions of the sphere of §7.2, Exercise 4; prove that a hemisphere is a fundamental domain for this group, and that equivalent points on the boundary circle are the antipodal points. In §12 below (see Figure 12.22 and the following explanations), we explain how to represent the surface obtained by glueing these together, by analogy with the Klein bottle. This surface is called the *projective plane* .

§10. Classification of all 2-dimensional locally Euclidean geometries

In this section we prove the main result of our whole study: we classify all possible 2-dimensional locally Euclidean geometries, that is, geometries which are identical with the plane in sufficiently small regions. In §7 we proved that every uniformly discontinuous group of motions of the plane corresponds to some locally Euclidean geometry (in §7 we denoted this geometry by Σ_Γ). Here we will prove that all locally Euclidean geometries are obtained in this way (Theorem 1 of this section). Since in §8 we classified all uniformly discontinuous groups of motions of the plane, as a result we get a complete description of all locally Euclidean geometries (the Main Theorem of this section).

In the proof of Theorem 1 we will face difficulties of a new kind. The point is that up to now we have specified a geometry by means of an explicit construction (in §§3-5 in a very transparent way, and in §7 in a less transparent, but entirely concrete way, starting from the given group Γ). But now we have to start off from a completely general geometry Σ, satisfying only the small number of conditions in the definition of a locally Euclidean geometry, as formulated in §6, and prove that it is one of the geometries Σ_Γ which we constructed in §7. Thus we have to learn to argue in the framework of a general geometry, using only the properties which appear in the definition. Here we meet an example of a mathematical investigation, where from a few simple and very general properties of some kind of object, we can deduce quite concrete general properties, going as far as a complete classification of all of the objects. Of course, this requires a new kind of arguments in comparison with those which we used when the geometry was given by some concrete construction (for example as Σ_Γ). Because of their very novelty, these new arguments demand long explanations, and as a result, the proof of Theorem 1 turns out to be notably longer than the proof of any other theorem in this book (although it consists of several separate arguments, each of which no harder than those which have already appeared). The reader who finds the proof too complicated can quite well omit it (reading carefully the statement only of Theorem 1 and of the Main Theorem), and proceed to Chapter III; in this case he might, after reading the book, come back to the proof of Theorem 1, and attempt to work through it.

Theorem 1. Every locally Euclidean geometry Σ corresponds to a uniformly discontinuous group Γ of motions of the plane, and Σ can be obtained from Γ by the construction of §7; in other words, Σ can be superposed on Σ_Γ.

The main result of our whole study follows of course at once from this theorem and from §8, Theorem 7:

Main Theorem. There are exactly 5 types of locally Euclidean geometries, namely:

I	geometry on the plane;
II.a	geometry on the cylinder;
III.a	geometry on the torus;
II.b	geometry on the twisted cylinder;
III.b	geometry on the Klein bottle.

In other words, any locally Euclidean geometry can be superposed on a geometry of one of the five types listed.

10.1. Constructions in an arbitrary geometry. Before starting on the proof of Theorem 1, we consider certain general notions and constructions which are applicable in an arbitrary locally Euclidean geometry Σ. According to the definition given in §6, in such a geometry Σ, there exists a positive number r such that the disc $D(a, r)$ centred at any point a of Σ can be superposed on a disc D in the plane, with some centre A and radius r', that is $D = D(A, r')$; note that we obviously have $r' = r$, and that the point a goes to A under the superposition. A superposition of $D(a, r)$ on $D(A, r)$ is a map f of the first disc onto the whole of the second disc which preserves the distance between points; we can think of taking a point x of $D(a, r)$ into the point $f(x)$ in $D(A, r)$ as representing x on a chart having the form of the disc $D(A, r)$. In view of this, we will from now on refer to a disc $D(A, r)$ and a superposition of $D(a, r)$ onto $D(A, r)$ as a *chart* of $D(a, r)$. This is an exact chart, since by definition, f preserves distances between points. Thus the property that a geometry should be locally Euclidean can be restated more transparently: every disc of radius r has an exact chart.

The same principle of representation in the plane can also be applied to other figures in a geometry Σ; by a figure in a geometry, we always mean some set of points of the geometry, for which therefore the distance between points is defined. The notion of superposition of a figure of one geometry onto a figure of another geometry is defined in exactly the same way as the superposition of geometries, namely as a map of the set of points of the first figure onto the set of points of the second which preserves distances. (Strictly speaking, the notion of superposition of figures of two geometries appeared much earlier, in fact in the definition of locally Euclidean geometry in §6. The point is that we do not know if a spherical neighbourhood in a geometry is a geometry in its own right; thus the superposition of the disc on a spherical neighbourhood in §6 must be taken as a superposition of figures.) Two figures which can be superposed on one another will be said to be *congruent*; in Euclidean geometry this is how congruence (or equality) of figures in the plane was defined. This allows us to apply to figures in Σ the names disc, line segment, parallelogram (and the corresponding notation), whereby we understand that these figures are congruent to figures in the plane with the same names (and therefore enjoy all the known geometrical properties of figures in the plane involving only points of the figures themselves). We will continually be using this. For example, using some chart of a disc D(a, r) of a geometry Σ, we can uniquely construct the centre a of the disc, the line segment [bc] joining points b and c of the disc, radiuses and diameters, the perimeter circle, and so on, and these constructions will satisfy the same assertions as the corresponding figures in a disc of radius r in the plane; recall that our conventions mean that the line segment [bc] is the set of points of Σ which is congruent to a line segment [BC] in the plane, with B mapping to b and C to c.

In what follows, line segments will appear continually in the constructions we will carry out. It will therefore be important for us to know when there exists a line segment [bc] joining two points b and c of Σ, and when it will be unique. For this, note that if both b and c are contained in a disc D(a, r), then we can construct a line segment [bc] in the disc, and it will be unique inside the disc. This method of construction depends on the disc D(a, r) in which b and c are assumed to lie, and it is not excluded that using some other disc containing b and c, we could join b and c by some other line segment. Nevertheless, in one extremely important case, we can prove that in Σ there exists a unique line segment [bc] joining two points b and c of Σ; let's prove this if $|bc| \leq r$. In fact, the points m of any line segment [bc], just as points of a line segment in the plane, satisfy the condition

$$|bm| + |mc| = |bc| \leq r. \tag{1}$$

Hence $|bm| \leq r$, and the points m of any line segment [bc] are inside the disc D(b, r). But since $|bc| \leq r$, we know that in D(b, r) there is just one line segment [bc] joining b and c. This proves what we wanted. Using the same technique, the reader will prove the following assertion, which we will require, as to the existence and uniqueness of the extension of a line segment in Σ . Suppose given points b

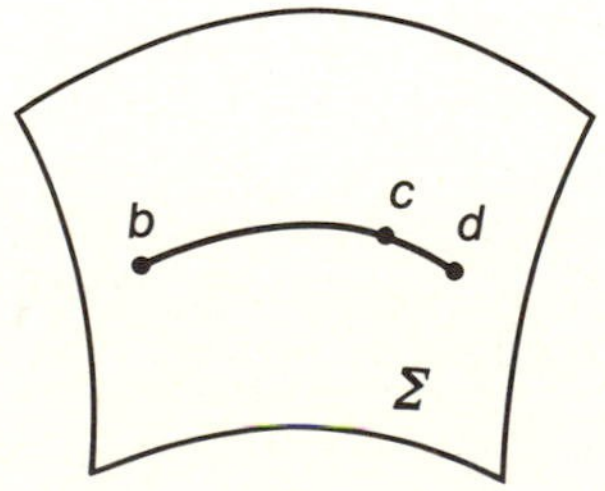

Figure 10.1

and c of Σ and a number $k > 0$ such that $|bc| < k \leq r$. As we have just proved, in Σ there is a unique line segment [bc]. Prove that in Σ, the segment [bc] can be extended in a unique way to a line segment [bd] of length k, that is, there exists one and only one line segment [bd] which contains [bc] and has length equal to k (Figure 10.1).

Before we finish these preparatory remarks, we note an interesting and extremely important phenomenon related to our definitions of figures in an arbitrary geometry. The point is that any figure in the plane (as in any other geometry) which is congruent to a disc, line segment, parallelogram, circle ... must now, according to our conventions, also be called a disc, line segment, parallelogram, circle and so on. Now, is it true that a plane figure congruent to a disc, line segment, parallelogram, circle ..., actually is a disc, line segment, parallelogram, circle and so on, corresponding to the definitions of Euclidean geometry? For example, is it true that a plane figure which is congruent to a disc of radius R consists exactly of all the points of the plane whose distance from some point of the plane is not greater than the number R? This is in fact the case, and it follows from the homogeneity property of the plane, which we formulate in Lemma 1.

Lemma 1 (Homogeneity of the plane). Every congruence f of a plane figure Φ onto a plane figure Ψ can be extended to a motion F of the plane; that is, there exists a motion F of the plane such that

$$F(X) = f(X) \text{ for } X \text{ belonging to } \Phi.$$

If Φ is not contained in a line, then the motion F is unique.

We first make use of the lemma to prove the properties listed above, and then prove the lemma afterwards.

So let f be a congruence of the disc $\Phi = D(A, R)$ onto some plane figure Ψ, and F the motion of the plane given by Lemma 1; then Ψ consists of the points F(X) as X runs through D(A, R). Let's prove that $\Psi = D(F(A), R)$. Since F is a motion,

$$|F(X)F(A)| = |XA| \leq r,$$

if X is a point of D(A, R). This proves that Ψ is contained in D(F(A), R). Now suppose that Y is any point of D(F(A), R). Since a motion maps the plane onto the whole plane, we can find a point X for which $Y = F(X)$. Then X satisfies

$$|AX| = |F(A)F(X)| = |F(A)Y| \leq R,$$

and hence X lies in D(A, R). Hence for every point Y of D(F(A), R), we can find a points X of D(A, R) such that $F(X) = Y$. This proves that conversely

D(F(A), R) is contained in Ψ. Hence Ψ = D(F(A), R). Thus a figure Ψ congruent to a disc D(A, R) is itself a disc D(F(A), R), as required to prove.

The other well-known figures of a Euclidean geometry course (line segment, parallelogram, circle and so on) can be treated in an entirely similar way: (see §8.1, Exercise 1).

Proof of Lemma 1. We consider the case that Φ is not contained in a line. Let A, B and C be three non-collinear points of Φ. By §8.1, Lemma 3, there exists one and only one motion F such that F(A) = f(A), F(B) = f(B), and such that F(C) and f(C) lie in the same half-plane bounded by the line (f(A)f(B)). It follows from this that F(C) = f(C), since the distances from either of F(C) or f(C) to f(A) and f(B) are the same, equal to |AC| and |BC|.

Let's prove that F is the required motion. Since f(A) = F(A), f(B) = F(B) and f(C) = F(C), for any point X of Φ, the distances from either of F(X) or f(X) to f(A), f(B) and f(C) are the same, equal to |AX|, |BX| and |CX|. From this it follows that F(X) = f(X): for if F(X) ≠ f(X), the points f(A), f(B) and f(C) all lie on the perpendicular bisector of [f(X)F(X)], and hence are collinear; but since F is a motion and A, B and C are non-collinear, the three points F(A) = f(A), F(B) = f(B) and F(C) = f(C) are also non-collinear.

The case when all points of Φ are collinear can be treated similarly, but much more easily; see Exercise 4.

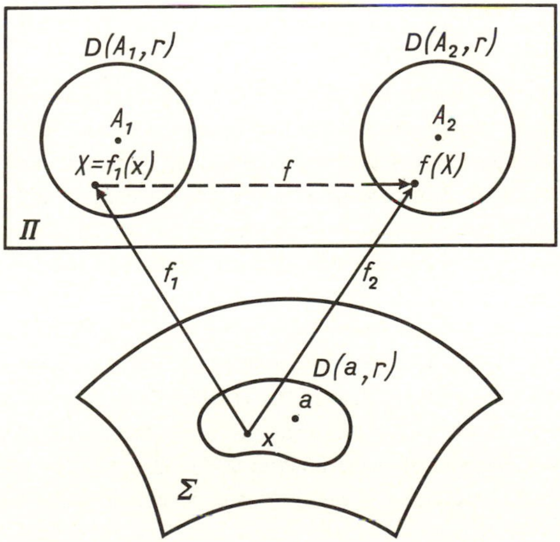

Figure 10.2

Corollary 1. Any two charts in the plane of the same disc $D(a, r)$ in Σ can be taken into one another by a motion of the plane.

The assertion means the following. Suppose that one of the charts is a map f_1 from $D(a, r)$ to a disc $D(A_1, r)$ in the plane, and the other a map f_2 to a disc $D(A_2, r)$; we can assume that $D(A_1, r)$ and $D(A_2, r)$ lie in the same plane. Then the assertion is that there exists a motion F of the plane such that $f_2(x) = F(f_1(x))$ for every point x of $D(a, r)$. In other words, we can make the two charts correspond by a motion of the plane which takes the two representatives of the same point into one another.

Proof (see Figure 10.2). We construct a congruence g of $D(A_1, r)$ onto $D(A_2, r)$ so that $f_2(x) = g(f_1(x)$ for any point x of $D(a, r)$; the existence of the required motion F will then follow from Lemma 1. For a point X of $D(A_1, r)$, we find a point x of $D(a, r)$ such that $f_1(x) = X$; this x exists since f_1 maps $D(a, r)$ onto the whole disc $D(A_1, r)$, and it is unique, since f_1 takes distinct points into distinct points: otherwise it would change the distances. Now set $g(X) = f_2(x)$. Now since f_1 and f_2 are both superpositions, g is obviously a congruence of $D(A_1, r)$ onto $D(A_2, r)$ as required (Figure 10.2). The corollary is proved.

Exercises

1. Prove that a line segment $[ab]$ of a geometry Σ will be a geometry in its own right if we measure distance between points of $[ab]$ as in Σ. Let $[ab]$ and $[bc]$ be line segments of a geometry Σ, joining points a, b and c, and write S for the union of $[ab]$ and $[bc]$. Prove that if we measure the distance between points of S as in Σ then S will be a geometry only if S is a line segment $[ac]$.

2. Let Σ be a locally Euclidean geometry, and r a positive number such that a disc of radius r of Σ can be superposed on a disc in the plane; let a and b be two points of Σ such that $|ab| < 2r$. Suppose that S is a line segment joining a and b, and c its midpoint; then prove that $S \subset D(c, r)$. Let S' be another line segment joining a and b. Prove that the points of the line segments S and S' sufficiently close to a coincide. [Hint: use the property of a line segment of Exercise 1.] Deduce from this that the points a and b can be joined by at most one line segment.

3. Let Σ be a locally Euclidean geometry, a and b two points of Σ, and suppose that $|ab| < r$. Prove that any piece S of the boundary of the disc $D(b, r)$ contained in $D(a, r)$ is represented as an arc of a circle in a chart of $D(a, r)$. [Hint: let X and Y be the ends of S; use the fact that in the disc $D(b, r)$, the angle XCY is a constant for all points C of S, expressing this as a relation between the distances $|XC|$, $|CY|$ and $|XY|$.]

4. Deduce from the hypotheses of Lemma 1 that if Φ is contained in a line then so is Ψ, and that provided Φ contains at least two points, there then exist exactly two extensions of the congruence f to a motion F.

5. Using Lemma 1 and Chasles' theorem in the plane, classify all motions of the line.

10.2. Coverings. We can now proceed to the proof of Theorem 1.

We are given a locally Euclidean geometry Σ, and we need to prove that Σ is one of the geometries Σ_Γ constructed from a uniformly discontinuous group Γ of motions of the plane. The proof breaks up into three stages.

I. We establish the relation between Σ and the plane Π.

II. Using this relation, we construct some uniformly discontinuous group Γ of motions of Π.

III. We prove that Σ can be superposed on the geometry Σ_Γ constructed from Γ.

We start on Stage I. First of all, we recall how to get from the plane at least to the simplest of the examples we considered, the geometry on a cylindrical surface of §3. We rolled up the plane Π around the cylinder Σ in such a way that sufficiently small discs of the plane were superposed on discs of the cylinder. This rolling up can be described in exact mathematical terms as a map φ from the plane Π to Σ. Also in the more general case of a geometry Σ_Γ considered in §8, we introduced a notion of equivalence of points of the plane Π (under the group Γ), and then defined a point a of the geometry Σ_Γ to be a set $\mathbf{A}$ of equivalent points of the plane. This can be expressed instead by saying that there is a map φ defined from the plane Π to the set of points of Σ, taking a point A of the plane to the point a of Σ defined by the set $\mathbf{A}$ of points equivalent to A. Just as in the case of the plane rolling up around the cylinder, this map φ has the important property that any disc of sufficiently small radius is superposed by φ onto a disc of Σ_Γ. Although this was not stated explicitly in §7, it was in fact proved there: the reader will easily convince himself of this if he looks again through the proof of the fact that Σ_Γ is locally Euclidean.

Thus it is plausible that in the case of an arbitrary locally Euclidean geometry Σ, we can establish its relation with the plane Π by constructing a map φ of Π onto Σ such that a disc of sufficiently small radius in the plane superposes onto a disc of Σ. This leads us to an important notion, enabling us to establish a relation not just between the plane and a geometry, but quite generally between two geometries.

Definition. Let Σ_1 and Σ_2 be any two geometries. A *covering* of Σ_2 by Σ_1 is a map φ from Σ_1 into Σ_2 which satisfies the two conditions:

(i) there exists a number $s > 0$ such that, for any point a of Σ_1, the map φ defines a superposition of the spherical neighbourhood D(a, s) of a in Σ_1 onto the spherical neighbourhood D(φ(a), s) of φ(a) in Σ_2;

(ii) φ maps Σ_1 onto the whole of Σ_2, that is, for any point x of Σ_2, there exists a point a of Σ_1 such that $\varphi(a) = x$.

We will denote a covering by the letter φ, or sometimes by the triple $(\Sigma_1, \Sigma_2, \varphi)$. By definition of a covering, two distinct points a and b of Σ_1 whose distance is less than s go into distinct points of Σ_2, since b is contained in the disc D(a, s), on which φ is a superposition, and hence $|\varphi(a)\varphi(b)| = |ab| \neq 0$.

However, for points at greater distances, this may not be the case. We will see later many examples of coverings under which distinct points go into the same point.

Let's give some examples of coverings which either occur often in real life, or which have already appeared in this book.

Example 0. Every superposition is obviously a covering. Thus a rotation, translation or glide reflection can be viewed as a covering from the plane to itself. Other examples of superpositions relating to various ways of specifying geometries were considered in §6. All of these are trivial examples of coverings, being just superpositions.

Example 1. The following example of a covering often occurs in life, and as the reader will notice, plays an important role in geometry in connection with

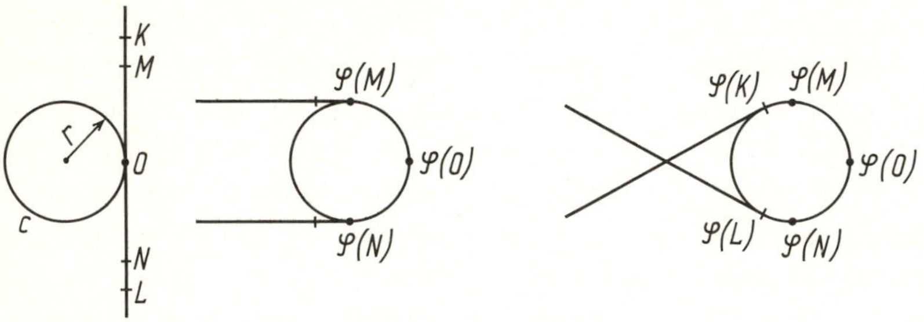

Figure 10.3

measuring angles in radians. Consider rolling up a line ℓ around a circle c (Figure 10.3). We can view this as winding up an infinitely long and thin thread on a very short cylindrical reel. Since the thread does not stretch, the distance between nearby points of the thread will be equal to that between the points of c to which they go under winding; that is, we have a covering of the circle by the line. Note that under this covering, points of ℓ at a distance of $2\pi rn$ will go into the same point of the

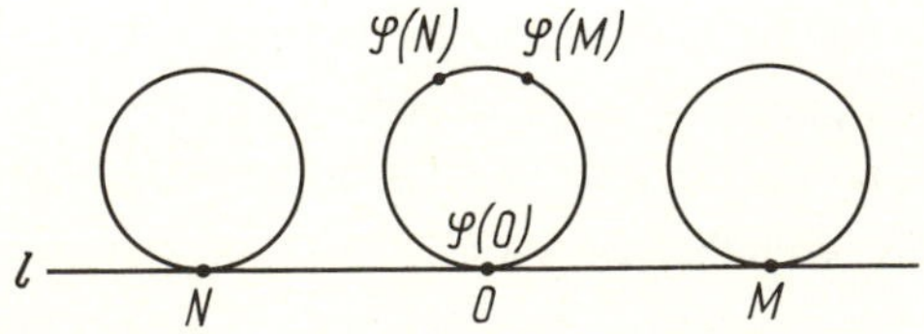

Figure 10.4

circle, where r is the radius of the circle c, and n is any integer. The same covering can be visualised as rolling a wheel c without slipping along a line ℓ (Figure 10.4); we then think of φ as taking a point M of the line ℓ into the point $\varphi(M)$ of the wheel c which touches it as it rolls.

The next two examples were already mentioned above, before we defined coverings, but we repeat them here to be able to compare them with the exact definition of covering.

Example 2. Similarly to Example 1, the 2-dimensional example is given by rolling up the plane around the cylinder, which by analogy with Example 1 we can visualise as rolling up a sheet of paper around a cylindrical axis, or rolling a cylinder without slipping along the plane (see Figures 10.3-4, which now depict a cross-section of this example). This example has already appeared during the study of geometry on a cylinder in §3. As in Example 1, under this covering distinct points of the plane will go into the same point of the cylinder.

Example 3. Consider a geometry Σ_Γ defined by a uniformly discontinuous group Γ of motions of the plane; we also think of the plane Π as being a geometry. To each point A of Π we assign the point a of Σ_Γ defined by the set **A** of points of the plane equivalent to A. We get a map φ_Γ from the points of Π into the set of points of Σ_Γ, with $\varphi_\Gamma(A) = a$. As we have mentioned already, in §7 it is actually proved that $(\Pi, \Sigma_\Gamma, \varphi_\Gamma)$ is a covering. Although the map φ_Γ was not explicitly defined there, we proved that the disc $D(A, s)$ in the plane is superposed by φ_Γ onto the disc $D(\varphi_\Gamma(A), s)$ of Σ_Γ if $s = d/4$, where d is the number involved in the definition of a uniformly discontinuous group Γ in §7.1. Note that as in Examples 1 and 2, distinct points of Π may go into the same point of Σ_Γ under the covering φ_Γ.

If we restrict attention to locally Euclidean geometries, then the definition of covering takes on a somewhat simpler form.

Lemma 2. Let Σ_1 and Σ_2 be two locally Euclidean geometries, and φ a map from Σ_1 to Σ_2 which satisfies the following condition: there exists some number $s > 0$ such that $|\varphi(a)\varphi(b)| = |ab|$ for any two points a and b of Σ_1 with $|ab| < s$. Then φ is a covering.

Proof. It follows from the condition in the statement of Lemma 2 that φ takes the disc $D(a, s/2)$ into the disc $D(\varphi(a), s/2)$, in such a way that distances between points are unchanged. What do we still need to check to prove that φ is a covering? To verify (i) in the definition of covering, we need to prove that for some $t > 0$, φ maps $D(a, t)$ onto the whole of $D(\varphi(a), t)$; and to prove (ii), we need to prove that φ maps Σ_1 onto the whole of Σ_2. We now prove these two assertions.

Proof of the first assertion. Recall that Σ_1 and Σ_2 are locally Euclidean geometries, and let r_1 and r_2 be the corresponding numbers, the radiuses of discs inside which these geometries are identical with that of the plane. We take $t > 0$

such that $t < s/2$, $t \le r_1$ and $t \le r_2$. With this choice of t, the disc $D(a, t)$ is contained in $D(a, r_1)$ and its image in the chart $D(A, r_1)$ which corresponds to $D(a, r_1)$ is a disc $D(A, t)$; thus $D(a, t)$ also has a chart. The same thing holds for $D(\varphi(a), t)$.

Let $D(A, t)$ and $D(B, t)$ be their charts; φ takes $D(a, t)$ into $D(\varphi(a), t)$, and preserves distances, since $t < s/2$, and φ preserves distances between

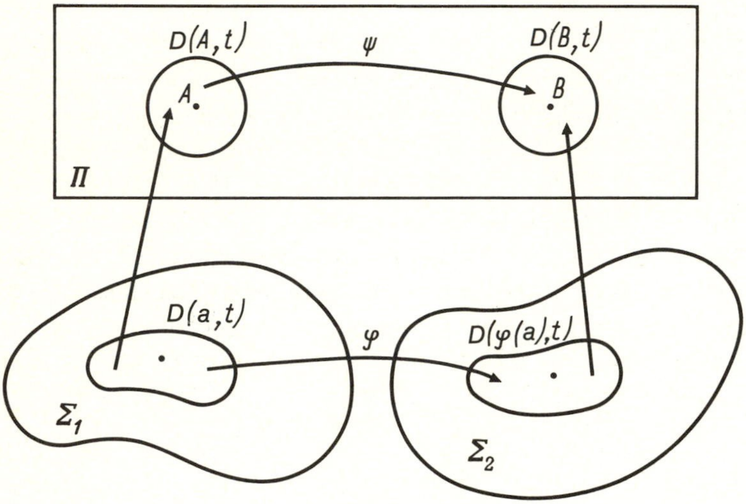

Figure 10.5

points of $D(a, s/2)$. On the charts $D(A, t)$ and $D(B, t)$ of these discs, φ is represented as a superposition ψ of the disc $D(A, t)$ onto some figure in $D(B, t)$ in the plane, with $\psi(A) = B$ (Figure 10.5). As we showed before the proof of Lemma 1 (using Lemma 1), in this situation, the figure must be the whole of $D(B, t)$, so that ψ superposes $D(A, t)$ onto the whole of $D(B, t)$; therefore φ superposes $D(a, t)$ onto the whole of $D(\varphi(a), t)$, as required.

Proof of the second assertion. Here we use for the first time the property (d) of the definition of geometry in §6.1. Let a be some point of Σ_1 and x any point of Σ_2. By property (d) of the definition of geometry (applied with $\alpha = t$ and any β), there exist points $b_1, b_2, \ldots, b_n$ of Σ_2 such that $b_1 = \varphi(a)$, $b_n = x$, and $|b_i b_{i+1}| < t$ for each i. By what we have proved, φ maps $D(a, t)$ in Σ_1 onto the whole of $D(\varphi(a), t)$ in Σ_2. Now since b_2 lies in $D(\varphi(a), t)$, there exists some point a_2 of Σ_1 such that $\varphi(a_2) = b_2$. Applying the same argument to b_2, we then find a point a_3 such that $\varphi(a_3) = b_3$, and so on. After n steps we construct a point y of Σ_1

such that $\varphi(y) = x$. The lemma is proved.

Note that in the case that both Σ_1 and Σ_2 are just the Euclidean plane, and φ is a map which does not change the distance between any pair of points, we deduce from Lemma 2 a new assertion of plane geometry: in the definition of a motion, we do not need to require that it maps the plane onto the whole plane. We recommend the reader to think through the proof of Lemma 2, and based on this, to find a proof of this assertion within the framework of plane geometry which is as simple as possible.

Remark. In discussing a covering $(\Sigma_1, \Sigma_2, \varphi)$ of locally Euclidean geometries, three numbers occur: r_1 (related to Σ_1), r_2 (related to Σ_2), and s (related to the map φ). We are obviously allowed to replace any of these by any smaller number, and it will have the same property. Hence from now on we will replace all three of these by some number smaller than all three of them, which we will denote by r.

Exercises

1. Let $(\Sigma_1, \Sigma_2, \varphi)$ be a covering of locally Euclidean geometries. Prove that for $s < r$, the inverse image in Σ_1 of a disc $D(a, s)$ in Σ_2 (that is, the set of all points b in Σ_1 such that $\varphi(b)$ is in $D(a, s)$) consists of disjoint discs of radius s.
2. Prove that winding a closed thread of length nL around a reel of perimeter L defines a covering of geometries in which each point of Σ_2 (the reel) is covered by n points of Σ_1 (the thread). Consider the analogous examples for the cylinder and the torus.
3. Suppose that a uniformly discontinuous group Γ of motions of the plane contains a group Γ'. We assign to a set of points of the plane equivalent under Γ' the set of points equivalent to these under Γ. Prove that this defines a covering φ of Σ_Γ by the geometry $\Sigma_{\Gamma'}$. How can the example of the previous exercise be included in this construction?
4. Construct a covering of the twisted cylinder by the cylinder.
5. Let $(\Sigma_1, \Sigma_2, \varphi)$ be a covering of locally Euclidean geometries. Suppose that for some point a of Σ_2, there is just a finite number n of points of Σ_1 which go to a. Prove that the same thing holds for every point of Σ_2.
6. Does there exist a covering of the plane by spherical geometry? Prove that if Σ is a geometry which admits a covering by a locally Euclidean geometry, then Σ is locally Euclidean, and conversely.

10.3. Construction of the covering. Stage I of the proof of Theorem 1, following the plan set out at the beginning of §10.2, consists of the proof of the next assertion.

Theorem 1a. For any locally Euclidean geometry Σ, there exists a covering $\varphi: \Pi \to \Sigma$ of Σ by the plane Π.

We construct a certain map φ from the set of points of Π to the set of points of Σ, and subsequently prove that φ is a covering. The idea of the construction is extremely natural: we must use the fact that our geometry is locally identical to the plane, so that every disc of radius r has a chart. We fix a point P of the plane Π, some point p of the geometry Σ, the disc D(p, r) of Σ centred at p, and a chart of D(p, r) which we can take to be a disc D(P, r) of the plane centred at P. By definition of a chart, we have a superposition f of D(p, r) onto D(P, r); this f has an inverse, which we denote by φ_0: this is the map which assigns to each point of the chart the point of Σ which it represents. Obviously φ_0 is a superposition of D(P, r) onto D(p, r). This already defines for us a map of the disc D(P, r) onto D(p, r). We will construct φ as an *extension* of φ_0, that is, as a map φ such that $\varphi(A) = \varphi_0(A)$ for any A in D(P, r).

For this, we use the following lemma.

Lemma 3 (Extension Lemma). Suppose given in the plane a line segment [PB] of length 2r with midpoint A, and a superposition φ_0 of the line segment [PA] onto a line segment [pa] of Σ (here r is the number involved in the definition of a locally Euclidean geometry Σ). Then there exists a superposition φ_1 of [PB] onto some line segment [pb] of Σ which is an extension of φ_0, so that $\varphi_1(X) = \varphi_0(X)$ for any point X in [PA].

Proof. We consider the disc D(a, r) in Σ. Note that the line segment [pa] is a radius of D(a, r), because $|pa| = r$, and [pa] is the unique line segment in Σ joining p and a (since $|pa| \le r$). Now since [pa] is a radius of D(a, r), we can extend it to a diameter [pb] as in Figure 10.6. We define φ_1 on the line segment [PA] by the condition $\varphi_1(X) = \varphi_0(X)$, and on [AB] as the superposition of [AB] onto the line segment [ab] of the same length, which takes A to a and B to b. It remains for usto check that φ_1 is a superposition of [PB] onto [pb]. This becomes obvious if

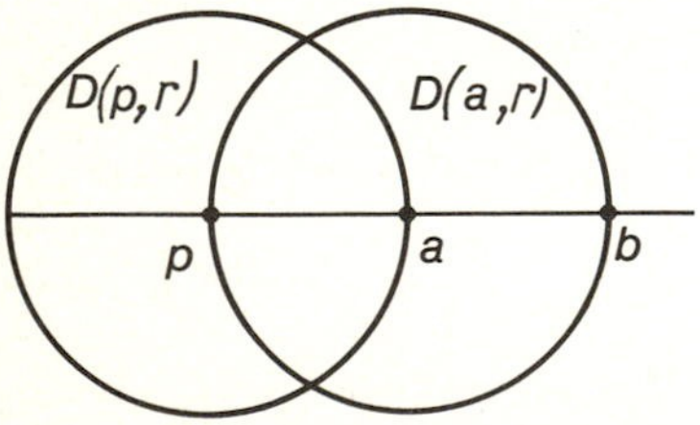

Figure 10.6

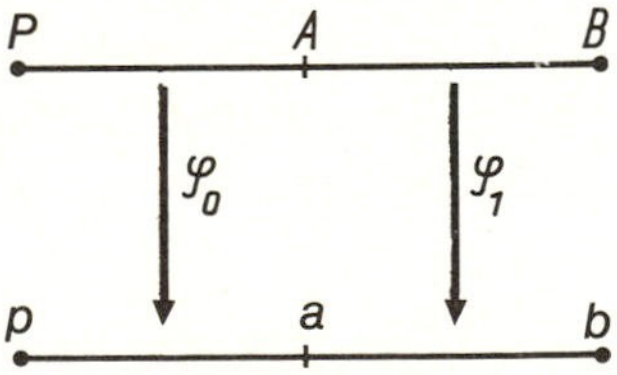

Figure 10.7

we use a chart of D(a, r) to represent [pb] on the plane. We have two line segments [PB] and [pb], and points A on [PB] and a on [pb]. The line segment [PA] superposes on [pa] and [AB] onto [ab]. Clearly then [PB] superposes onto [pb] (Figure 10.7). The lemma is proved.

To construct the covering, we now draws rays from the point P of the plane, and define the map φ separately on each individual ray. To do this, divide each ray into equal segments of length r by points P, A_1, A_2, ..., with $|PA_1| = ... = |A_nA_{n+1}| = r$. On the segment PA_1, which is contained in the disc D(P, r) we define φ to be equal to φ_0. By the preceding lemma, we can extend this map to the segment $[A_1A_2]$; using the lemma again, we can extend it to $[A_2A_3]$. Proceeding in the same way, we eventually get an extension to the whole ray. By Lemma 3, our map φ will be a superposition of the line segment $[A_nA_{n+2}]$ onto the line segment $[\varphi(A_n)\varphi(A_{n+2})]$ of Σ. Since any line segment of length r lying in the ray is contained in some segment $[A_nA_{n+2}]$, we can say that the map of the ray constructed in this way is a superposition for any line segment of length $\leq r$ contained in the ray.

The rays we have constructed cover the whole plane, and each point of the plane lies on just one ray, so that the map defined on all the rays gives a unified well-defined map φ from the plane Π to Σ, in such a way that

(A) φ defines a superposition of the disc D(P, r) onto the disc D(p, r), with $p = \varphi(P)$;

(B) φ defines a superposition of every line segment [AB] of length $\leq r$ lying in a ray of Π emanating from P onto the line segment [ab] of Σ, where $\varphi(A) = a$ and $\varphi(B) = b$.

This completes the construction of our map φ, and it remains for us to prove that it is actually a covering. For this, let's prove that $|\varphi(A)\varphi(B)| = |AB|$ for any points A and B of Π with $|AB| \leq r/2$. By Lemma 2, Theorem 1a already follows from this.

Let A and B be two points of Π with $|AB| \leq r/2$. We consider three possible cases. (a) If A and B lie on a ray emanating from P then by property (B) of φ we have $|\varphi(A)\varphi(B)| = |AB|$. (b) If A and B lie together with P on one line, but on opposite sides of P, then $|AP| \leq r/2$ and $|BP| \leq r/2$, so that A and B belong to the disc D(P, r/2), which is contained in D(P, r); in this case $|\varphi(A)\varphi(B)| = |AB|$ by property (A) of φ. There remains the main case: (c) A and B are not collinear with P. For this, we make the following construction: choose a sufficiently large natural number k such that $k\cdot(r/2)$ is greater than or equal to both of $|PA|$ and $|PB|$. Now subdivide each of the rays PA and PB into k equal intervals by points P, $M_1, M_2, ..., M_k = A$ and P, $N_1, N_2, ..., N_k = B$. This divides up the triangle PAB into k-1 trapeziums and a triangle PM_1N_1; we consider this first piece as a degenerate trapezium with $M_0 = N_0 = P$ (see Figure 10.8). Notice that under the assumption $|AB| \leq r/2$, all the sides of these k trapeziums have length $\leq r/2$.

We need to prove that $|\varphi(A)\varphi(B)| = |AB|$; however, for our purposes it is convenient to prove more. Namely, we will prove that the distance between any two of the points A, B, M_{k-1} and N_{k-1} is preserved by the map φ; so that, not only $|\varphi(A)\varphi(B)| = |AB|$, but also $|\varphi(A)\varphi(M_{k-1})| = |AM_{k-1}|$, and so on for all six of the pairs (A, B), (A, M_{k-1}), (A, N_{k-1}), (B, M_{k-1}), (B, N_{k-1}) and (M_{k-1}, N_{k-1}). In other words, we prove that φ defines a congruence of the figures

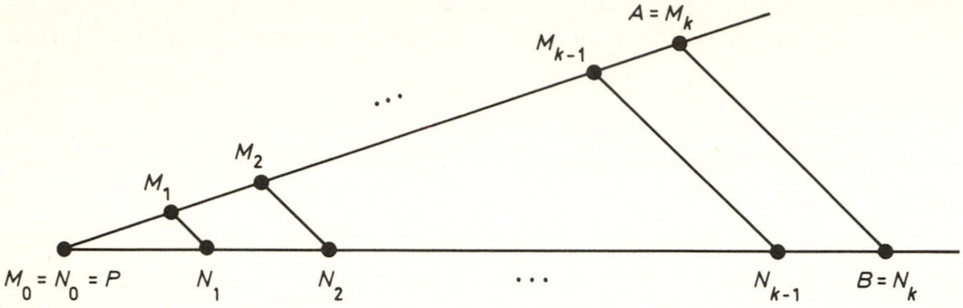

Figure 10.8

$ABN_{k-1}M_{k-1}$ and $\varphi(A)\varphi(B)\varphi(N_{k-1})\varphi(M_{k-1})$ (each figure consisting of four points).

The idea of the proof is to prove that for each i, the map φ defines a congruence of the figures $M_{i-1}N_{i-1}N_iM_i$ and $\varphi(M_{i-1})\varphi(N_{i-1})\varphi(N_i)\varphi(M_i)$. We prove this fact by the method of mathematical induction on i, that is going from the ith trapezium to the $(i+1)$st, starting from the very first, which is the triangle PM_1N_1. For $i = k$ we get the required result on $ABN_{k-1}M_{k-1}$.

The first step is obvious. Indeed, by construction $|PM_1| \le r/2$ and $|PN_1| \le r/2$, so that M_1 and N_1 are contained in the disc $D(P, r/2)$. From the property (A) of φ it follows that

$$|\varphi(P)\varphi(M_1)| = |PM_1|, \ |\varphi(P)\varphi(N_1)| = |PN_1| \text{ and } |\varphi(M_1)\varphi(N_1)| = |M_1N_1|,$$

which are the equalities we require.

Now suppose that for the points M_{i-1}, N_{i-1}, N_i and M_i our equalities are known, and let's prove them for M_i, N_i, N_{i+1}, M_{i+1}. Obviously (the reader can check this easily) the diagonals of any of the trapeziums $M_iN_iN_{i+1}M_{i+1}$ have length not exceeding that of the final trapezium, which is less than r. It follows from this that the trapeziums $M_{i-1}N_{i-1}N_iM_i$ and $M_iN_iN_{i+1}M_{i+1}$ are both contained in the disc $D(N_i, r)$ of Π (Figure 10.9, left).

We apply φ to the line segments $[M_{i-1}M_{i+1}]$ and $[N_{i-1}N_{i+1}]$; by property (B) of the map φ, each of these is superposed on some line segment of Σ. Set

$$\varphi(M_{i-1}) = m_{i-1}, \ \varphi(M_i) = m_i, \ \varphi(M_{i+1}) = m_{i+1},$$

and

$$\varphi(N_{i-1}) = n_{i-1}, \ \varphi(N_i) = n_i, \ \varphi(N_{i+1}) = n_{i+1}.$$

First of all, we verify that all of these points are contained in the disc $D(n_i, r)$ of Σ. This is obvious for the points n_i, n_{i-1}, m_{i-1} and m_i, since by the inductive hypothesis, the distances between them are equal to those between the points of Π which mapped to them under φ, so that

$$|n_i n_{i-1}| = |N_i N_{i-1}|,\ |n_i m_{i-1}| = |N_i M_{i-1}| \text{ and } |n_i m_i| = |N_i M_i|,$$

and as we have seen, each of the four points N_i, N_{i-1}, M_{i-1} and M_i are contained in $D(N_i, r)$. Our assertion is also obvious for the point n_{i+1}, since under

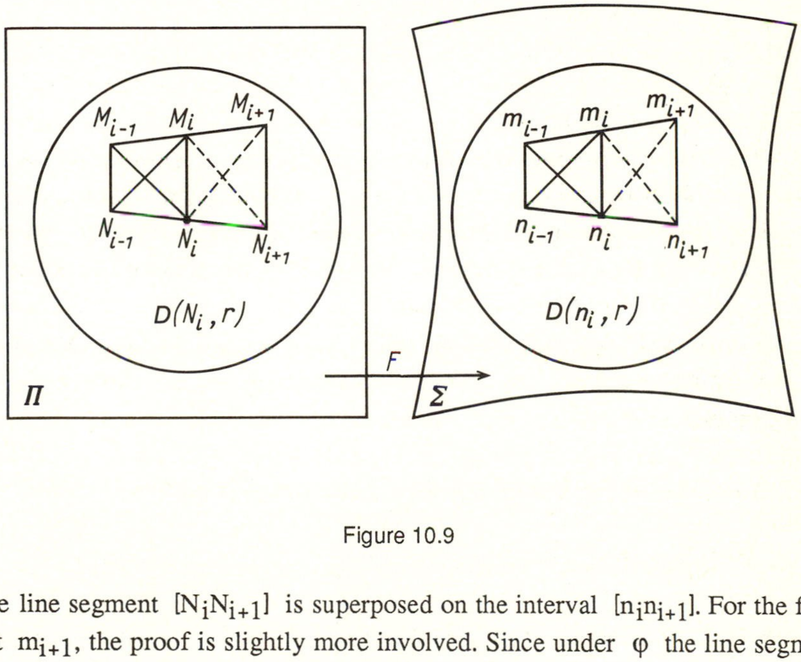

Figure 10.9

φ the line segment $[N_i N_{i+1}]$ is superposed on the interval $[n_i n_{i+1}]$. For the final point m_{i+1}, the proof is slightly more involved. Since under φ the line segment $[M_i M_{i+1}]$ is superposed on $[m_i m_{i+1}]$, we have $|M_i M_{i+1}| = |m_i m_{i+1}|$; in addition, as we have already verified, $|N_i M_i| = |n_i m_i|$. Finally,

$$|n_i m_{i+1}| \le |n_i m_i| + |m_i m_{i+1}|.$$

From these relations we get that

$$|n_i m_{i+1}| \le |N_i M_i| + |M_i M_{i+1}| \le r/2 + r/2 = r.$$

Thus we have also proved the required assertion for m_{i+1}. Now recall that the geometry Σ is locally Euclidean. More precisely, the disc $D(n_i, r)$ of Σ has a chart which can be assumed to lie in Π; using this chart we identify it with a disc in Π, and assume that the points $n_{i-1}, n_i, n_{i+1}, m_{i-1}, m_i$ and m_{i+1} lie in Π. Thus we get the following simple picture (see Figure 10.9). In the plane we have two figures, consisting of points $N_{i-1}, N_i, N_{i+1}, M_{i-1}, M_i$ and M_{i+1}, with the line segments $[N_{i-1}N_{i+1}]$ and $[M_{i-1}M_{i+1}]$ containing N_i and M_i on the one hand, and of points $n_{i-1}, n_i, n_{i+1}, m_{i-1}, m_i$ and m_{i+1}, with the line segments $[n_{i-1}n_{i+1}]$ and $[m_{i-1}m_{i+1}]$ containing n_i and m_i on the other (Figure 10.9). Under this, all the line segments drawn on the left-hand side of Figure 10.9 have the same lengths as those on the right with corresponding endpoints:

$$|M_{i-1}N_{i-1}| = |m_{i-1}n_{i-1}|, \ldots, |N_iN_{i+1}| = |n_in_{i+1}|.$$

The assertion we need to prove is that the corresponding line segments drawn as dotted lines in Figure 10.9 also have the same length:

$$|M_iN_{i+1}| = |m_in_{i+1}|, \quad |M_{i+1}N_i| = |m_{i+1}n_i| \quad \text{and} \quad |M_{i+1}N_{i+1}| = |m_{i+1}n_{i+1}|.$$

This will of course be proved if we can show that the figures themselves are congruent; this is almost obvious, and can for instance be checked as follows.

Since the distances between all the points of the figures $n_{i-1}m_{i-1}m_in_i$ and $N_{i-1}M_{i-1}M_iN_i$ (made up of four points each) are equal, the two figures are congruent, so that by Lemma 1 there is a motion F of the plane which takes N_{i-1} into n_{i-1}, N_i into n_i, M_i into m_i, and M_{i-1} into m_{i-1}; now F takes the ray $M_{i-1}M_i$ into $m_{i-1}m_i$ and hence the point M_{i+1} on the ray $M_{i-1}M_i$ into the point m_{i+1} on the ray $m_{i-1}m_i$, since the distances $|M_{i-1}M_{i+1}|$ and $|m_{i-1}m_{i+1}|$ of these points from the starting-point of the rays is the same. In the same way one proves that F take N_{i+1} into n_{i+1}.

Since the motion F preserves distances it follows that

$$|m_im_{i+1}| = |M_iM_{i+1}|, \quad |m_in_i| = |M_iN_i|, \quad \ldots, \quad |n_in_{i+1}| = |N_iN_{i+1}|.$$

This proves the assertion that we needed for the ith step of the induction. Thus we have proved that

$$|\varphi(A)\varphi(B)| = |AB| \text{ whenever } |AB| \le r/2.$$

This completes the proof of Theorem 1a.

Exercise

1. Prove the following analogue of the Extension Lemma. Suppose given a superposition φ of the disc D(P, r) in the plane onto a disc of a locally Euclidean geometry Σ; then for any point A of D(P, r), φ can be extended to the disc D(A, r) such that it is also a superposition on this disc.

10.4. Construction of the group. We proceed to Stage II of the proof of Theorem 1, that is, to the construction of the group Γ. To explain the idea of this construction, we return again to Example 3, given in §10.3 in connection with the notion of covering; that is, suppose that some uniformly discontinuous group Γ of motions of Π is given, and consider the geometry Σ_Γ defined by Γ, and the corresponding cover $(\Pi, \Sigma_\Gamma, \varphi_\Gamma)$. In §7, we introduced the definition of equivalence of points in the plane and defined a point a of the geometry Σ_Γ as a set **A** of equivalent points. Hence under φ_Γ, all equivalent points (and only these) go into a single point, that is, $\varphi_\Gamma(A) = \varphi_\Gamma(B)$ if and only if A and B are equivalent. In the general case, when we have an arbitrary covering, this example suggests that we should go round the opposite way, taking the relations $\varphi(A) = \varphi(B)$ as the definition of equivalence.

Definition 1. If $(\Sigma_1, \Sigma_2, \varphi)$ is any covering, then points A and B of Σ_1 are said to be *equivalent with respect to* φ if they go into the same point of Σ_2, that is, if $\varphi(A) = \varphi(B)$. Thus a set of equivalent points is just the set of all points of Σ_1 which go into some given point of Σ_2.

Now recall how we defined equivalence in §7: there we had a given group Γ, and we defined points A and B of the plane Π to be equivalent if Γ contains a motion F such that $F(A) = B$. Together with §7.1, Theorem 1, this suggests that in the case of an arbitrary covering $\varphi: \Pi \to \Sigma$ of a geometry Σ by the plane, (where we have a notion of equivalence, as just defined, but as yet no group Γ), we should go round the opposite way, and define the group Γ, basing ourselves on the relation $F(A) = B$ for equivalent points A and B:

Definition 2. Given a covering (Π, Σ, φ), the *covering group* of (Π, Σ, φ) is the set Γ of all motions F of the plane Π which take every point of Π into an equivalent point.

If we recall that A and B are by definition equivalent with respect to φ if and only if $\varphi(A) = \varphi(B)$, we can rewrite this condition on F in the following form: $\varphi(F(A)) = \varphi(A)$ for every point A of Π. In turn, by the definition of composition of maps, this condition means that the maps φF and φ are equal; thus we can express the definition more compactly: Γ consists of all motions F for which $\varphi F = \varphi$.

Theorem 1b. The covering group Γ of a covering (Π, Σ, φ) is a uniformly discontinuous group of motions of the plane.

First of all, we must justify using the word 'group' in the definition of the set Γ. If F and G are motions belonging to Γ, then $\varphi FG = \varphi G = \varphi$, since $\varphi F = \varphi$ and $\varphi G = \varphi$; moreover, using $\varphi = \varphi F$ and $FF^{-1} = E$, gives $\varphi F^{-1} = \varphi FF^{-1} = \varphi E = \varphi$. Hence Γ is a group. Let's prove that Γ is uniformly discontinuous. Suppose that $F(X) \neq X$ for some motion F belonging to Γ and for some point X of Π; we prove that then $|F(X)X| > r$. Indeed, if we suppose that $|F(X)X| \leq r$ then F(X) lies in the disc D(X, r); but the covering φ is a superposition on this disc, so that

$$|XF(X)| = |\varphi(X)\varphi(F(X))| = |\varphi(X)\varphi(X)| = 0,$$

that is, $X = F(X)$. We get a contradiction. This proves that Γ is a uniformly discontinuous group. The theorem is proved.

We require one final property of the covering group Γ of (Π, Σ, φ). By definition of this group, for any motion F in Γ and any point A of Π, the point F(A) is equivalent to A. We now prove that we get in this way all points equivalent to A. From this, it will follow that the notion of equivalence with respect to the covering φ defined above agrees exactly with the equivalence defined by the group Γ, as introduced in §7.

Theorem 1c. Let (Π, Σ, φ) be any covering and Γ its covering group. For any two points A and A′ of Π which are equivalent with respect to φ, there exists a motion F in Γ such that $F(A) = A'$.

Proof. It is easy to propose a plausible candidate for the required motion F. Recall that by definition, A and A′ are equivalent if $\varphi(A) = \varphi(A')$. Write a for the point $\varphi(A) = \varphi(A')$ of Σ. By definition of a covering, it follows that φ defines superpositions f_1 of D(A, r) onto D(a, r), and f_2 of D(A′, r) onto the same disc. In other words, we have two charts of D(a, r). By the Corollary of Lemma 1, there exists a motion F of Π which takes one chart into the other, that is, such that $f_1(B) = f_2(F(B))$ for all points B of D(A, r). This motion F is our candidate; it only remains for us to prove that it really serves our purpose, in other words, that it belongs to Γ.

By definition of Γ, this means that we need

$$(\varphi F)(B) = \varphi(B)$$

for every point B of Π. Note that by the very definition of F, it follows that this relation holds when B is a point of D(A, r). Indeed, by definition, for points of this disc we have $f_1(B) = (f_2F)(B)$. But φ is equal to f_1 in D(A, r), and to f_2 in D(A′, r). Replacing f_1 and f_2 by φ gives the required relation.

Now we try to extend the relation we have obtained from points of D(A, r) to all points of Π.

The assertion which we need to prove can be formulated in another way, if we remark that $\psi = \varphi F$ is obviously also a covering of Σ by Π (since φ is a covering, and F is a motion of the plane).

Lemma 4 (Criterion for equality of two coverings). Let φ and ψ be two coverings of a geometry Σ by the plane Π; if φ and ψ agree on the points of some disc in the plane, then they agree everywhere, that is $\varphi = \psi$.

Let B be any point of Π. We choose points $A_0 = A, A_1, ..., A_{n-1}, A_n = B$ on the line segment [AB], such that the distance between any adjacent pair of points A_i and A_{i+1} is less than r. Then each point A_i is the centre of a disc $D(A_i, r)$, and these discs are so closely spaced that the centre A_{i+1} of the next disc $D(A_{i+1}, r)$ is contained in $D(A_i, r)$ (Figure 10.10). We know that φ and ψ agree

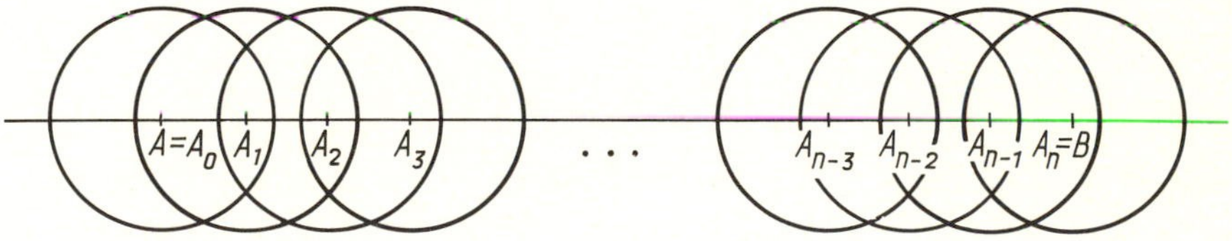

Figure 10.10

on the first disc $D(A_0, r)$. From this, we will show how to deduce that φ and ψ agree on the following disc $D(A_1, r)$, then on the following disc $D(A_2, r)$, ..., and so on until we get to $D(A_n, r)$. If we prove that φ and ψ agree on $D(A_n, r)$, then we can deduce that they agree at the centre of this disc; but $A_n = B$, and from this we get that $\varphi(B) = \psi(B)$. Since B is any point of the plane, this means that φ and ψ are equal.

Clearly, we only have to deal with the passage from any one disc to the next, for example from $D(A_0, r)$ to $D(A_1, r)$. We are given that the maps φ and ψ agree on $D(A_0, r)$, and in particular $\varphi(A_0) = \psi(A_0)$; denote this point of Σ by a_0. The map φ is a superposition of $D(A_0, r)$ onto the disc $D(a_0, r)$ in Σ, and ψ agrees with φ on $D(A_0, r)$. In particular, $\varphi(A_1) = \psi(A_1)$ (recall that the A_i are so closely spaced that A_1 is contained in $D(A_0, r)$). We denote the point $\varphi(A_1) = \psi(A_1)$ by a_1. By definition of a covering, φ defines a superposition of $D(A_1, r)$ onto $D(a_1, r)$, and the same holds for ψ; that is, φ and ψ define two superpositions of the same disc $D(A_1, r)$ onto the same disc $D(a_1, r)$. Moreover, by what we have already proved, φ and ψ agree on the part of the disc $D(A_1, r)$

which is in contained in $D(A_0, r)$ (Figure 10.11). Let's prove that this implies that φ and ψ agree on the whole disc $D(A_1, r)$. In fact since Σ is locally Euclidean, we can superpose $D(a_1, r)$ onto some disc $D(B, r)$ of the plane Π. Then our problem translates into the following question of plane geometry: given two discs $D(A_1, r)$ and $D(B, r)$ in Π, and two maps of one disc onto the other which are

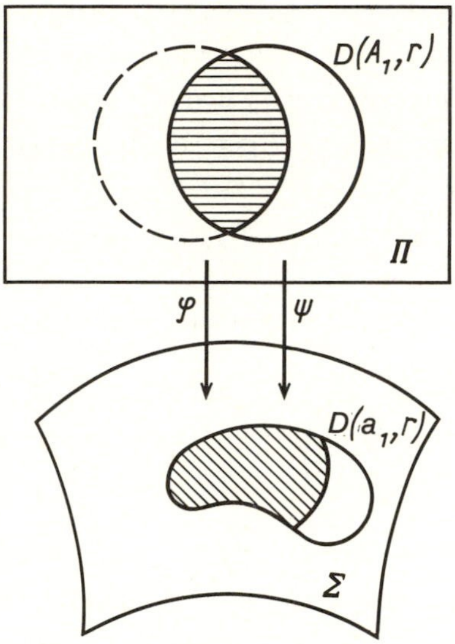

Figure 10.11

superpositions and which agree in the shaded area of $D(A_1, r)$ (as shown in Figure 10.12); then we need to prove that these maps agree on the whole disc. This is now very easy.

According to Lemma 1, each of the superpositions is defined by some motion of the plane Π; write G_1 and G_2 for these motions. They both take $D(A_1, r)$ to $D(B, r)$ and they agree in the shaded area of $D(A_1, r)$. Then obviously, this shaded region is not contained in a line, so that by Lemma 1, $G_1 = G_2$, as required. This completes the proof of Theorem 1c.

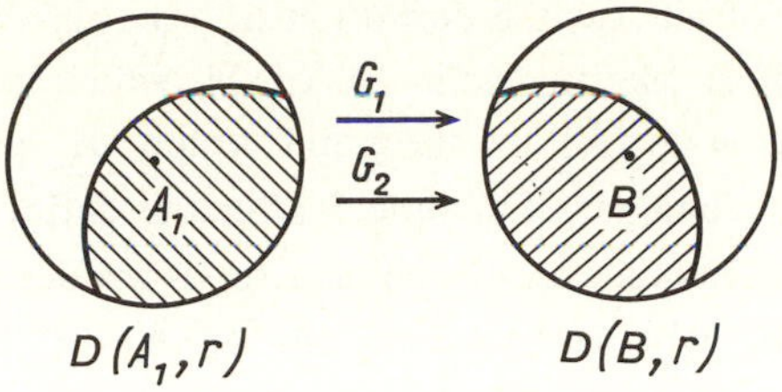

Figure 10.12

Exercises

1. Generalise Lemma 4 to the case when instead of Π we have an arbitrary locally Euclidean geometry.

2. Give an example of a locally Euclidean geometry for which the assertion of Lemma 1 is false.

3. Give an example of a covering $(\Sigma_1, \Sigma_2, \varphi)$ of locally Euclidean geometries for which the assertion of Theorem 1c is false. Nevertheless, it holds for certain coverings (see §13.1, Exercise 4).

10.5. Conclusion of the proof of Theorem 1. Finally we can proceed directly to the proof of the assertion made in Theorem 1. Let Σ be an arbitrary locally Euclidean geometry.

We construct the covering (Π, Σ, φ) of Σ by the plane Π, the existence of which is established in Theorem 1a. Using §10.4, Definition 1, we introduce on the plane the notion of equivalence with respect to φ. Using Definition 2, and starting from the notion of equivalence just constructed, we define a group Γ. According to Theorem 1b, Γ is a uniformly discontinuous group of motions of the plane. As in §7, we construct a geometry Σ_Γ corresponding to Γ. We now prove that the original geometry Σ can be superposed on this geometry Σ_Γ. This will then prove Theorem 1.

For this, we define first of all a map from the set of points of Σ into the set of points of Σ_Γ.

Take any point a of Σ. Let $\mathbf{A}$ be the set of all points in the plane Π which go to a under the covering φ (the set $\mathbf{A}$ is usually called the *inverse image* of a in Π). According to Theorem 1c, $\mathbf{A}$ is exactly a set of points of Π which are equivalent under Γ. This is just the type of set which determines a point of the geometry Σ_Γ; let a' be the point of Σ_Γ specified by $\mathbf{A}$.

Denote by $\psi\colon \Sigma \to \Sigma_\Gamma$ the map of Σ into Σ_Γ we have just constructed (ψ

is given by $a \mapsto \mathbf{A} \mapsto a'$). Obviously the only remaining thing for us to check to complete the proof of Theorem 1 is the next assertion: ψ is a superposition of Σ onto Σ_Γ.

We remark first of all that the definition of ψ can be reformulated as follows: for a point a of Σ, take any point A of Π which goes to a under the covering φ, and let $a' = \varphi_\Gamma(A)$ be the point which A goes into under the covering $\varphi_\Gamma: \Pi \to \Sigma_\Gamma$ (compare Example 3 after the definition of covering in §10.2). Then because of Theorem 1c, this point a′ does not depend on the point A of Π we chose, provided only that $\varphi(A) = a$ and $\varphi_\Gamma(A) = a'$ (Figure 10.13).

Let's check that ψ maps the set of points of Σ onto the whole set of points of Σ_Γ. This is completely obvious: if a′ is any point of Σ_Γ and A is a point of Π which goes to a′ under the covering $\varphi_\Gamma: \Pi \to \Sigma_\Gamma$, then letting $a = \varphi(A)$, we have $\psi(a) = a'$ by the above. But moreover, ψ is a one-to-one correspondence from the set of points of Σ to the set of points of Σ_Γ; this means that two points a

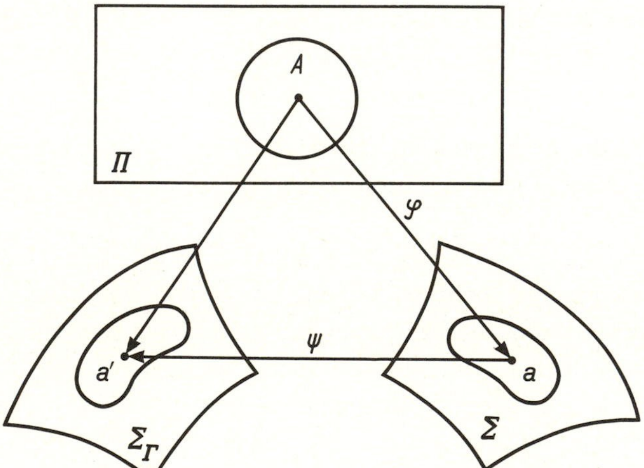

Figure 10.13

and b of Σ go into the same point of Σ_Γ (that is, $\psi(a) = \psi(b)$) only if $a = b$. This is also very simple to see: let $a = \varphi(A)$ and $b = \varphi(B)$, where A and B are points of Π. By definition, $\psi(a) = \psi(b)$ means that the points A and B go into the same point of Σ_Γ under the covering $\Pi \to \Sigma_\Gamma$. By definition of Σ_Γ, this can only happen if A and B are equivalent under Γ. But then by definition of a covering group, since Γ is the covering group of (Π, Σ, φ), we have $\varphi(A) = \varphi(B)$, that is,

$a = b$.

One further property is just as simple to show, namely that ψ is a covering of Σ_Γ by Σ. For the proof, take points A, a and a' as shown in Figure 10.13; recall that for some positive number r the maps $D(A, r) \to D(a, r)$ and $D(A, r) \to D(a', r)$ are superpositions (since $\Pi \to \Sigma$ and $\Pi \to \Sigma_\Gamma$ are coverings). If we construct the map ψ for points b of $D(a, r)$ by choosing the point B such that $\varphi(B) = b$ in $D(A, r)$, then we get at once that ψ defines a superposition of $D(a, r)$ onto $D(a', r)$.

We thus have the following situation: two locally Euclidean geometries Σ and $\Sigma' = \Sigma_\Gamma$, and a map ψ of Σ into Σ' which is both a one-to-one correspondence and a covering.

We prove that in these circumstances, ψ is always a superposition of Σ on Σ', and that furthermore, this is even true for arbitrary geometries, not necessarily locally Euclidean. This shows that it is not by chance that in all our examples of coverings which were not superpositions, the covering was not a one-to-one correspondence (see the examples in §8.2, and the exercises to that section).

Lemma 5. Let ψ be a covering of a geometry Σ' by a geometry Σ, and suppose that ψ is a one-to-one correspondence; then ψ is a superposition of geometries.

Proof. We need to prove the equality $|ab| = |\psi(a)\psi(b)|$ for points a and b of Σ. For this, it is enough to prove that

$$|ab| \geq |\psi(a)\psi(b)|. \tag{2}$$

Indeed, since ψ is a one-to-one correspondence, it has an inverse map ψ^{-1}, with the property that $\psi^{-1}(a') = a$ if $\psi(a) = a'$. Obviously ψ^{-1} is also a one-to-one correspondence. Moreover, it is also a covering of Σ by Σ'. Indeed, if a' is a point of Σ' and $a = \psi^{-1}(a')$, that is $a' = \psi(a)$, then ψ is a superposition of a spherical neighbourhood $D(a, r)$ onto a spherical neighbourhood $D(a', r)$, and therefore the inverse map ψ^{-1} is a superposition of $D(a', r)$ onto $D(a, r)$. Thus Σ and Σ' play a symmetrical role.

If (2) is proved for all coverings (Σ, Σ', ψ) satisfying the properties we have indicated, then it will also hold for the covering $(\Sigma', \Sigma, \psi^{-1})$ and this at once implies the opposite inequality to (2), and hence the required equality.

Proof of (2). Let r be the number involved in the definition of the covering ψ, and let β be any positive integer. Then by condition (d) in the definition of geometry, there exist points $a_1, \ldots, a_n$ such that $a_1 = a$, $a_n = b$, $|a_i a_{i+1}| < r$ for all i, and

$$\sum |a_i a_{i+1}| \leq |ab| + \beta. \tag{3}$$

Then by the choice of r we have

$$|a_i a_{i+1}| = |\psi(a_i)\psi(a_{i+1})|,$$

so that

$$\sum |a_i a_{i+1}| = \sum |\psi(a_i)\psi(a_{i+1})|,$$

and as we have seen in §6.1,

$$\sum |\psi(a_i)\psi(a_{i+1})| \geq |\psi(a)\psi(b)|$$

(this is a general property of any chain $\psi(a)$, $\psi(a_2)$, ..., $\psi(a_{n-1})$, $\psi(b)$). Putting this together with (3) gives

$$|\psi(a)\psi(b)| \leq \sum |\psi(a_i)\psi(a_{i+1})| = \sum |a_i a_{i+1}| \leq |ab| + \beta.$$

Since this inequality holds for any $\beta > 0$, this implies (2). This completes the proof of Theorem 1.

Exercises

1. Think through the proof of Theorem 1 for a 1-dimensional locally Euclidean geometry. What simplifications appear? What geometries does one get in consequence?
2. Try to carry through the proof of Theorem 1 we have given to the case of locally spherical geometries, that is, geometries which are identical in sufficiently small regions with that of the sphere. [Hint: difficulties arise in the proof of Theorem 1a, because rays emanating from one point intersect again at the antipodal point.]

Chapter III
Generalisations and applications

§11. 3-dimensional locally Euclidean geometries

11.1. Motions of 3-space. In the preceding chapters we have classified all 2-dimensional locally Euclidean geometries, that is, the worlds in which all properties of Euclidean geometry in the plane are satisfied in sufficiently small regions. An inhabitant of such a world who always remains within some distance r of a fixed point (home, for example) could not detect in his world any contradictions to Euclidean plane geometry. But the real space in which we live is 3-dimensional. Thus, it is of course more interesting to describe the 3-dimensional locally Euclidean geometries, that is, the worlds in which all properties of Euclidean geometry in 3-space are satisfied in sufficiently small regions; we can think of the description of the 2-dimensional geometries as just a model for this more interesting problem. In this section, we will concern ourselves with the description and some of the properties of 3-dimensional locally Euclidean geometries.

First of all, the two main theorems, §7, Theorem 2 and §10, Theorem 1, carry over to 3-dimensional geometries without any changes in their proofs, provided we replace the plane by 3-space and the disc by the ball throughout. In other words, just as in the 2-dimensional case, the description of locally Euclidean geometries in 3 dimensions is equivalent to the description of uniformly discontinuous groups of motions of 3-space. These groups can be classified in a similar way to our classification in §8 of uniformly discontinuous groups of motions of the plane. The arguments are in many ways similar to those which we got to know in §8. But for 3-space, the number of possible types of groups is quite a lot larger, and the arguments turn out to be considerably longer and more complicated. We therefore restrict ourselves to drawing up a complete list of such groups (there turn out to be 18 types), accompanying the list with some comments. In addition, similarly to the 2-dimensional case, we determine which of the resulting 18 types of 3-dimensional locally Euclidean geometries are bounded, and which are *orientable*. This last property generalises to 3-dimensions the property of 2-dimensional locally Euclidean geometries, appearing in §§4 and 9, that right and left are distinguishable.

We start with examples of motions of 3-space: twist, glide reflection and rotary reflection; these replace translation, rotation and glide reflection in the plane. They are obtained as composites of the well-known simple motions of space: translation $T_{\mathbf{a}}$ in a vector $\mathbf{a}$, rotation $R_{\ell,\varphi}$ through an angle φ about an axis ℓ, and reflection S^{Π} in a plane Π.

The *twist* $R_{\ell,\varphi}{}^{\mathbf{a}}$ with axis ℓ, angle φ and vector $\mathbf{a}$ is by definition the composite

$$R_{\ell,\varphi}{}^{\mathbf{a}} = T_{\mathbf{a}} R_{\ell,\varphi}$$

of a rotation $R_{\ell,\varphi}$ through an angle φ about an axis ℓ, and a translation $T_{\mathbf{a}}$ in a vector $\mathbf{a}$ parallel to ℓ. It is characterised by the fact that lines parallel to ℓ are rotated through the angle φ about ℓ, and planes perpendicular to ℓ are translated in the vector $\mathbf{a}$ (Figure 11.1).

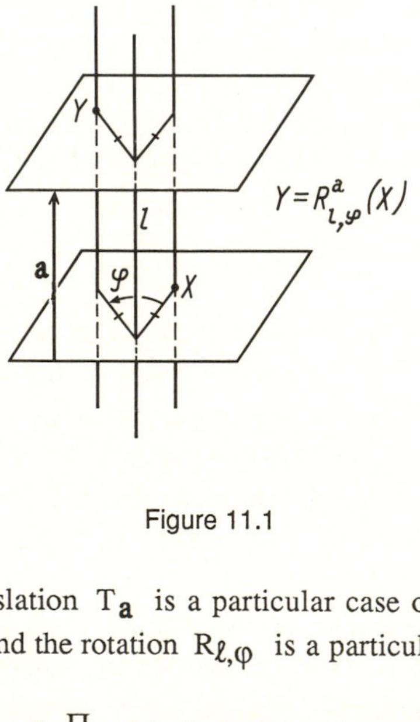

Figure 11.1

Note that the translation $T_{\mathbf{a}}$ is a particular case of a twist, with angle of rotation φ equal to 0; and the rotation $R_{\ell,\varphi}$ is a particular case, with translation vector $\mathbf{a}$ equal to $\mathbf{0}$.

The *glide reflection* $S_{\mathbf{a}}{}^{\Pi}$ with plane Π and vector $\mathbf{a}$ is by definition the composite

$$S_{\mathbf{a}}{}^{\Pi} = T_{\mathbf{a}} S^{\Pi}$$

of a reflection in a plane Π and a translation in a vector $\mathbf{a}$ parallel to Π. This is characterised by the fact that lines perpendicular to Π are all translated in the same vector $\mathbf{a}$, and planes parallel to Π are taken into their reflections in Π

(Figure 11.2).

Note that the reflection S^{Π} in a plane Π is a particular case of a glide reflection with translation vector **a** equal to **0**.

The *rotary reflection* $S_{\ell,\varphi}{}^{\Pi}$ with plane Π, axis ℓ perpendicular to Π and rotation angle φ is by definition the composite

$$S_{\ell,\varphi}{}^{\Pi} = R_{\ell,\varphi}S^{\Pi}$$

of a reflection in a plane Π and a rotation through an angle φ around the axis ℓ. It is characterised by the fact that lines perpendicular to Π are rotated through an angle φ around ℓ, and planes parallel to Π are taken into their reflections in Π (Figure 11.3).

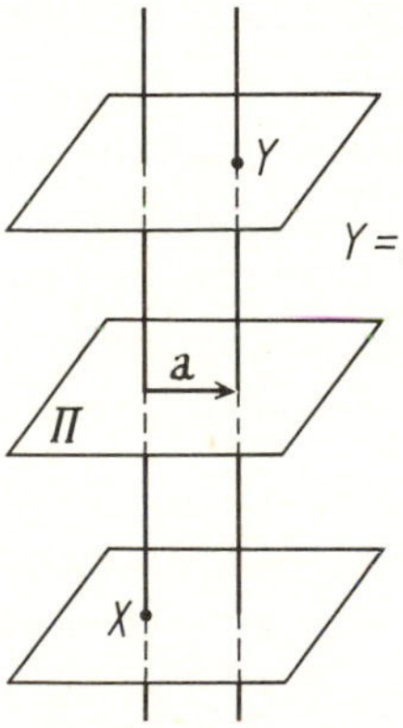

Figure 11.2

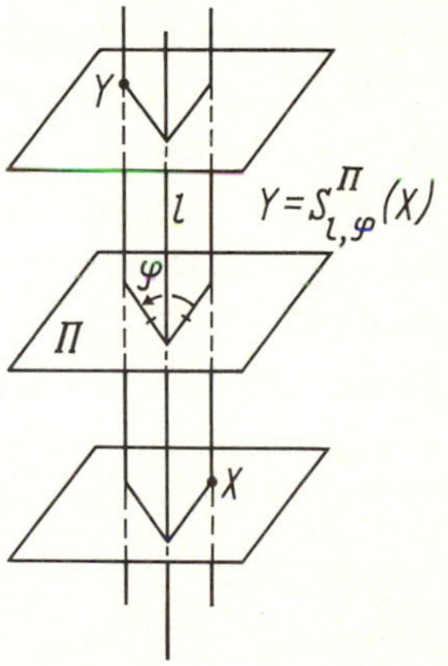

Figure 11.3

Note that an important particular case of a rotary reflection is the central reflection in a point: a rotary reflection $S_{\ell,\pi}{}^{\Pi}$ with angle π is a central reflection in O, the point of intersection of Π and ℓ.

We would like to draw to the reader's attention one property which all these types of motions have in common. As one sees from the definitions, each of them is related to a net of parallel lines and a perpendicular pencil of parallel planes in 3-space, which the motion preserves (recall that a net of parallel lines or a pencil of parallel planes in 3-space is respectively a set of all lines parallel to a given line, or of all planes parallel to a given plane). In the notation $R_{\ell,\varphi}{}^{\mathbf{a}}$, $S_{\mathbf{a}}{}^{\Pi}$, $S_{\ell,\varphi}{}^{\Pi}$ we are using, the subscript indicates what is happening to the net of parallel lines, and the superscript what is happening to the pencil of parallel planes. Since every point A of 3-space is a point of intersection of exactly one line m of the net and one plane M of the pencil, the image of A is the point of intersection of the image of the line m and of the plane M, so that the motion of 3-space is entirely determined.

Similar to, but rather more complicated than Chasles' Theorem in the plane (see §8), one can prove the next result.

Theorem 1 (Chasles' theorem in 3-space). Every motion of 3-space is a twist, a glide reflection or a rotary reflection.

A proof can be found in any of the following: H.S.M. Coxeter, Introduction to geometry, J.Wiley, New York, 1969; E.G. Rees, Notes on geometry, Springer, Berlin-New York, 1983, or J. Hadamard, Leçons de géométrie élémentaire, vol.2, Géométrie dans l'espace, A. Colin, Paris, 1949.

The reader can prove this theorem for himself, using the hints given in Exercise 1: the main difficulty consists in showing that every motion of 3-space takes a certain net of parallel lines into itself; the perpendicular pencil of parallel planes is then also automatically taken into itself.

As in the case of the plane, not every motion can appear in a uniformly discontinuous group of motions of 3-space. Indeed, §8, Proposition 1, which of course continues to hold in 3-space, forbids any motion other than the identity from having a fixed point. Thus a uniformly discontinuous group of motions of 3-space cannot contain a rotation $R_{\ell,\varphi}$ through a non-zero angle φ around an axis ℓ, a reflection S^{Π} or any rotary reflection $S_{\ell,\varphi}{}^{\Pi}$, since these motions have respectively the points of the axis ℓ, the plane Π and the point of intersection of ℓ and Π as fixed points. Thus the non-identity elements of a uniformly discontinuous group of motions of 3-space can only be twists $R_{\ell,\varphi}{}^{\mathbf{a}}$ and glide reflections $S_{\mathbf{a}}{}^{\Pi}$, and in either case the translation vector $\mathbf{a}$ must be non-zero.

Exercises

1. (a) State and prove the analogues in 3-space of §8, Lemma 3 and §10, Lemma 1. (b) Prove that the motions of spherical geometry are induced by motions of 3-space taking the centre of a sphere to itself. State and prove the analogue of §8, Lemma 3 for motions of spherical geometry. (c) State and prove the analogues of §8, Lemmas 1 and 2, and of Chasles' theorem for spherical geometry. [Hint: see §8.1, Exercise 3.] (d) Prove that every motion of 3-space takes some net of parallel lines into itself. [Hint: replace F by the composite $T_{\mathbf{a}}F$ which fixes a point O; here $T_{\mathbf{a}}$ is a translation in $\mathbf{a} = \overrightarrow{F(O)O}$; then use the results of (b) and (c) above.] (e) Prove Chasles' theorem in 3-space. [Hint: observe that a net of parallel lines and a pencil of parallel planes, with the usual definition of distance between lines and planes, form geometries: the first is the plane, and the second the line; then use Chasles' theorem on the plane and the line.]

2. Verify that the composite $T_{\mathbf{b}}R_{\ell,\pi}{}^{\mathbf{a}}$ of a twist with angle π and a translation in a vector $\mathbf{b}$ perpendicular to the axis ℓ is a twist $R_{\ell',\pi}{}^{\mathbf{a}}$ with axis $\ell' = T_{\mathbf{b}/2}(\ell)$ parallel to ℓ, and the

same angle π and vector **a**.

3. Verify that the composite $T_b S_a^{\Pi}$ of a glide reflection with plane Π and a translation in a vector **b** perpendicular to Π is a glide reflection $S_a^{\Pi'}$ with plane $\Pi' = T_{b/2}(\Pi)$ parallel to Π and with the same translation vector **a**.

11.2. Uniformly discontinuous groups in 3-space: generalities. As in the case of groups in the plane, the classification of uniformly discontinuous groups of motions of 3-space is based on the analysis of the translations belonging to these groups. Obviously, if Γ is a uniformly discontinuous group then the translations in Γ also form a uniformly discontinuous group Γ'. As in the case of the plane, it is not hard to see which groups Γ' can appear; it turns out that there can be four types:

Type	General form of the translations of Γ'
I	T_0
II	$T_{n\mathbf{a}}$, where $\mathbf{a} \neq \mathbf{0}$ is some vector, and n is any integer
III	$T_{m\mathbf{a}+n\mathbf{b}}$, where **a** and **b** are certain non-collinear vectors, and m, n are any integers
IV	$T_{\ell\mathbf{a}+m\mathbf{b}+n\mathbf{c}}$, where **a**, **b** and **c** are certain non-coplanar vectors, and ℓ, m, n are any integers

If we lay off from some point the endpoints of all the vectors **x** for which the translation $T_{\mathbf{x}}$ is in Γ' then we get: for Type I, a single point; for Type II, a series of points on a line such that the distance between any two successive points

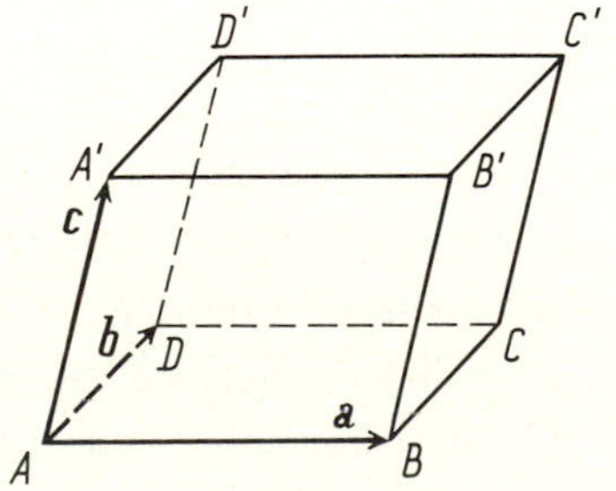

Figure 11.4

is equal; for Type III, a lattice in a plane; the set of points obtained for Type IV is called a *space lattice*. A space lattice is obtained from a plane lattice formed by the endpoints of vectors $\ell\mathbf{a} + m\mathbf{b}$ (contained in the plane Π spanned by $\mathbf{a}$ and $\mathbf{b}$) by translations in vectors $\mathbf{c}, 2\mathbf{c}, 3\mathbf{c}, \ldots, -\mathbf{c}, -2\mathbf{c}, -3\mathbf{c}, \ldots$ If Δ is the parallelepiped ABCDA′B′C′D′ for which the edges out of A are the vectors $\mathbf{a}, \mathbf{b}$ and $\mathbf{c}$ (Figure 11.4), then the whole of space is filled with parallelepipeds equal and parallel to Δ, touching one another along their faces.

It is easy to verify that groups of Type I-IV actually are uniformly discontinuous. For example, in the case of Type IV, the points $T_{\mathbf{x}}(P)$ run through a space lattice, one of whose vertices is P. If $\mathbf{x} = \ell\mathbf{a}+m\mathbf{b}+n\mathbf{c}$ with $n \neq 0$ then $T_{\ell\mathbf{a}+m\mathbf{b}}(P)$ lies in the plane Π which contains $\mathbf{a}$ and $\mathbf{b}$ laid off from P; the point $T_{\mathbf{x}}(P)$ is contained in the plane $\Pi' = T_{n\mathbf{c}}(\Pi)$ parallel to Π, and obtained by translating Π in $n\mathbf{c}$. The distance between parallel planes Π and Π' is not less than the distance between the planes Π and $\Pi_1 = T_{\mathbf{c}}(\Pi)$ (Figure 11.5). If the

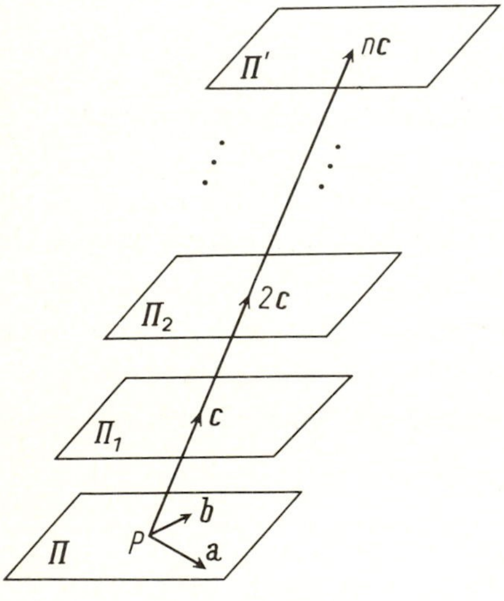

Figure 11.5

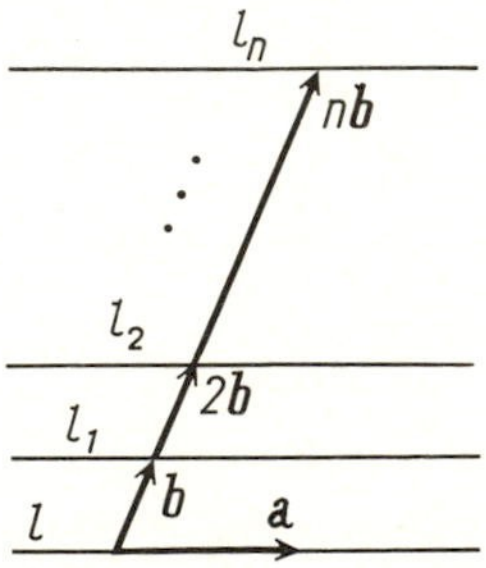

Figure 11.6

distance between Π and Π_1 equals d, then $|PT_{\mathbf{x}}(P)| \geq d$. If on the other hand $n = 0$, then $T_{\mathbf{x}}(P)$ is contained in the plane Π, and we obtain a similar plane picture, which can be dealt with in an entirely similar way (Figure 11.6).

To relate a group Γ' consisting only of translations to the whole group Γ containing it, we need to introduce another idea.

By definition, a motion F acts on points of 3-space, taking A into some point F(A). Notice that the action of F can be extended to vectors, by applying it to the origin and endpoint of a vector; that is, if $\mathbf{x} = AB^{\rightarrow}$, then by definition we set $F(\mathbf{x}) = F(A)F(B)^{\rightarrow}$. However, we must recall that the same vector may be defined by different pairs of points (A, B) and (C, D): $AB^{\rightarrow} = CD^{\rightarrow}$ if $|AB| = |CD|$ and AB and CD are parallel rays; we must check that in this case, the vectors $F(A)F(B)^{\rightarrow}$ and $F(C)F(D)^{\rightarrow}$ given by our rule will be equal. But this follows at once from the fact that under a motion, distances are preserved, and parallel rays go into parallel rays.

We needed the preceding remarks in order to state an important property of translations. The reader is probably already familiar with the fact that motions do not commute, that is, in general $FG \neq GF$. It turns out that a translation $T_{\mathbf{x}}$ 'commutes after correction' with any other motion F. More precisely, we have the relation

$$FT_{\mathbf{x}} = T_{F(\mathbf{x})}F. \tag{1}$$

The truth of this equality follows at once by definition of $F(\mathbf{x})$; it is obvious from Figure 11.7, which shows that for any point P, $F(T_{\mathbf{x}}(P))$ can be obtained from F(P) by translating in $F(\mathbf{x}) = F(P)F(T_{\mathbf{x}}(P))^{\rightarrow}$. This is the assertion of (1).

It follows from (1) that $T_{F(\mathbf{x})} = FT_{\mathbf{x}}F^{-1}$, and since by assumption F, $T_{\mathbf{x}}$ and F^{-1} belong to Γ, it follows that $T_{F(\mathbf{x})}$ also belongs to Γ; since it is a translation, it belongs to Γ'. In other words, the set of all vectors $\mathbf{x}$ such that the translation $T_{\mathbf{x}}$ belongs to Γ' must go into itself under the action of any motion F in Γ. This can be expressed in a different way: lay off all such vectors from some point O of space; as we have seen, the endpoints of the vectors then form a certain

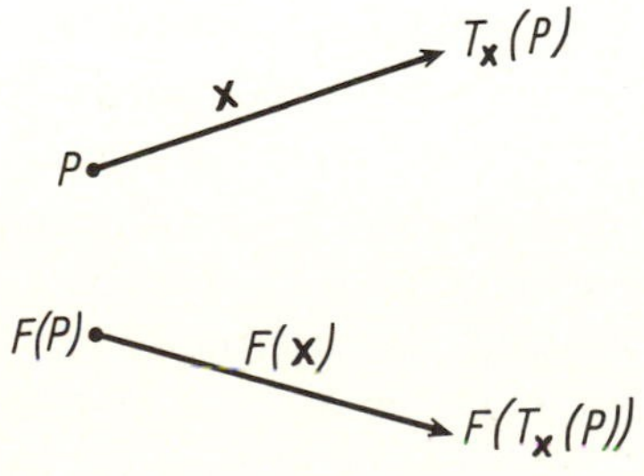

Figure 11.7

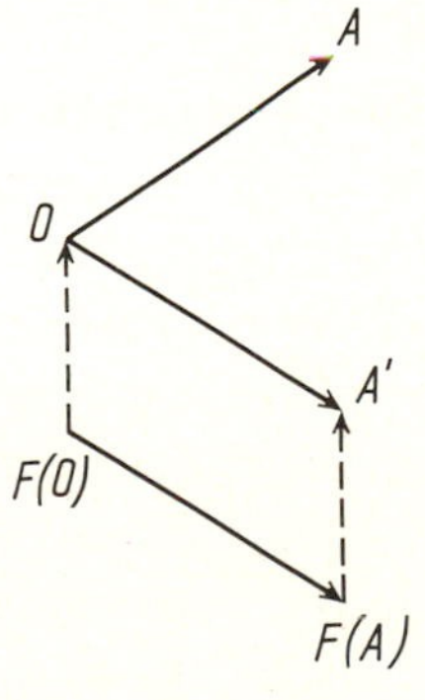

Figure 11.8

lattice Ω. If A is a point of Ω then the vector $\vec{OA}$ belongs to Γ'. Hence, as we have seen, Γ' also contains $F(\vec{OA})$, for any motion F in Γ. We also lay off $F(\vec{OA})$ from O; for this, we need to translate its endpoint F(A) by the vector $\vec{F(O)O}$ (Figure 11.8), taking it to some point A′. The assertion we have proved means that A′ must again belong to Ω. Thus Ω goes to itself if we apply to it any motion F of Γ, and then translate in $\vec{F(O)O}$. In other words, Ω is taken to itself by the motion $T_{F(O)O}F$, where here and in what follows we use T_{AB} to denote the translation in the vector $\vec{AB}$. The motion $T_{F(O)O}F$ obviously also fixes O. A motion of this kind, taking a lattice into itself and fixing one of its points is called a *symmetry* of a lattice. For example, the square lattice in the plane has symmetries consisting of rotations of the plane about some lattice point through angles of 0, $\pi/2$, π and $3\pi/2$, as well as reflections in the axes ℓ_1, ℓ_2, ℓ_3 and ℓ_4 (Figure 11.9).

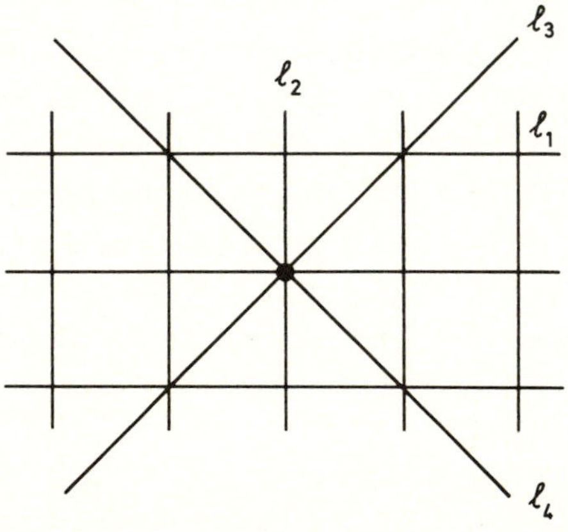

Figure 11.9

Thus we can say that knowing the translation group Γ' already imposes strong restrictions on the possibilities for the whole group Γ: its motions must be composites of symmetries of the lattice Ω corresponding to Γ' and translations.

We have already met this idea in §8 in the course of the analysis of uniformly discontinuous groups of Type III.b in the plane. In this case Γ' consists of translations in vectors of the form $m\mathbf{a} + n\mathbf{b}$, and the corresponding lattice Ω is obtained by laying these vectors off from some point O. Motions of Γ not belonging to Γ' can only be glide reflections, that is, are composed of a reflection and a translation. But an arbitrary lattice Ω will not in general go into itself under any reflection. In order for such a symmetry to exist, we choose Ω to be a rectangular lattice, that is, the vectors $\mathbf{a}$ and $\mathbf{b}$ to be perpendicular. Then for example, the reflection S in the line containing $\mathbf{a}$ will take Ω into itself. We choose the glide reflection $F = T_{\mathbf{a}/2}S$, composed of this reflection and of the translation $T_{\mathbf{a}/2}$, such that the square of F, a translation, belongs to Γ' (in the present instance, it equals $T_{\mathbf{a}}$). This is how we arrive at the group of Type III.b described in §8.3, Theorem 6.

We have discussed these ideas so that the description of uniformly discontinuous groups which we are now going to give does not seem too artifical. We divide up all the groups into four types according to the structure of the translation group Γ' in Γ (we already had a division of these into Type I, II, III and IV at the beginning of §11.2).

Type of Γ	**Condition on Γ and Γ'**
I	$\Gamma = \{E\}$
II	$\Gamma \neq \{E\}$ and Γ' belongs either to Type I or Type II
III	Γ' belongs to Type III
IV	Γ' belongs to Type IV

This classification is justified in part by the fact that the groups corresponding to bounded geometries are exactly the groups of Type IV. It is easy to see that geometries corresponding to groups of Type IV are bounded. Indeed, if the group contains translations $T_{\mathbf{a}}$, $T_{\mathbf{b}}$ and $T_{\mathbf{c}}$ in three non-coplanar vectors $\mathbf{a}$, $\mathbf{b}$ and $\mathbf{c}$, then it also contains the composite $T_{\mathbf{a}}^{\ell}T_{\mathbf{b}}^{m}T_{\mathbf{c}}^{n}$ where ℓ, m and n are any integers, that is, any translation $T_{\mathbf{v}}$ in a vector $\mathbf{v}$ of the form $\ell\mathbf{a} + m\mathbf{b} + n\mathbf{c}$. Let P be any point of space and $\mathbf{h}$ the vector $\overrightarrow{OP}$. We decompose $\mathbf{h}$ as a linear combination of the three non-coplanar vectors $\mathbf{a}$, $\mathbf{b}$ and $\mathbf{c}$, so that $\mathbf{h} = x\mathbf{a} + y\mathbf{b} + z\mathbf{c}$, where x, y and z are real numbers. We can write these numbers in the form $x = \ell + x'$, $y = m + y'$ and $z = n + z'$, where ℓ, m and n are integers and $0 \le x', y', z' < 1$. It follows that

$$\mathbf{h} = \ell\mathbf{a} + m\mathbf{b} + n\mathbf{c} + \mathbf{h}', \text{ where } \mathbf{h}' = x'\mathbf{a} + y'\mathbf{b} + z'\mathbf{c}. \qquad (2)$$

If the origin of the vector $\mathbf{h}'$ is at A then its endpoint P is contained in the parallelepiped $\Delta = ABCDA'B'C'D'$ (see Figure 11.4) which has vectors $\mathbf{a}$, $\mathbf{b}$ and $\mathbf{c}$ as its edges out of A: $\mathbf{a} = \overrightarrow{AB}$, $\mathbf{b} = \overrightarrow{AD}$, $\mathbf{c} = \overrightarrow{AA'}$. Since, as we have seen, the translation in $\ell\mathbf{a} + m\mathbf{b} + n\mathbf{c}$ belongs to our group Γ, (2) shows that every point P of space is equivalent under Γ to some point P' contained in the parallelepiped Δ. It follows from this that the distance between any two points of the geometry Σ corresponding to Γ does not exceed the distance between two points of Δ, so is bounded by some fixed length (the length of the longest diagonal of Δ).

Conversely, the geometries corresponding to groups of Type I-III are unbounded. This can be read off from the list of these groups (Table I below), by checking separately for the groups of each type. We now turn our attention to groups of Type I-III.

Exercises

1. Let Ω be a 3-space lattice which is general, so that it does not have any symmetries other than central reflections in its points. Prove that if Γ is a uniformly discontinuous group for which the translation group Γ' is represented by Ω then $\Gamma = \Gamma'$. [Hint: prove that a motion of 3-space which is the composite of a central reflection in a point of O and a translation in a vector **a** has a fixed point.]

2. Prove that there exist two uniformly discontinuous groups of motion of spherical geometry: the group consisting of the identity only, and the group with two elements, consisting of the identity and the motion taking each point of the sphere into the antipodal points. [Hint: see §11.1, Exercise 1.]

11.3. Uniformly discontinuous groups in 3-space: classification. We now proceed to describe the groups Γ of Type I, II and III. For this, we start from a lattice Ω consisting either of O only, or spanning a line or a plane, form the group Γ' of translations in Ω, and then add in some motions to Γ'; these must be of the form $T_{\mathbf{a}}F$, where F is some symmetry of Ω and $T_{\mathbf{a}}$ is a translation. The choice of the translations $T_{\mathbf{a}}$ is subject to two conditions: (i) we must get a uniformly discontinuous group, and (ii) this group should not contain translations other than those in Γ'. We now show how to do this in a specific example.

In the plane Oxy, consider vectors **a**, **b** pointing along the x and y-axes; let Γ' be the group of translations in vectors $m\mathbf{a} + n\mathbf{b}$, where m, n are any integers. Since the corresponding lattice Ω is rectangular, it goes into itself under reflection of the plane in the x-axis; and this plane reflection is in turn obtained by a rotation through an angle π about the x-axis $Ox = \ell$. Composing this rotation with a translation in $\mathbf{a}/2$ gives the twist $R = R_{\ell,\pi}{}^{\mathbf{a}/2}$, which satisfies

$$R^2 = T_{\mathbf{a}}; \tag{3}$$

that is, R^2 belongs to Γ'.

In addition, reflecting 3-space in the plane $\Pi = Oxy$ induces the identity map of Π to itself, and hence the identity symmetry of Ω. Composing this reflection with a translation in $\mathbf{b}/2$ gives the glide reflection $S = S_{\mathbf{b}/2}{}^{\Pi}$, which satisfies

$$S^2 = T_{\mathbf{b}}; \tag{4}$$

that is, S^2 belongs to Γ'. We now set $\Gamma = \langle R, S \rangle$ for the group generated by R and S, and show how to verify in this example that Γ is a uniformly discontinuous group, and that the corresponding geometry is unbounded.

Firstly, since $R^2 = T_{\mathbf{a}}$ and $S^2 = T_{\mathbf{b}}$ both belong to Γ, the whole of $\Gamma' = < T_{\mathbf{a}}, T_{\mathbf{b}} >$ is in Γ. Recall that, as explained in §11.2, both R and S 'commute after correction' with translations in Γ'. In our case, (1) reduces to the relations

$$RT_{\mathbf{h}} = T_{\sigma(\mathbf{h})}R, \tag{5}$$

$$ST_{\mathbf{h}} = T_{\mathbf{h}}S, \tag{6}$$

where σ is the reflection of the plane Oxy in the axis Ox. Finally, the reader can easily verify that R and S also 'commute after correction' with one another, namely, one has the formula

$$SR = RST_{-\mathbf{b}} \tag{7}$$

(one could for example see what happens to the coordinates (x, y, z) of any point under the motions on the left and right-hand side of (7)). Thus any composite of the elements R and S taken in any order can be transformed, using the 'commutation rules' (5 - 7), to take R and S to the left, and R to the left of S; if R^2 or S^2 appears in this process, we replace them as in (3) and (4). Thus any such motion (so any element of Γ) can be written in the form $T_{\mathbf{h}}$ or $RT_{\mathbf{h}}$ or $ST_{\mathbf{h}}$ or $RST_{\mathbf{h}}$, where $\mathbf{h}$ is a vector of Ω.

As an example, we can transform the composite

$$SRS^{-1}R^{-1}SR$$

into

$$SRST_{-\mathbf{b}}RT_{-\mathbf{a}}SR$$

by using formulas (3) and (4) to give $R^{-1} = RT_{-\mathbf{a}}$ and $S^{-1} = ST_{-\mathbf{b}}$. Moreover, using (7), this composite can be written

$$RST_{-\mathbf{b}}ST_{-\mathbf{b}}RT_{-\mathbf{a}}SR.$$

Then using (6) and (4), we get

$$RT_{-\mathbf{b}}T_{\mathbf{b}}T_{-\mathbf{b}}RT_{-\mathbf{a}}SR, \quad \text{or} \quad RT_{-\mathbf{b}}RT_{-\mathbf{a}}SR.$$

Now using (5) and (3), we represent him as

TABLE 1

Uniformly discontinuous groups of motions of 3-space of Types I, II and III

I		$\Gamma = \langle E \rangle$	
II.1	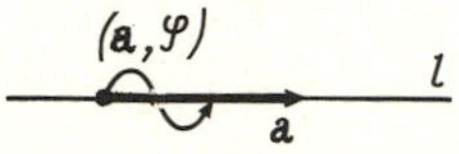	$\Gamma = \langle R^{\mathbf{a}}_{\ell,\varphi} \rangle$	with $\mathbf{a} \neq 0$
II.2	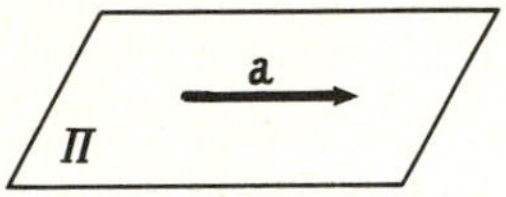	$\Gamma = \langle S^{\Pi}_{\mathbf{a}} \rangle$	with $\mathbf{a} \neq 0$
III.1	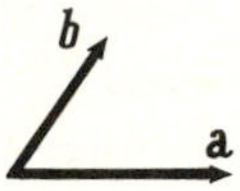	$\Gamma = \langle T_{\mathbf{a}}, T_{\mathbf{b}} \rangle$	with **a**, **b** not parallel
III.2	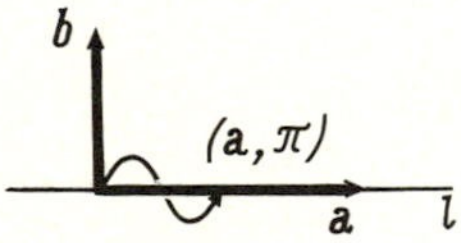	$\Gamma = \langle R^{\mathbf{a}}_{\ell,\pi}, T_{\mathbf{b}} \rangle$	with **a**, **b** perpendicular
III.3	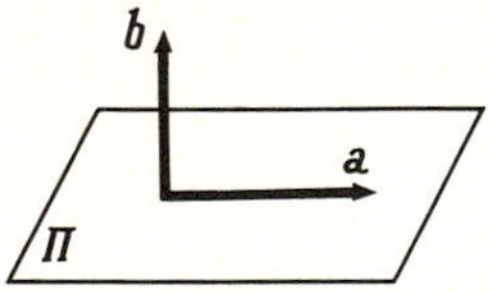	$\Gamma = \langle S^{\Pi}_{\mathbf{a}}, T_{\mathbf{b}} \rangle$	with **b** perpendicular to Π
III.4	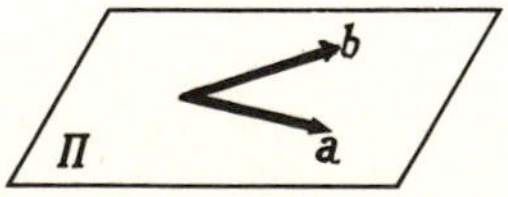	$\Gamma = \langle S^{\Pi}_{\mathbf{a}}, S^{\Pi}_{\mathbf{b}} \rangle$	with **a**, **b** not parallel
III.5	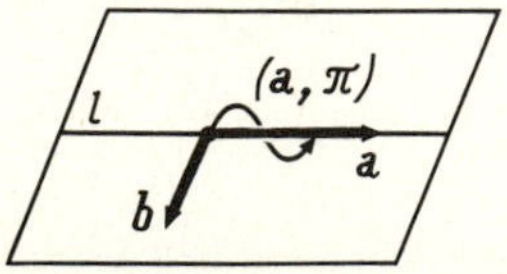	$\Gamma = \langle R^{\mathbf{a}}_{\ell,\pi}, S^{\Pi}_{\mathbf{b}} \rangle$	with **a**, **b** perpendicular

$$T_{\mathbf{a}}T_{\mathbf{b}}T_{-\mathbf{a}}SR, \qquad \text{or} \qquad T_{\mathbf{b}}SR,$$

which by (5) and (6) can be transformed to $SRT_{-\mathbf{b}}$; and finally by (7) we get

$$RST_{-2\mathbf{b}}.$$

Now it is easy to prove the required properties of Γ. Let's prove the relation $|AG(A)| \geq r > 0$ for $G \neq E$, which will imply that Γ is a uniformly discontinuous group. Suppose for example, that G is a motion in Γ of the form $RT_{\mathbf{h}}$, where $\mathbf{h} = m\mathbf{a} + n\mathbf{b}$; then the first coordinate x' of $G(A)$ is obtained from the first coordinate x of A by adding $(2m+1)\alpha$, where α is the length of $\mathbf{a}/2$. Hence $|x' - x| \geq \alpha$, and therefore $|AG(A)| \geq \alpha$. The cases when G is of the form $ST_{\mathbf{h}}$, $RST_{\mathbf{h}}$ or $T_{\mathbf{h}}$ are treated similarly.

We now check that the geometry Σ obtained from Γ is unbounded. We must find points a_1 and a_2 of Σ which are an arbitrarily large distance apart, say having $|a_1a_2| > N$. To do this, we need to find points A_1 and A_2 of 3-space such that, for all G in Γ, the distance from A_1 to the equivalent point $G(A_2)$ is greater than N. Note that for all G in Γ, the z-coordinates of A_2 and $G(A_2)$ are either equal or differ just by a minus sign. Hence if A_1 and A_2 have positive coordinates z_1 and z_2, and $|z_1 - z_2| > N$, then the z-coordinates of A_1 and $G(A_2)$ will differ by at least N, and hence $|A_1G(A_2)| > N$.

The other types of group given in Table 1 are all similar to the example we have treated, which is Type III.5 of the table.

Exercise

1. Why does Table 1 not contain the following group?

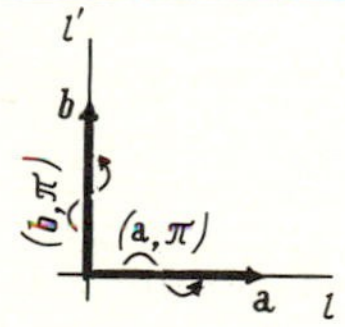	$\Gamma = \langle R_{\ell,\pi}^{\mathbf{a}}, R_{\ell',\pi}^{\mathbf{b}} \rangle$	with **a**, **b** perpendicular

We now proceed to groups of Type IV. Type IV.1 is the translation group which has already appeared, consisting of translations $T_{\mathbf{x}}$ with $\mathbf{x}$ of the form $\mathbf{x} = \ell\mathbf{a} + m\mathbf{b} + n\mathbf{c}$, where $\mathbf{a}$, $\mathbf{b}$ and $\mathbf{c}$ are three non-coplanar vectors, and ℓ, m and n are any integers.

The majority of the remaining groups can be obtained from symmetries of plane lattices, but used in a slightly different way from before; however, first of all, we should describe all symmetries of plane lattices. It turns out that these can be listed completely:

	Type of symmetry	Lattice
(1)	identity	any
(2)	rotation about a lattice point through π (central reflection)	any
(3)	rotation through $\pi/2$ or $3\pi/2$	square
(4)	rotation through $\pi/3, 2\pi/3,$ $4\pi/3$ or $5\pi/3$	equilateral triangle
(5)	reflection in a line through two lattice points	rectangular or 'centred rectangular', as Figure 8.16.

The fact that this list is complete is not too complicated to prove, and the reader might like to have a go at it himself as an exercise (the proof will be given in §14.1).

Let Ω be a plane lattice, having a symmetry of type (2), (3) or (4), which is a rotation through an angle φ (= π, $\pi/2$, $3\pi/2$, $\pi/3$, $2\pi/3$, $4\pi/3$ or $5\pi/3$), write $\mathbf{c}$ for any non-zero vector orthogonal to the plane of Ω, and let R be the twist $R_{\ell,\varphi}^{\mathbf{c}}$, where ℓ is a line through a lattice point and parallel to $\mathbf{c}$. If Ω is obtained by laying off vectors $m\mathbf{a} + n\mathbf{b}$ from some point, set $\Gamma = \langle R, T_{\mathbf{a}}, T_{\mathbf{b}} \rangle$. As above, we have a rule for 'commutation after correction':

$$T_{\mathbf{h}}R = RT_{R(\mathbf{h})}, \tag{8}$$

where $\mathbf{h}$ is a vector of Ω. In exactly the same way as above, using (8), we see that every motion of Γ can be written in the form

$$F = R^k T_{\mathbf{h}} \tag{9}$$

where k is any integer, and $\mathbf{h}$ any vector of Ω; here R^k means the composite of R with itself k times, that is, $R^0 = E$, $R^k = R \cdots R$ (k times), and $R^{-k} = (R^k)^{-1}$. From this, we see easily that Γ is uniformly discontinuous. Suppose that Ω lies in the plane Oxy, and that $\mathbf{c}$ points along the z-axis and has length γ. Then a motion F with $k \neq 0$ changes the z-coordinate of any point by $k\gamma$, so that $|PF(P)| \geq \gamma$. On the other hand if $k = 0$ then F changes the x and y-coordinates just as if we had the plane translation group corresponding to Ω, which is a uniformly discontinuous group. For example, suppose that we have a lattice with a symmetry of type (3). Then R^4 is a translation in $4\mathbf{c}$ and the corresponding translation group Γ' is $\langle T_{\mathbf{a}}, T_{\mathbf{b}}, T_{4\mathbf{c}} \rangle$.

This procedure gives the groups of Type IV.2, IV.3, IV.4 and IV.5 in Table 2; the symmetry of type (4) leads to groups of Type IV.4 and IV.5, since we can set $\varphi = \pi/3$ or $\varphi = 2\pi/3$.

If Ω has a symmetry of type (5), let Π denote the plane through the axis of reflection and perpendicular to the plane containing Ω (if Ω lies in the plane Oxy, with Ox the axis of the reflection, then Π is the plane Oxz). Let **c** be any vector lying in Π; we set $S = S_{\mathbf{c}}^{\Pi}$ and $\Gamma = \langle T_{\mathbf{a}}, T_{\mathbf{b}}, S \rangle$; entirely similar arguments show that Γ is a uniformly discontinuous group and $\Gamma' = \langle T_{\mathbf{a}}, T_{\mathbf{b}}, T_{2\mathbf{c}} \rangle$. We arrive at groups of Type IV.6 or IV.7, corresponding to the two choices of lattice having a symmetry of type (5).

Finally, the remaining three types of group are constructed using variations of these ideas. We start with a group which has already appeared, of Type III.5. The motions $R = R_{\ell,\pi}^{\mathbf{a}}$ and $S = S_{\mathbf{b}}^{\Pi}$ which generate it both take **c** into **-c**, where **c** is the vector orthogonal to **a** and **b**; tt follows from this that they preserve the lattice Ω generated by the vectors **2a**, **2b** and **c**. Thus if we set $\Gamma = \langle R, S, T_{\mathbf{c}} \rangle$, then Γ contains the translation group $\Gamma' = \langle T_{2\mathbf{a}}, T_{2\mathbf{b}}, T_{\mathbf{c}} \rangle$, and we have the following rules for 'commutation after correction':

$$T_{\mathbf{h}}R = RT_{R(\mathbf{h})}, \quad T_{\mathbf{h}}S = ST_{S(\mathbf{h})},$$

where **h** is any vector such that $T_{\mathbf{h}}$ is in Γ'. Then all the arguments which we used to study groups of Type III.5 turn out to be applicable, and using them we prove, exactly as before, that Γ is a uniformly discontinuous group. We thus arrive at groups of Type IV.8.

If in the preceding group we do not take all the elements, but just some subset which forms a group, then we of course again get a uniformly discontinuous group. This construction gives Type IV.9: we set $R_1 = T_{\mathbf{c}}R$ and $\Gamma = \langle R_1, S \rangle$. Then obviously $S^2 = T_{2\mathbf{b}}$, and hence $T_{2\mathbf{b}}$ belongs to Γ. A simple calculation using (8) shows that $R_1{}^2 = T_{2\mathbf{a}}$, so that $T_{2\mathbf{a}}$ belongs to Γ. Finally, it it easy to deduce from (7) that $R_1S = SR_1T_{-2\mathbf{b}+2\mathbf{c}}$, so that Γ contains $T_{-2\mathbf{b}+2\mathbf{c}} = R_1{}^{-1}S^{-1}R_1S$, and therefore also $T_{2\mathbf{c}} = T_{2\mathbf{b}}T_{-2\mathbf{b}+2\mathbf{c}}$. This proves that Γ is a group of Type IV. It is easy to see that R_1 is the twist $R_{\ell',\pi}{}^{\mathbf{a}}$, whose axis ℓ' is parallel to Ox and is obtained from Ox by a translation in $\mathbf{c}/2$. (Check this for yourself as an exercise.)

The final group requires a highly specialised choice of generating motions. This is related to the 'missing' group of Type III, mentioned in the exercise after Table 1, which is not in the list, although it suggests itself naturally. It is the group

$$\Gamma^* = \langle R, R_1 \rangle, \text{ with } R = R_{\ell,\pi}^{\mathbf{a}} \text{ and } R_1 = R_{m,\pi}^{\mathbf{b}},$$

where ℓ is the x-axis Ox, m the y-axis Oy, and **a**, **b** are vectors pointing along Ox and Oy. We hope that the reader has done this exercise, and discovered that Γ^* is not in the list, since it contains the motion RR_1, which is a rotation through an angle π about an axis parallel to Oz and passing through the centre of the

TABLE 2

Uniformly discontinuous groups of motions of 3-space of Type IV

IV.1	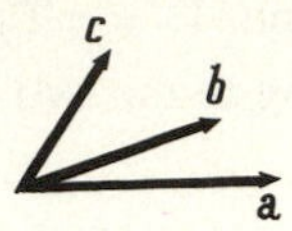	$\Gamma = \langle T_{\mathbf{a}}, T_{\mathbf{b}}, T_{\mathbf{c}} \rangle$
IV.2	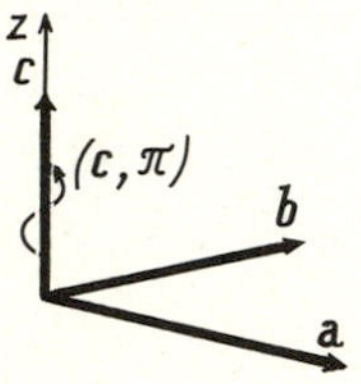	$\Gamma = \langle T_{\mathbf{a}}, T_{\mathbf{b}}, R^{\mathbf{c}}_{\ell,\pi} \rangle$ ℓ = Oz, **c** points along Oz, **a** and **b** lie in Oxy
IV.3	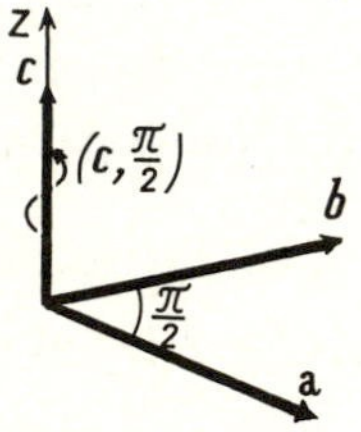	$\Gamma = \langle T_{\mathbf{a}}, T_{\mathbf{b}}, R^{\mathbf{c}}_{\ell,\pi/2} \rangle$ ℓ = Oz, **c** points along Oz, **a** and **b** lie in Oxy, satisfy $\lvert\mathbf{a}\rvert = \lvert\mathbf{b}\rvert$ and make an angle of $\pi/2$
IV.4	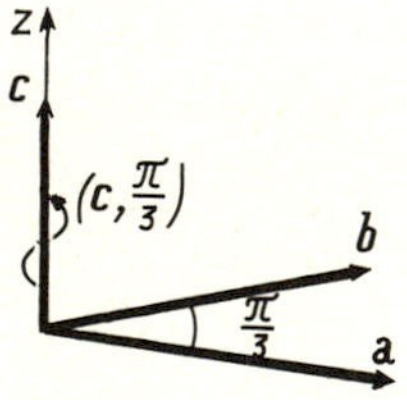	$\Gamma = \langle T_{\mathbf{a}}, T_{\mathbf{b}}, R^{\mathbf{c}}_{\ell,\pi/3} \rangle$ as IV.3, but **a** and **b** make an angle of $\pi/3$
IV.5	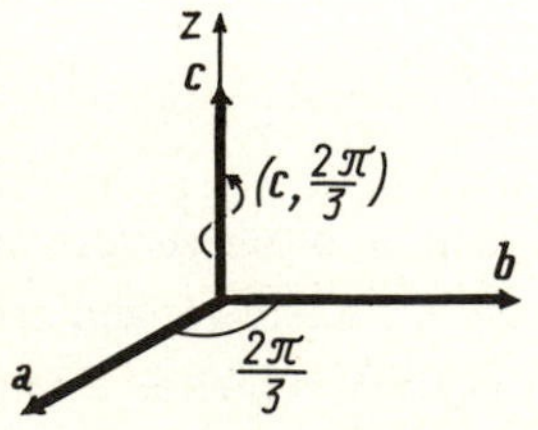	$\Gamma = \langle T_{\mathbf{a}}, T_{\mathbf{b}}, R^{\mathbf{c}}_{\ell,2\pi/3} \rangle$ as IV.3, but **a** and **b** make an angle of $2\pi/3$

IV.6	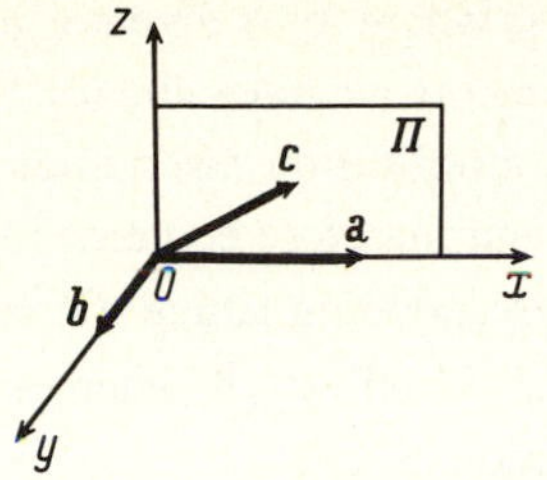	$\Gamma = \langle T_{\mathbf{a}}, T_{\mathbf{b}}, S_{\mathbf{c}}^{\Pi} \rangle$ Π = Oxz, **a** points along Ox, **b** along Oy, **c** lies in Oxz
IV.7	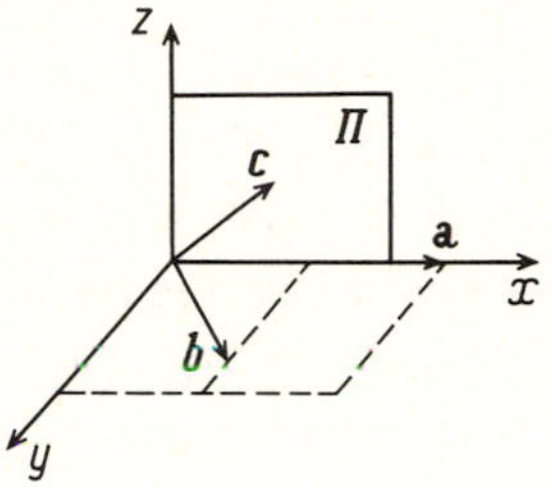	$\Gamma = \langle T_{\mathbf{a}}, T_{\mathbf{b}}, S_{\mathbf{c}}^{\Pi} \rangle$ Π = Oxz, **a** points along Ox, **b** lies in Oxy and has projection **a**/2 to Ox, **c** lies in Oxz
IV.8	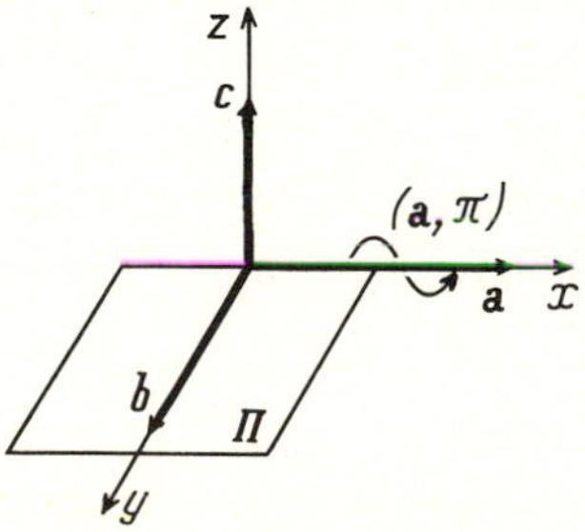	$\Gamma = \langle R_{\ell,\pi}^{\mathbf{a}}, S_{\mathbf{b}}^{\Pi}, T_{\mathbf{c}} \rangle$ ℓ = Ox, Π = Oxy, **a** points along Ox, **b** along Oy, **c** along Oz
IV.9	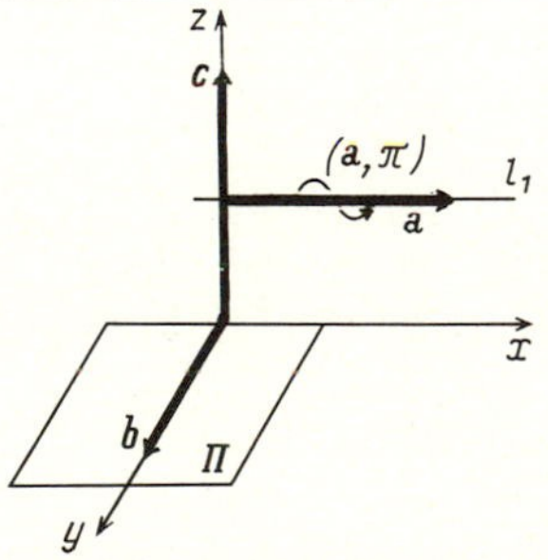	$\Gamma = \langle S_{\mathbf{b}}^{\Pi}, R_1 \rangle$, $R_1 = T_{\mathbf{c}} R_{\ell,\pi}^{\mathbf{a}}$ notation is as in IV.8, and ℓ_1 is the translate of Ox by **c**/2; $R_1 = R_{\ell_1,\pi}^{\mathbf{a}}$
IV.10	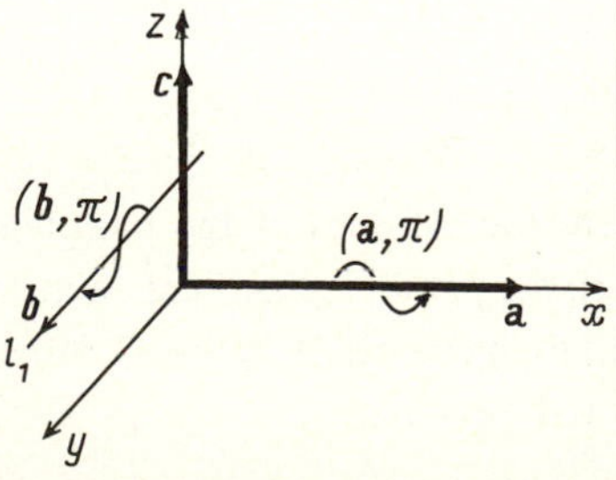	$\Gamma = \langle R, R' \rangle$, where $R = R_{\ell,\pi}^{\mathbf{a}}$ and $R' = R_{\ell',\pi}^{\mathbf{b}}$; **a**, **b**, **c** point along Ox, Oy, Oz ℓ = Ox, ℓ' is translate of Oy by **c**/2

rectangle made up by **a** and **b**; thus Γ^* is not a uniformly discontinuous group. We try to cure this defect by setting $R' = T_{\mathbf{c}}R_1$, where **c** is a vector perpendicular to **a** and **b**. If we carry out carefully the arguments we have already been through several times, we see that $\Gamma = \langle R, R' \rangle$ is a uniformly discontinuous group; it contains the translations $T_{2\mathbf{a}}$, $T_{2\mathbf{b}}$ and $T_{2\mathbf{c}}$, and therefore contains the translation group $\Gamma' = \langle T_{2\mathbf{a}}, T_{2\mathbf{b}}, T_{2\mathbf{c}} \rangle$, so that the corresponding lattice Ω is made up of rectangular parallelepipeds. It is easy to see that $R' = R_{m',\pi}{}^{\mathbf{b}}$, where m' is the axis parallel to Oy, obtained from Oy by a translation in $\mathbf{c}/2$.

This completes the classification of uniformly discontinuous groups of motion of 3-space (Table 2).

How should we imagine the 3-dimensional geometries corresponding to these groups? Consider for example the group $\Gamma = \langle T_{\mathbf{a}}, T_{\mathbf{b}}, T_{\mathbf{c}} \rangle$ of Type IV.1. Obviously, a fundamental domain (defined in §9) of Γ is given by a parallelepiped with the generating vectors **a**, **b** and **c** as edges (see Figure 11.4), and the equivalent points on its boundary are the points of the opposite faces of the parallelepiped which can be obtained from one another by translation along the edges which intersect the face, that is in the corresponding generating vector **a**, **b** or **c**. The geometry corresponding to Γ consists of points of the parallelepiped with equivalent points of the boundary considered to be the same, that is, with opposite faces glued together in equivalent points. Ball neighbourhoods of equivalent boundary points of the parallelepiped glue together into a single ball, which can be identified with a ball neighbourhood in the geometry of the

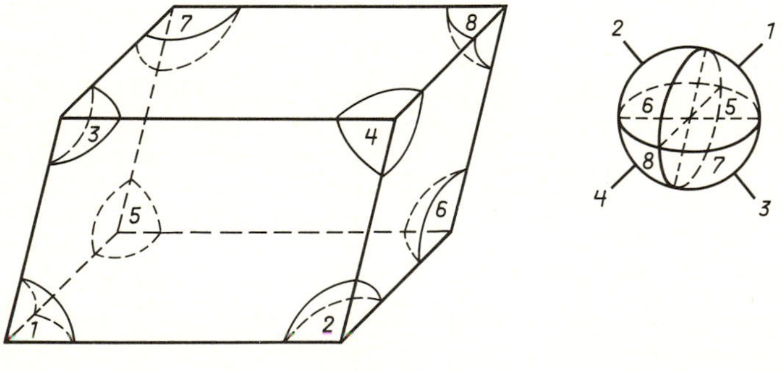

Figure 11.10

corresponding point. For example, all 8 of the vertices of the parallelepiped are equivalent, and their ball neighbourhoods in the parallelepiped are sectors of a ball, which glue together into an ordinary ball in the geometry (Figure 11.10); this ball is cut up by three planes through its centre into 8 sectors.

The distance between two points is exactly as in §§3-5, visualised in terms of a 'jet service' or defined by 'tracks'. By analogy with the 2-dimensional case, the resulting geometry is called the 3-*dimensional torus* .

The situation is the same with all the 3-dimensional geometries corresponding to the groups we have classified. All of their differences from the 2-dimensional case come from the fact that we are glueing polyhedra rather than polygons, and the glueing takes place along faces rather than edges. Of course the result of the glueing is much harder to visualise in this case, requiring more stretching of our spatial imagination. However, in many cases this can be avoided, if we stick to the following point of view: since our geometries are obtained from groups Γ, any question which we are interested in concerning them can somehow or other be reformulated in terms of the groups Γ; the remaining problem is to find this formulation. We have already encountered one such question and its answer: we know that the geometry is bounded if and only if Γ contains translations in non-coplanar vectors.

We conclude this section with a discussion of a similar question and answer.

Exercise

1. Find fundamental domains for the groups of Types IV.2-5. Indicate the equivalent points on the boundaries. If you're keen, do the same for groups of Types IV.6-10, III.1-5 and I, II.2.

11.4. Orientability of the geometries. We know from previous sections (§4 and §9) that among the 2-dimensional locally Euclidean geometries, there are some in which right and left can be distinguished (for example, the plane itself), and also some in which they are indistinguishable (for example, the twisted cylinder or the Klein bottle). Thus a number of natural questions present themselves: what is the

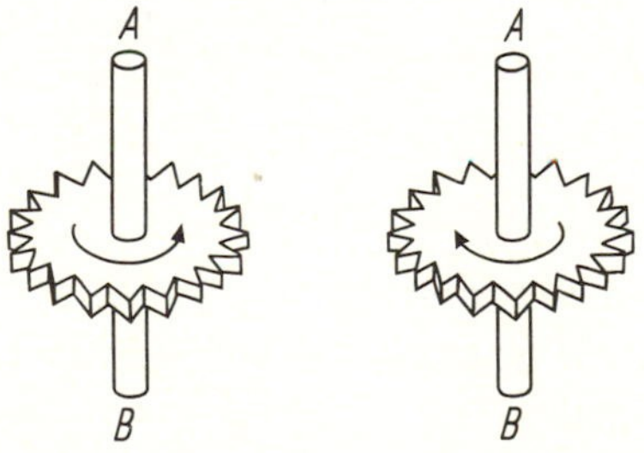

Figure 11.11

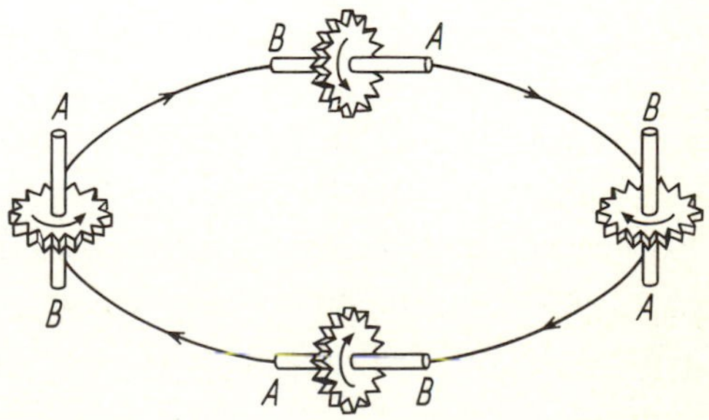

Figure 11.12

meaning of right and left in 3-space? What does it mean to be able to distinguish right and left in 3-space? For which of the 18 types of geometry corresponding to the 18 types of uniformly discontinuous groups described above can right and left be distinguished, and for which are they indistinguishable? The answers to the first two of these questions are frequently encountered in real life.

Consider a revolving motion, for example that of a cogwheel revolving around an axle. This revolution can take place in either of two opposite senses (Figure 11.11); this is right and left in 3-space.

Suppose now that the axle, together with the continually revolving cogwheel, moves around in space. The second question asks whether, if the axle returns to its original position, the sense of motion of the cogwheel can change (Figure 11.12).

From everyday experience we know that in the space we live in, the sense of motion cannot change. We convince ourselves continually of this from the example

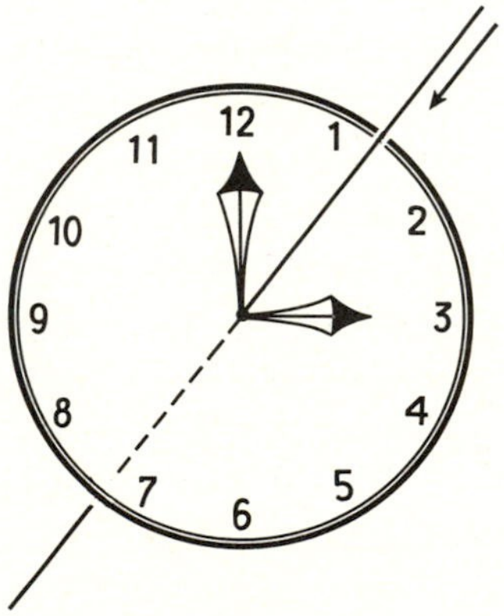

Figure 11.13

of a clock: however we travel in 3-space, the sense of revolution of the hands of a clock (viewed from the face side) does not change (Figure 11.13). This therefore means that right and left are distinguishable in 3-space.

Another method of distinguishing right and left in space is related to the existence of two different types of helixes (Figure 11.14). If a point on the surface of a cylinder moves upwards, going round the cylinder in the anticlockwise sense (or moves downwards, going round the cylinder clockwise) then we get one type of helix; if motion around the cylinder is in the opposite sense, then we get the other. We say that the first of these is a right-handed helix (like a corkscrew), and the second a left-handed helix. To remember which is which, we can grasp a cylindrical object with the right hand: the fingers indicate the anticlockwise

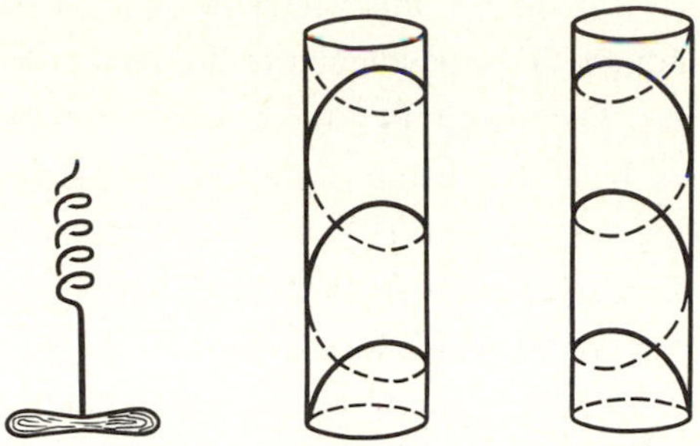

Figure 11.14

direction around the cylinder, while the thumb points upwards.

It is not possible to turn a right-handed helix into a left-handed one by a continuous motion in space. But the mirror-image, or reflection in a plane, of a right-handed helix is a left-handed one. It is an interesting fact that Mother Nature, for reasons which we do not understand, has chosen betwen these two possibilities: all molecules of amino acids, which make up proteins, have the form of a helix, and all of them are left-handed.

The opposite does happen in certain 3-dimensional locally Euclidean geometries: there exist geometries in which right and left are indistinguishable. Such geometries are said by mathematicians to be *non-orientable* ; those in which right and left are distinguishable are said to be *orientable* . The problem as to which of the 18 types of 3-dimensional locally Euclidean geometries are orientable, and which not, is the subject-matter of this section.

First of all we consider 3-space, which as we have said above, we know by experience to be orientable. Now let's prove this.

Theorem 1. 3-space is orientable.

Proof. Suppose that at time t, a clock P is in a position P_t, and moves continuously in space for t in the interval $0 \le t \le 1$, in such a way that at $t = 1$, it returns to the point of departure, but going in the opposite sense. (Here strictly speaking, we could think of the clock as being framed by a tetrahedron $OABC$ with edges OA, OB and OC of the same length and perpendicular to one another; the centre of the clock is held at O, the clock-face in the plane (OBC) so that it is visible from A, and the clock shows 3 o'clock, with the hour hand pointing along OB, and the minute hand along OC.) Now for each t with $0 \le t \le 1$, we translate the clock P_t to move its centre back to the centre O of P_0, and denote by P_t' the new position of the clock which we thus obtain. We get the same situation as before: the clock moves continuously, occupying the position P_t' at time t, but now the centre O remains fixed; and again, P_0' and P_1' are at the same point, but go in opposite senses. Now choose a point B on the boundary circle of the clock, and set $\mathbf{e} = OB^{\rightarrow}$. At time t this vector occupies a position $\mathbf{e}_t$. We rotate space about an axis through O perpendicular to the vectors $\mathbf{e}_0$ and $\mathbf{e}_t$ so that the vector $\mathbf{e}_t$ goes into $\mathbf{e}_0$. Denote by P_t'' the new position of the clock which we thus obtain. Now we have the same situation as before: the clock moves continuously, occupying the position P_t'' at time t, but now in addition to the centre O, the point B also remains fixed throughout the motion, so that P_t'' is obtained from P_0'' by a motion which leaves fixed all the points of the axis (OB). Finally, we can choose the other point C on the clock, which does not lie on (OB), and in a similar way arrange that for all t, the points O, B and C of the clock P_t''' are left fixed throughout the motion, and again P_1''' occupies the same position as P_0''', but goes in the opposite sense. Thus we have arrived at the situation that the clock moves continuously in space, occupying the position P_t''' at time t for $0 \le t \le 1$, such that all points of their plane (that is, all points of the clock-face) are left fixed throughout the motion; and when they arrive at P_1''', the clock is going in the opposite sense.

But such a motion is clearly impossible: under our assumptions, at any instants t and u the clocks P_t''' and P_u''' must either coincide or be mirror reflections of one another. Indeed, imagine that we have an arrow OA rigidly

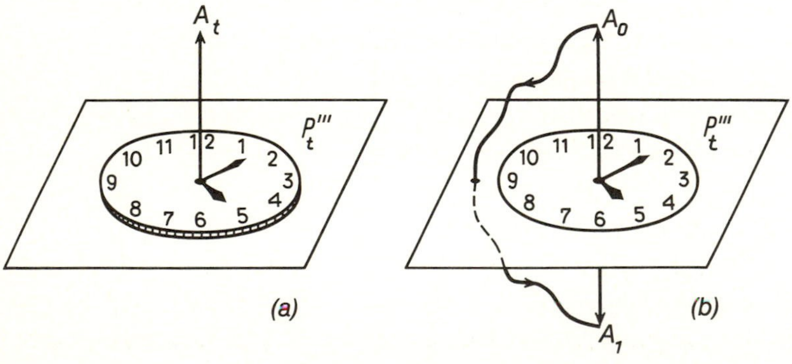

Figure 11.15

fixed perpendicularly to the clock on the face side, as in Figure 11.15, (a). As the clock moves continuously, so does the endpoint A_t of OA, so that it describes a continuous curve joining the symmetrically opposite positions A_0 and A_1 of the arrow at times $t = 0$ and $t = 1$ (Figure 11.15, (b)). But under our conditions on the motion of the clock, this curve consists of only two points, A_0 and A_1. There cannot be any such continuous curve. We get a contradiction, which proves the theorem.

The orientability of 3-space just proved allows us to divide up all clocks of space into two kinds, such that clocks of one kind go in one sense, and those of the other kind in the opposite sense. Here when we say that two clocks P and Q go in the same sense, we mean that after moving the face of P continuously to lay it over that of Q, the hands turn out to go in the same sense. If however, after such a continuous motion the hands of P and Q turn out to go in opposite senses, we say that the clocks themselves are going in opposite senses. Now clocks P and Q cannot go both in the same sense and in the opposite sense.

Indeed, suppose that moving P along the curve c to lay it over Q they turn out to go in the same sense, and along c′, in the opposite sense (Figure 11.16). Then on moving P back along c in the opposite direction, and then along c′, it

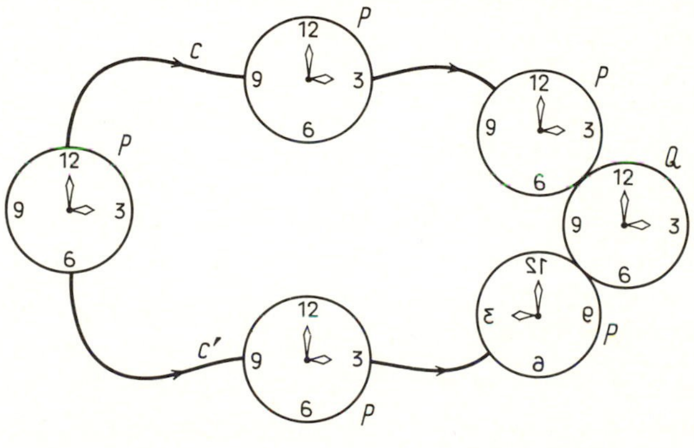

Figure 11.16

moves around a closed curve, and goes in the opposite sense at the end; this contradicts Theorem 1.

The inhabitants of the planet Earth have agreed a long time ago on the sense of rotation of their clocks, thus introducing a unit of sense of revolution. Thanks to this, for any revolving motion in the space we live in, we can say in which sense it takes place: clockwise or anticlockwise. This is especially important in engineering, where most components have a 'handedness' and most mechanisms - for example clocks themselves, or propellers, can in view of their construction only revolve in one sense.

Thanks to the existence of two kinds of clocks in 3-space, all motions of 3-space can be divided into two kinds of motions: motions of the first kind preserve the sense of rotation of a clock, and motions of the second kind reverse it. Thus, it is not hard to see that a twist is of the first kind, and a glide or rotary reflection is of the second. Corresponding to this, all groups of motions of 3-space can be divided into two kinds: a group of the first kind contains only motions of the first kind, and a group of the second kind contains at least one motion of the second kind. Among the 18 types of uniformly discontinuous groups of motion of 3-space, those of Type II.2, III.3-5 and IV.6-9 are of the second kind, since in each case one of the generators is a glide reflection. The remaining groups of Type I, II.1, III.1-2, IV.1-5 and IV.10 are of the first kind, since all of their generators are of the first kind (and since the inverse of a motion of the first kind, or the composite of two motions of the first kind is again of the first kind).

Now we proceed to arbitrary 3-dimensional locally Euclidean geometries. The question which we are interested in is to know which of the geometries corresponding to these groups are orientable; to answer this, we will now prove Theorem 2 below. It might appear that the statement of Theorem 2 includes the case of 3-space itself, which we have already considered in Theorem 1. But this is not the case, since the division of motions into two kinds, used in the statement of Theorem 2, assumes the orientability of 3-space, that is, it assumes Theorem 1. Thus without Theorem 1, Theorem 2 would be meaningless.

Theorem 2. The geometry corresponding to a uniformly discontinuous group Γ of motions of 3-space is orientable if Γ is a group of the first kind, and non-orientable if Γ is a group of the second kind.

We denote 3-space by Σ, and the geometry corresponding to a uniformly discontinuous group Γ of motions of 3-space by Σ_Γ. A point a of Σ_Γ is a set **A** of points of Σ equivalent under Γ. If A is any point of Σ, the set **A** of all points equivalent to A defines a point a of Σ_Γ. The map f which takes a point A of 3-space to the point a of Σ_Γ is called the covering of Σ_Γ by Σ.

Suppose that Γ is a group of the second kind, that is, it contains glide reflections, and let F be one of these. We take a point A of Σ lying on the plane of symmetry of F, set B = F(A), and consider the line segment L joining A and B. Under the covering f, A and B go into the same point a of Σ_Γ, and L into a closed curve ℓ passing through a. Recall the basic property of the covering f: for a ball B(A, r) in Σ of sufficiently small radius r, f defines a superposition of B(A, r) onto the ball B(a, r) in Σ_Γ (it is clear that everything we said about discs D(A, r) and D(a, r) in §§9-10 carries over word-for-word for balls in our case).

Inside the ball B(A, r) we put a clock P_0, in such a way that it lies in the plane of symmetry of F, with its centre at A. We consider the corresponding clock p_0 in B(a, r) under the superposition of B(A, r) onto B(a, r). Taking P_0 as the starting point of a clock P, we move it parallel so that its centre moves along the line segment L. Then to each position P of the clock, by the covering property of f, there will be a corresponding position of the clock p in Σ_Γ corresponding to P; this clock p will move in such a way that its centre describes the curve ℓ. As a result of the motion along ℓ, the centre of p returns to its previous position. What now is the sense of the clock?

To answer this question precisely, we need to recall that the position p_0 of the clock p corresponds to the position P_0 of P under the superposition of B(A, r) onto B(a, r). After the motion along L the clock P takes a position P_1 with centre B and is contained in B(B, r). The corresponding position p_1 of the clock p will be obtained by applying the map f to P_1 (Figure 11.17), that is $p_1 = f(P_1)$. By definition of the covering f, the maps f and fF^{-1} are equal. Hence $p_1 = f(F^{-1}(P_1))$, that is, p_0 and p_1 are obtained by the same map f from the position of P_0 and $F^{-1}(P_1)$ in the ball B(A, r). On the other hand, the position $F^{-1}(P_1)$ of the clock is obtained from the position P_0 by a translation and a

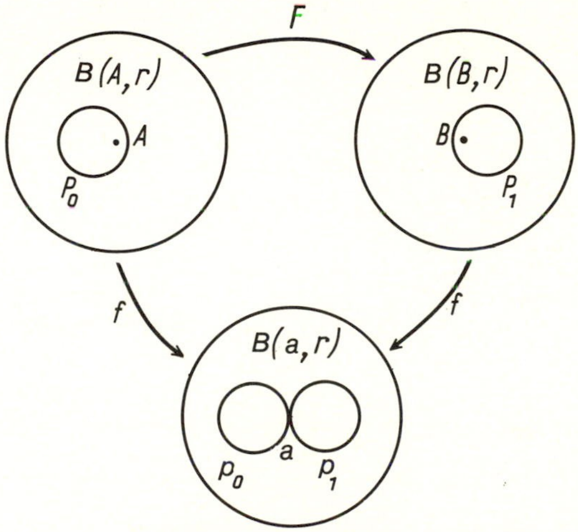

Figure 11.17

reflection, so that in these positions, P goes in opposite senses. Hence also p in the two positions p_0 and p_1 will go in opposite senses.

The reader can easily convince himself that the arguments we have just carried out are just the same as those by which we showed in §4 and §9 that in geometry on the twisted cylinder and on the Klein bottle, right and left are indistinguishable, but in a more general situation. Now let's prove that the same thing cannot happen if Γ is a group of the first kind.

Let ℓ be a closed path in Σ starting and ending at a. We show first of all that there exists a curve L in Σ which maps to ℓ under the covering f of Σ_Γ by 3-space Σ (this is true for an arbitrary covering).

Divide up the curve ℓ from its starting point a to its endpoint b (since we are dealing with a closed curve, a = b, but the argument works for any curve) by points $a_0 = a, a_1, a_2, \dots, a_n, a_{n+1} = b$ into curves $\ell_i = a_i a_{i+1}$ for $i = 0, 1, \dots, n$, of length less than s (Figure 11.18) Then since each curve $\ell_i = a_i a_{i+1}$ has length less than s, it will lie in a ball $B(a_i, s)$. We choose a point A in Σ for which $f(A) = a$. Since f defines a superposition of $B(A, s)$ in Σ onto the ball $B(a, s)$ in Σ_Γ, there is some curve $L_0 = AA_1$ of $B(A, s)$, starting at A and ending at some point A_1, which maps onto ℓ_0, that is, $f(L_0) = \ell_0$. Then the endpoint A_1 of L_0 goes to the endpoint a_1 of ℓ_0, that is, $f(A_1) = a_1$. Repeating the same argument with a replaced by a_1, a_1 by a_2 and A by A_1, we construct some curve $L_1 = A_1A_2$ in $B(A_1, s)$ which covers $\ell_1 = a_1a_2$ under f, that is, $f(L_1) = \ell_1$ (see Figure 11.18). Since again the endpoint A_2 of L_1 is taken by f to the endpoint a_2 of ℓ_2, we can continue this process. As a result, we get curves $L_0 = AA_1$, $L_i = A_iA_{i+1}$ for $i = 1, \ldots, n-1$, and $L_n = A_nB$, which under f cover $\ell_0 = aa_1$, $\ell_i = a_ia_{i+1}$, and $\ell_n = a_nb$ respectively. Since each curve L_{i+1} starts off at the endpoint A_{i+1} of its predecessor, we have constructed a composite curve $L_0, L_1, \ldots, L_n$ in Σ, which we will denote from now on by L, for which $f(L) = \ell$; as a consequence we get

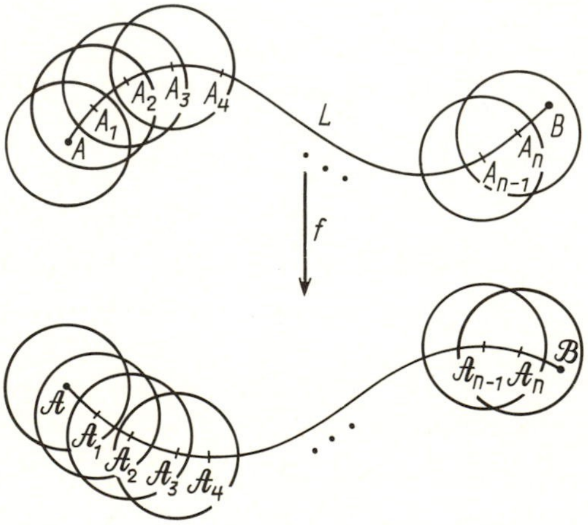

Figure 11.18

that the endpoints B and b of L and ℓ satisfy $f(B) = b$ (see Figure 11.18). This proves the required relation between curves of Σ_Γ and those of Σ.

Since our curve ℓ is closed (that is, $a = b$), the starting point and endpoint of L must be equivalent under the group Γ, that is $B = F(A)$ where F is a motion in Γ. Now we can argue exactly as in the first part of the proof, since using the covering f, the motion of p along the curve ℓ of Σ_Γ can be considered as the corresponding motion of P along the curve L from A to B. In the same notation as in the first part of the proof, the question reduces to verifying that the clock P at positions P_0 and $F^{-1}(P_1)$ in $B(A, r)$ go in the same sense.

We have arrived at a problem very similar to that which we solved in the proof of Theorem 1 for the case of 3-space. The difference is just that now the clock is not moving along a closed curve, since in general $B \neq A$. This situation can easily be reduced to the one we have already considered. For this, we recall that the group Γ is of the first kind, and therefore the motion F^{-1} is also of the first kind, that is, is a twist. Let $F^{-1} = R_{\ell,a}{}^{\varphi}$; then we can obtain the result of this motion by a continuous motion, which at time τ takes X into $R_{\ell,\tau a}{}^{\tau\varphi}(X)$ for $0 \le \tau \le 1$. Then for $\tau = 0$ we get the point X, and for $\tau = 1$ the point $F^{-1}(X)$. Suppose that under this, the point B traces out a trajectory L' (L' is the piece of the helix $R_{\ell,\tau a}{}^{\tau\varphi}(B)$ with $0 \le \tau \le 1$). Thus we move the clock P continuously along the curve L' from the position P_1 at the point B to the position $F^{-1}(P_1)$ at A. Consider the motion of the clock P firstly along the curve L, and then along L'. Then it will move continuously along the closed curve $L + L'$, and return to the point A in position $F^{-1}(P)$. By Theorem 1 in 3-space Σ, on its return to A the clock will be going in the same sense. This proves that the clocks P and $F^{-1}(P)$ go in the same sense. The theorem is proved.

We see that the uniformly discontinuous groups in 3-space break up as shown in Table 3 into 4 types, according to whether the corresponding geometries are bounded or orientable.

Table 3

Geometries	Orientability	How many	Type of groups
bounded	orientable	6	IV.1, 2, 3, 4, 5, 10
	non-orientable	4	IV.6, 7, 8, 9
unbounded	orientable	4	I; II.1, III.1, 2
	non-orientable	4	II.2; III.3, 4, 5

Exercises

1. Prove that no continuous movement of a circle in the plane can reverse a sense of rotation marked on it.

2. State and prove the analogues of the theorems of this section for the case of a 2-dimensional locally Euclidean geometry.

3. Which of the locally spherical geometries are orientable (see §11.2, Exercise 2)?

4. Prove that if there exists a covering of a geometry Σ' by a geometry Σ (where Σ and Σ' are both 2 or 3-dimensional locally Euclidean geometries), and if Σ is non-orientable, then Σ' is also non-orientable.

5. Suppose that Σ is a 2 or 3-dimensional locally Euclidean geometry, and is non-orientable. Prove that Σ then has a covering by a geometry $\Sigma^{\sim}$ which is orientable, and such that in the covering, the inverse image in $\Sigma^{\sim}$ of each point of Σ consists of exactly 2 points. Which are the geometries $\Sigma^{\sim}$ covering in this way the non-orientable geometries which we know?

§12. Crystallographic groups and discrete groups

12.1. Symmetry groups. The description of 2 and 3-dimensional locally Euclidean geometries has led us in the previous sections to groups of motions of the plane and of 3-space which are uniformly discontinuous. The appearance of groups of motion here is not at all obvious, in fact perhaps rather unexpected; thus in §10 we had to work quite hard to show that the description of such geometries reduces to the description of groups of motions of the plane or of 3-space which are uniformly discontinuous.

However, in many cases the appearance of groups of motions is natural from the start. Suppose given some figure Φ in the plane or in 3-space, and consider the symmetries of Φ, that is, the motions of the plane or of 3-space which take Φ onto itself. Here it is obvious that if F and G are two symmetries of Φ then their composite FG is also a symmetry of Φ: indeed, $FG(\Phi) = F(G(\Phi)) = \Phi$, since $G(\Phi) = \Phi$ and $F(\Phi) = \Phi$; also obvious is the fact that if F is a symmetry of Φ then so is F^{-1}: indeed, $F^{-1}(\Phi) = F^{-1}(F(\Phi)) = E(\Phi) = \Phi$, since $\Phi = F(\Phi)$ and $FF^{-1} = E$. Thus we can say that the set Γ of symmetries of a figure Φ satisfies Properties I and II of §7, that is, Γ is a group of motions.

The word 'symmetry' we have introduced is explained by the fact that the more symmetries a figure has, then the more symmetrical, or regular, it is in the naive sense of the word. By specifying various symmetrical figures Φ, we can get in this way a whole variety of groups of motions of the plane or 3-space. Consider a few examples.

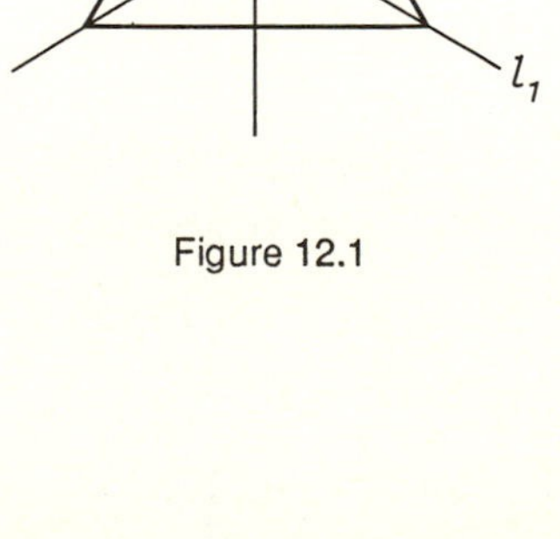

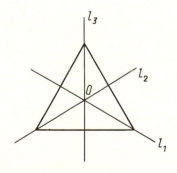

Figure 12.1

Example 1. The group of symmetries of a plane equilateral triangle (Figure 12.1) consists of the 6 motions

$$\{E, R_O{}^{120°}, R_O{}^{240°}, S_{\ell_1}, S_{\ell_2}, S_{\ell_3}\},$$

three rotations about the centre O of the triangle, and three reflections in axes ℓ_1, ℓ_2 and ℓ_3, the three heights of the triangle (Figure 12.1).

Example 2. Consider convex quadrilaterals in the plane. All quadrilaterals in the plane can be divided into 7 types, depending on their symmetry group Γ; each of these plays an important role in plane geometry (see Figure 12.2).

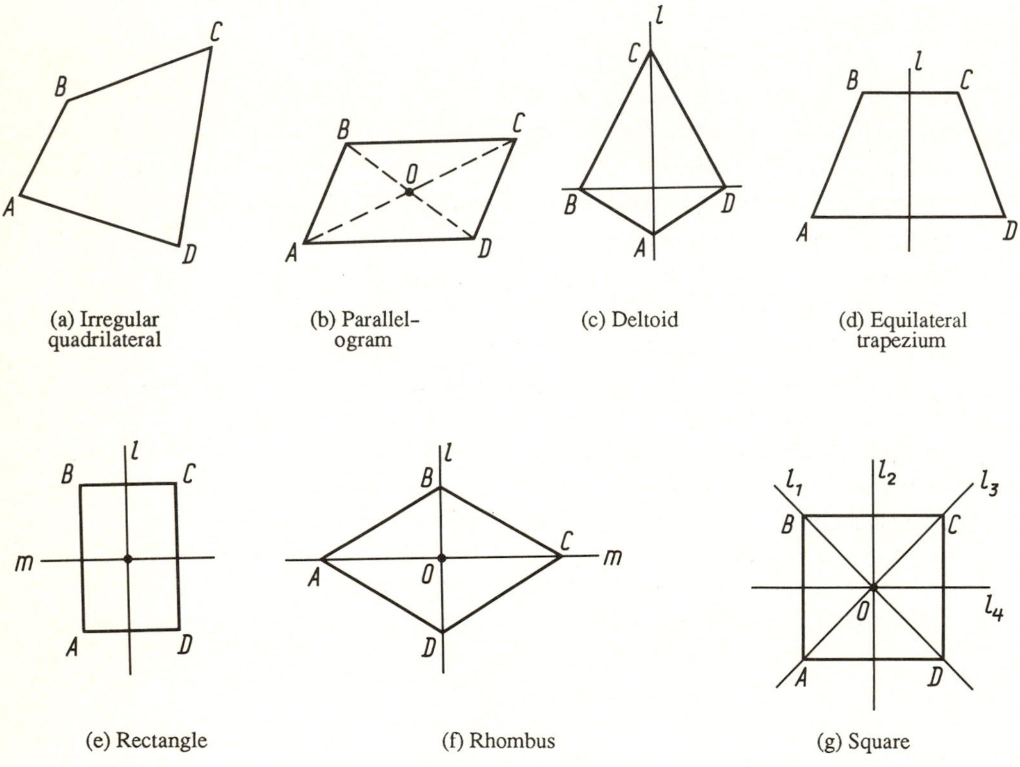

Figure 12.2

(a) irregular quadrilateral: $\Gamma = \{E\}$; (b) parallelogram: $\Gamma = \{E, R_O{}^{180°}\}$;
(c) deltoid: $\Gamma = \{E, S_\ell\}$; (d) equilateral trapezium: $\Gamma = \{E, S_\ell\}$;
(e) rectangle: $\Gamma = \{E, R_O{}^{180°}, S_\ell, S_m\}$; (f) rhombus $\Gamma = \{E, R_O{}^{180°}, S_\ell, S_m\}$;
(g) square: $\Gamma = \{E, R_O{}^{90°}, R_O{}^{180°}, R_O{}^{270°}, S_{\ell_1}, S_{\ell_2}, S_{\ell_3}, S_{\ell_4}\}$.

This example suggests the idea that certain kinds of figures can in a natural way be classified by their symmetry groups.

Example 3. A figure may of course have an infinite symmetry group. For example, the symmetries of a line in the plane consist of translations along it, glide reflections with respect to it, reflections in lines perpendicular to it, and central reflections in its points. The symmetries of a circle consist of rotations around its centre and reflections in lines through its centre. The reader can check as an exercise that in both cases there are no further symmetries.

Example 4. If we try to construct figures in the plane whose symmetry groups are uniformly discontinuous groups of Type III.a and III.b, we get patterns which repeat infinitely often in two directions (see Figure 12.3, (a) and (b), which illustrated such patterns made up of letters).

A single glance at Figure 12.3 is enough to see the difference between these patterns: the pattern for a group of Type III.b is much more specific and symmetrical, which is not surprising, since a group of Type III.b has a very special group of translations, and in addition contains glide reflections.

The notion of symmetry is applicable not just to geometrical figures. For example, one can speak of more or less symmetrical algebraic expressions, understanding by this (in complete analogy with the geometrical situation) that our expression keeps the same shape under a bigger or smaller number of substitutions of the quantities appearing in it. Thus among cubic polynomials in three variables x, y and z, the polynomials

$$x^3 + y^3 + z^3$$

is not altered by any permutation of x, y and z. The polynomial

$$x^2y - xy^2 - x^2z + xz^2 + y^2z - yz^2$$

has a more delicate symmetry: it is not altered if we make a cyclic permutation of x, y and z, that is, if we replace x by y, y by z, and z by x, or the other way round: x by z, z by y and y by x. But under other permutations of x, y and z (other than the identity, of course), the expression changes sign. The quadratic polynomial

$$x^2 + y^2 + 2z^2$$

is unaltered if we interchange x and y, and the polynomial

$$x^2 + 2y^2 + 3z^2$$

is only left unaltered by the identity permutation, that is, if we don't change any of x, y and z. The theory of 'algebraic symmetries' is the basis of the theory of equations of higher degree and its generalisations in Galois theory.

Finally, relations of a more general character can have symmetries, by which one understands transformations (what kind depends on the concrete situation) which do not change these relations. For example, it is very important to know the changes of coordinates which preserve

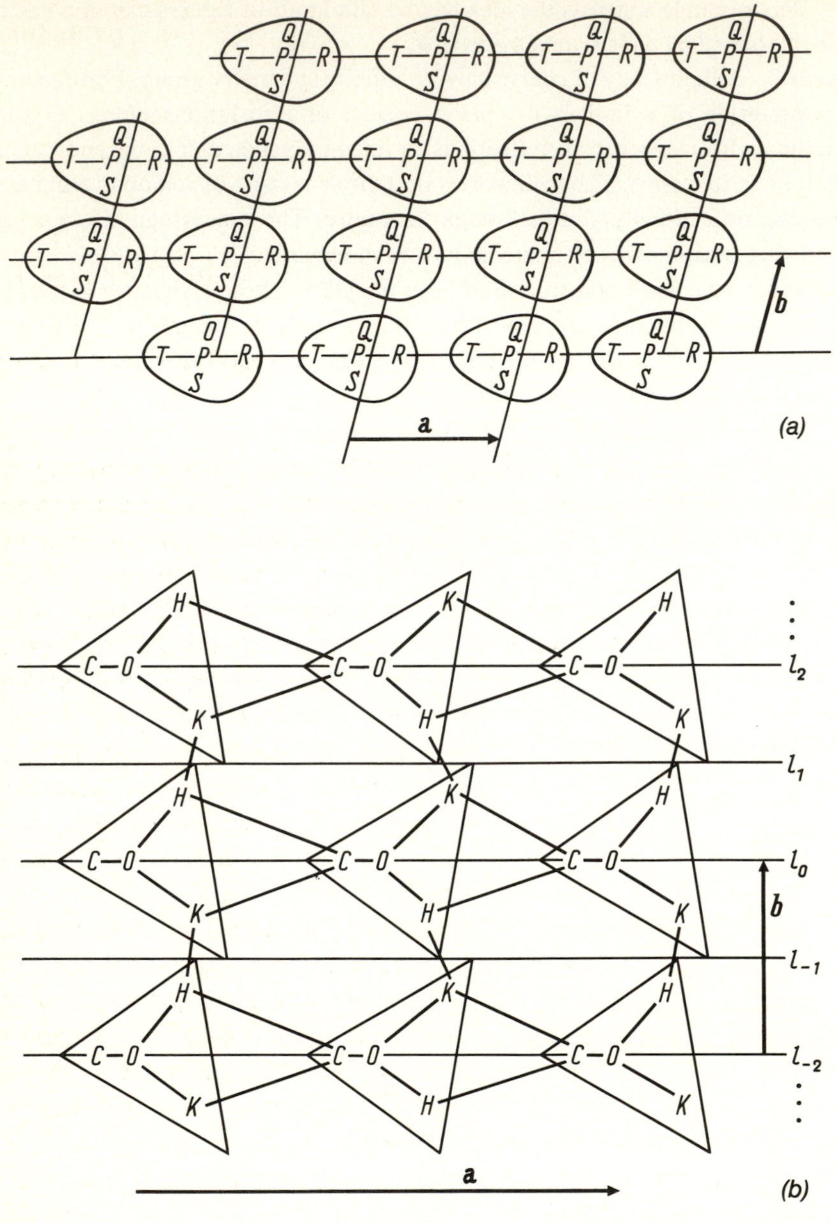

Figure 12.3

one or other physical law. Thus the laws of mechanics do not change on passing from one inertial coordinate system to another. For the case of motion along a line c with coordinate x these transformations have the form

$$x' = x + vt, \quad t' = t$$

in the classical mechanics of Gallileo and Newton. In the mechanics of the special theory of relativity, they are of the form

$$x' = \frac{x + vt}{\sqrt{1 - v^2c^2}} \quad \text{and} \quad t' = \frac{t + vx/c^2}{\sqrt{1 - v^2c^2}},$$

where c is the speed of light in a vacuum.

Certain symmetries of physical laws are similar to the symmetries of figures which we considered in Examples 1 and 2. For example, all phenomena will be unaffected if we replace positive and negative electrical charges, and interchange right and left, and vice versa. The reader can find out more about the symmetries of physical laws from R. Feynman, The Feynman lectures on physics, vol.III, chap.17.

The general notion of symmetry is the subject of the book H. Weyl, Symmetry, Princeton University Press, Princeton, 1952. In the remainder of this section we will talk about one concrete application of the idea of symmetry, to crystallography.

Exercises

1. Can a bounded figure have two central reflections in distinct centres as symmetries? Two rotations about distinct centres?

2. Prove that among all tetrahedrons, the regular tetrahedron has the biggest symmetry group.

3. Classify rectangular parallelipipeds by their symmetry groups.

4. Find the symmetry group of a tetrahedron whose edges are given by diagonals of the faces of an irregular rectangular parallelipiped (it has 4 motions). Prove that every tetrahedron with this symmetry group is constructed in this way.

5. Give an example of a figure Φ in the plane for which there exists a motion F taking Φ to itself, but not onto itself; such an F is then not a symmetry of Φ.

6. Construct figures whose symmetry groups are uniformly discontinuous groups of Types II.a and II.b. Note that the resulting patterns, in contrast to those for groups III.a and III.b, only repeat in one direction. Explain this.

12.2. Crystals and crystallographic groups. Symmetry groups, and classification by means of these groups have a particularly important role in *crystallography* , the branch of physics which studies the properties of crystals. Here we should bear in mind that in nature, the bounded size of crystals is explained by the fact that the growth of the crystal is stopped by some external circumstances. Theoretically though, they could grow to infinity. Therefore in crystallography, one supposes that nothing prevents the crystal from growing indefinitely, and that it occupies the whole of space. From this point of view, a crystal has two properties: (a) on the one hand, it is *discrete*, that it, it consists of elements of different nature, atoms, molecules, and so on, which can arrange themselves relative to one another in space in various ways; (b) on the other hand, the properties of a crystal are repeated in space: there exists a ball B of a sufficiently large radius, so that for every point X of space, there is a point Y in B such that all the physical properties of the crystal at X repeat those at Y. In the words of Hermann Weyl (Symmetry, p.121), we can say that crystals are solid ornaments found in Nature rather than in Art.

To get away from the physics, we restate properties (a) and (b) in a mathematical way. For this, note that by the homogeneity of space (see §10.1, Lemma 1), points X and Y of space for which the physical properties of the crystal are the same (in what follows we will call such points equivalent points) are characterised by the fact that there exists a symmetry F of the crystal taking X into Y. Indeed, let F be the correspondence between points of the crystal which preserves all the physical properties of the crystal and takes X to Y (this is the precise meaning of the fact that the physical properties of the substance are the same at X and Y). The distance $|AB|$ between points A and B of the crystal is some physical property of this pair of points, so that it is preserved by F. Thus F is a motion which preserves all the physical properties of the crystal, that is, it is a symmetry. (Note that by a symmetry of a crystal we always mean a symmetry of the crystalline substance thought of as extending to fill the whole of space.)

If we denote by Γ the group of symmetries of the crystal, we can therefore restate properties (a) and (b) as follows:

(1) For every point A of space, there is a number $d(A) > 0$, depending on A, such that, for every motion F in Γ with $F(A) \neq A$, the inequality $|AF(A)| \geq d(A)$ holds.

(2) There exists a ball B in space such that, for every point X of space, there is some motion F in Γ for which F(X) belongs to B.

In view of the fact that groups of motion of space satisfying (1) and (2) first arose in crystallography as the symmetry groups of crystals, and play an extremely important role there, they are called *crystallographic groups* . Of course, we can also consider groups of motion of the plane satisfying (1) and (2), (replacing the ball by a disc in (2)), and these will again be called crystallographic groups. These also have physical meaning, for example as the symmetry groups of suitable cross-sections of crystals. Example 4 of §12.1 shows another application of plane crystallographic groups. They can be used in the classification of repeating patterns in the plane, called *ornaments* (or *wallpaper patterns* in the English literature). Thus these groups are not just important in science, but also in art (see the books H. Weyl, Symmetry, Princeton University Press, Princeton, 1952, and A.B. Shubnikov and V.A. Koptsik, Symmetry in science and art, Nauka, Moscow, 1972, English translation Plenum, New York, 1974). To run ahead of ourselves, we note that there exist 17 different types of plane crystallographic groups; at the end of this section, in Figure 12.23 we give 17 ornaments with symmetry groups of each of the 17 types. Thus every plane crystallographic group is a symmetry group of one of these, and every plane ornament is 'similar' to one of these. Ornaments of all 17 types were for the first time gathered together in one place by the Arabs in the 13th century in the form of the wall decorations and paintings of the Cathedral of Alhambra in Granada, Spain; although apparently all 17 types of ornaments were known to the ancient Egyptians, that is, some time before the appearance of the notion of a group (19th century A.D.), and a fortiori before the classification of plane crystallographic groups.

Since in previous sections we have studied uniformly discontinuous groups of motions of the plane and 3-space, it is natural to determine which of these are crystallographic. The answer is obvious: exactly those for which the corresponding geometry is bounded. Indeed, property (1) is satisfied quite generally for every uniformly discontinuous group (see §7, Property III), and in fact the constant $d(A)$ can even be chosen independently of the point A. From (2), it follows obviously that the distance between any two points of the geometry Σ_Γ corresponding to Γ does not exceed $2R$, where R is the radius of B. Conversely, if the distance between any two points of Σ_Γ is less than some $R > 0$, then for any point X of space, the ball $B = B(A, R)$ of radius R centred at any point A of space contains a point X' equivalent to X, since otherwise $|ax| \geq R$, where a is the point of Σ_Γ defined by the set of points $\mathbf{A}$ equivalent to A, and x is defined by X in the same way.

Thus uniformly discontinuous groups of motion provide us with two types of crystallographic groups of motions of the plane (Type III.a and III.b, see §8), and 10 types of those of 3-space (Type IV.1–10, see §11.3). As imaginary plane crystals, having uniformly discontinuous crystallographic groups of Type III.a and III.b, we can consider crystals with chemical formulas PQRST and OCKH whose molecules have the form, and relative positions illustrated in Figure 12.3, (a) and (b); of course, the letters P, Q, ... we are using here are not in any way related to the notation for elements in Mendeleev's periodic table of the elements. However, for the spatial groups of Type IV.1–10, one can exhibit genuine existing crystals with these symmetry groups.

The simplest of these, the symmetry group of Type IV.2 is exhibited by potassium hydroxide KOH. This hard crystal is a very strong alkali, and is widely known as caustic potash. We now describe the structure of KOH; this description is obtained in crystallography by illuminating the substance at different angles, usually with X-rays or other radiation, and decyphering the resulting diffraction patterns on photographic plates. In the notation of §11, the symmetry group of KOH is of the form

$$\Gamma = \langle T_{\mathbf{a}}, T_{\mathbf{b}}, R_{\ell,\pi}^{\mathbf{c}} \rangle,$$

where the vectors $\mathbf{a}, \mathbf{b}$ and $\mathbf{c}$ have lengths $|\mathbf{a}| = 3.95$Å, $|\mathbf{b}| = 5.73$Å and $|\mathbf{c}| = 2.00$Å (here Å = Angstrom unit = 10^{-10} m.), the angle between $\mathbf{a}$ and $\mathbf{b}$ is $103°36'$, and $\mathbf{c}$ is perpendicular to $\mathbf{a}$ and $\mathbf{b}$ by definition of a group of Type IV.2 (Figure 12.4). One can even say how the potassium and oxygen atoms K and O are positioned relative to the group (the hydrogen atom H is not considered). In crystalline KOH, we can choose coordinates with origin Q and direction vectors $\mathbf{a}, \mathbf{b}, \mathbf{c}$, so that a point P has coordinates (x, y, z) if $\overrightarrow{QP} = x\mathbf{a} + y\mathbf{b} + z\mathbf{c}$; then points of the axis ℓ have coordinate $(0, 0, z)$, so Q is on the axis ℓ. The centres of the potassium and oxygen atoms K and O will be at points with coordinates

$$K(0.175, 0.288, 0) \text{ and } O(0.318, 0.770, 0). \tag{1}$$

The remaining potassium and oxygen atoms are obtained by applying motions in Γ to these. In the given coordinate system the generators of Γ act as follows:

$$T_{\mathbf{a}}: (x, y, z) \mapsto (x+1, y, z),$$
$$T_{\mathbf{b}}: (x, y, z) \mapsto (x, y+1, z), \tag{2}$$
$$R_{\ell,\pi}^{\mathbf{c}}: (x, y, z) \mapsto (-x, -y, z+1).$$

Owing to the small number of atoms in the molecule of KOH, and the simple nature of the group Γ of Type IV.2, we can even, using (1) and (2), draw a picture (see Figure 12.4) of the pattern in space formed by the atoms K and O in KOH; the picture also illustrates the packing of the molecules of KOH, by which

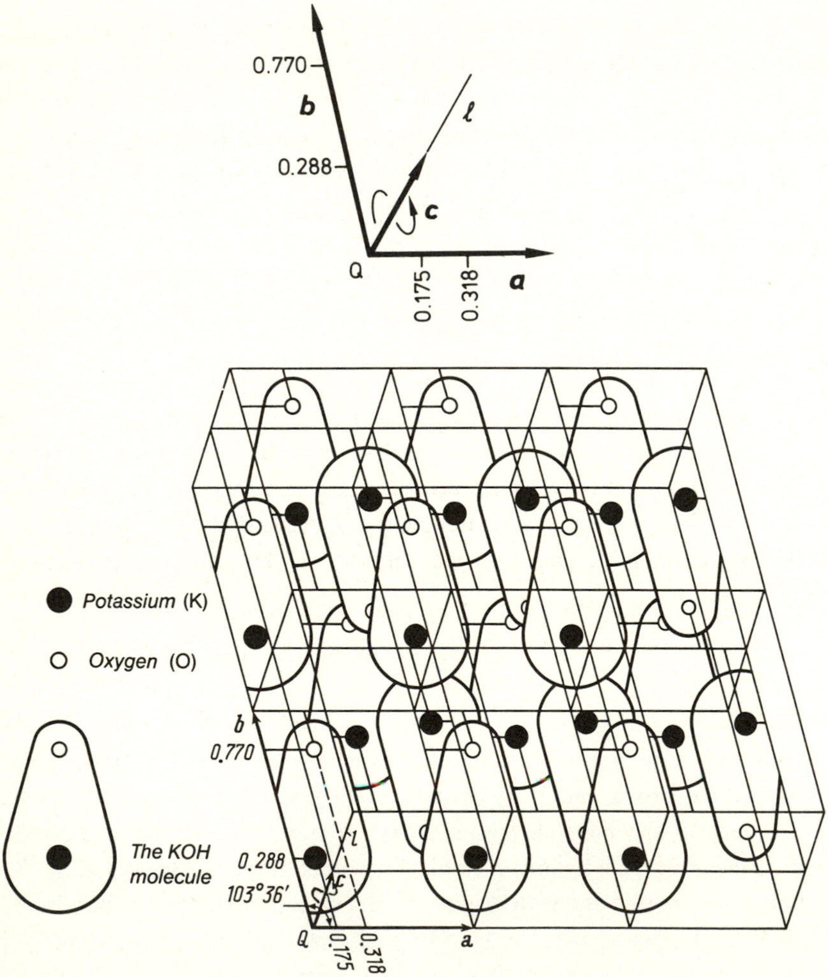

Figure 12.4. The structure of KOH

we mean that we take the closest pairs of atoms K and O (among all such pairs), and the portions of space closest to these. It is not hard to calculate that one molecule is formed by the pair (1), and all the remaining molecules are obtained from it by motions in Γ.

Using the symmetry group, we can describe in an entirely similar way the structure of other, even very complicated crystals. The myoglobin protein is at present one of the most complicated crystals whose structure has been decyphered. A molecule of myoglobin consists of approximately 1200 atoms, not counting hydrogen atoms. The symmetry group is of the same Type IV.2, but the scale is of course substantially larger

$$|\mathbf{a}| = 64.4\text{Å}, \ |\mathbf{b}| = 34.8\text{Å}, \ |\mathbf{c}| = 15.6\text{Å},$$

and the angle between **a** and **b** is equal to 105.5°. For a complete description of the structure of this protein we would have to give the coordinates of its 1200 atoms, similarly to (1).

There are rather a lot of crystals with the other symmetry groups of Type IV.1-10. Of these types of groups, that most commonly found in the world of crystals is the most complicated group IV.10; rather a lot of inorganic substances (but still less that 1%) belong to this group, and 13% of the organic substances; one such substance is given in Exercise 1. The simplest group IV.1 accounts for about 1% of inorganic, and 5% of organic substances. The group IV.2 which we considered above accounts for some (less than 1%) of inorganic, and 8% of organic substances. The remainder of the groups from IV.1-10 are much less common, although for each of these it is possible to give a crystal having this as symmetry group. The groups IV.1-10 are very common among substances of biological extraction. It is an interesting fact that among these, only groups of the first kind can appear, that is groups of Type IV.1-5 and IV.10. This is related to the fact that amino acids, from which the molecules of biological substances are built up, form helixes, which in nature are always left-handed (compare §11.4).

From the data we have given, the conclusion should be presenting itself to the reader that the uniformly discontinuous groups IV.1-10 appear rather rarely as symmetry groups of crystals. For example, why have we not so far introduced as an example a crystal well-known to everyone, such as rock salt (known in culinary science as table salt)?

The crystal rock salt NaCl consists of a cubical lattice, with vertices occupied alternately by atoms of chlorine and sodium (Figure 12.5). From the picture it can be seen that the symmetry group of the rock salt crystal contains reflections in planes (one such plane is shaded in Figure 12.5), and rotations around axes in multiples of 90° (with any of the lines shown in Figure 12.5 as axis). Hence this crystallographic group is not uniformly discontinuous, since in a uniformly discontinuous group, any motion other than the identity cannot have fixed points. For this reason, we cannot take rock salt (or many other well-known crystals) as

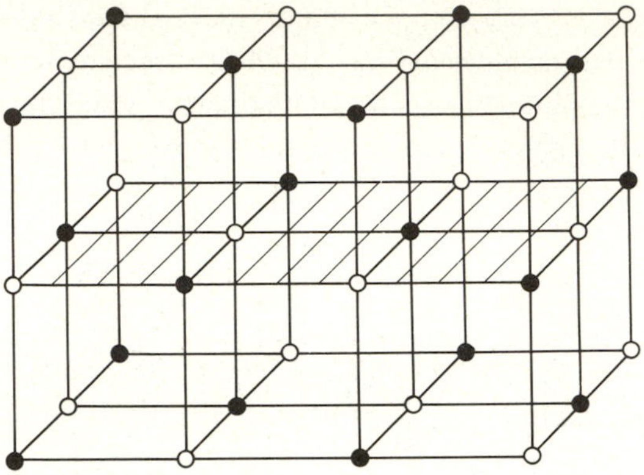

Figure 12.5. NaCl

examples of crystals with uniformly discontinuous crystallographic groups.

The example of the crystallographic group of salt shows that the 10 types of uniformly discontinuous groups form only a part, and as it turns out, only a small part of all types of crystallographic groups in space. This is not surprising, since as we have remarked, uniformly discontinuous groups satisfy a much stronger property than that required in the definition of crystallographic group: in property (1) the constant $d(A)$ can be chosen independently of A. A remarkable discovery made at the end of the last century was the classification by E.S. Fëdorov (1853-1919) in Russia and A. Schoenflies (1853-1928) in Germany of all crystallographic groups of motion of 3-space[1]. There turn out to be 219 types[2], and as we know, only 10 of these types give uniformly discontinuous groups.

We will not enumerate all these groups here; the interested reader will find their description in the specialist literature on crystallography (see for example

1 It is interesting to note that E.S.Fëdorov used a different, equivalent definition of a crystallographic group Γ: the group of translations in Γ must contain a translation in a shortest non-zero vector (or be uniformly discontinuous), and translations in non-coplanar vectors. It follows from this definition, just as for plane groups (see §8) that the translations in Γ are of the form $T_{k\mathbf{a}+\ell\mathbf{b}+m\mathbf{c}}$ where $\mathbf{a}$, $\mathbf{b}$ and $\mathbf{c}$ are three non-coplanar vectors, and k, ℓ, m are any integers.

2 The number of types of crystallographic groups is given as 230 in several textbooks; this comes from distinguishing groups which are conjugate by a motion of the second kind (that is, orientation reversing), but not by a motion of the first kind. In the present context, the number 219 is more natural.

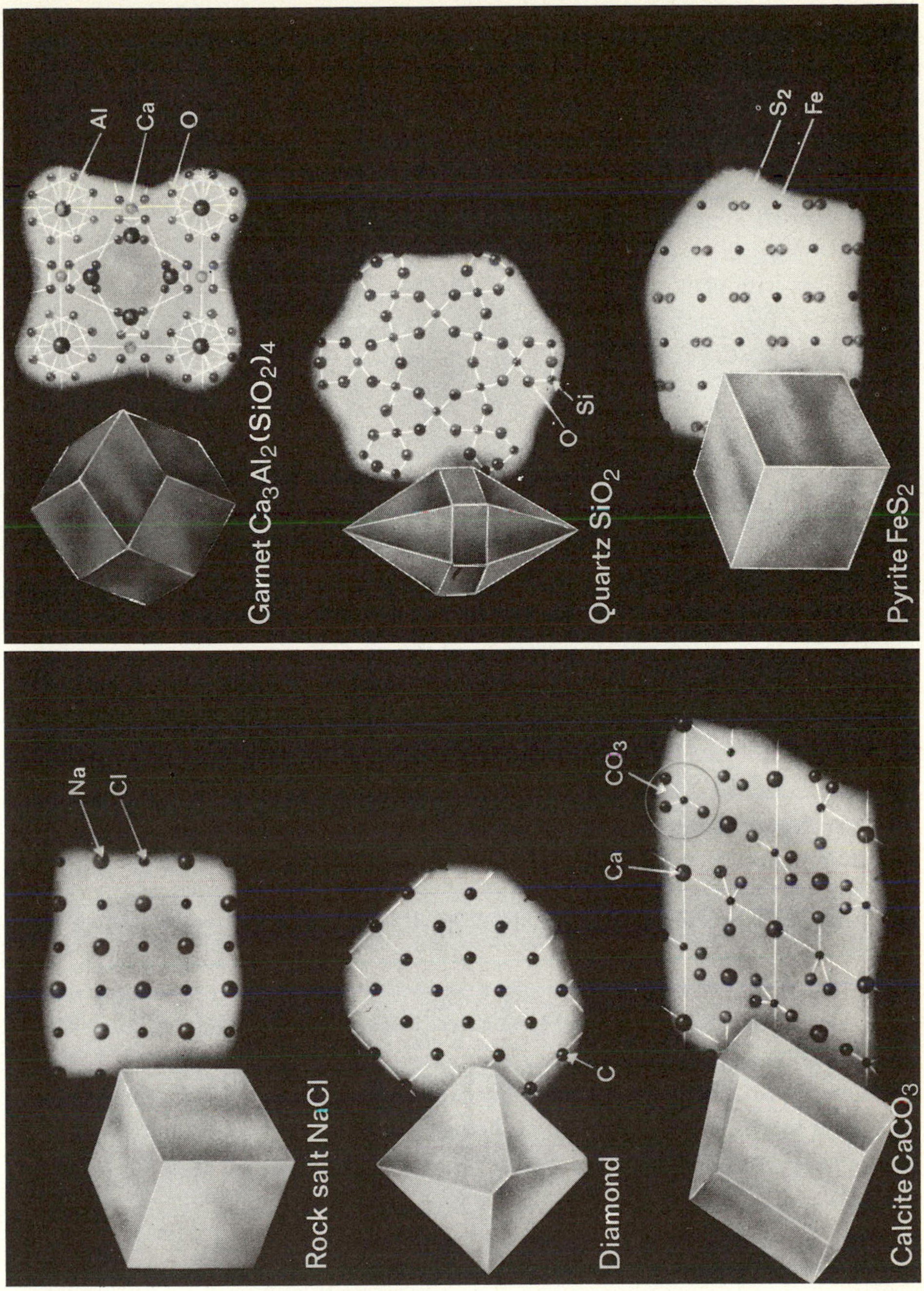
Garnet $Ca_3Al_2(SiO_2)_4$
Al
Ca
O
Quartz SiO_2
O
Si
Pyrite FeS_2
S_2
Fe
Rock salt NaCl
Na
Cl
Diamond
C
Calcite $CaCO_3$
Ca
CO_3

Figure 12.6

J.J. Burckhardt, Die Bewegungsgruppen der Kristallographie, Birkhäuser, Basel, 1957 or M.J. Buerger, Introduction to crystal geometry, McGraw-Hill, New York, 1971).

The reader will find pretty examples of crystals with particularly rich symmetry groups in Figure 12.6. It is an interesting fact that the symmetry of the atomic structure of crystals is reflected in the external form. Later we will classify (without proof) all plane crystallographic groups, which turn out to be far fewer than the space groups, just 17 types. We have already mentioned these in connections with plane patterns; the 17 patterns given at the end of the section (Figure 12.23) of each of the 17 types can be considered as a model of plane crystals with these symmetry groups.

Exercises

1. Hydroxylamine NH_2OH has the symmetry group IV.10, with (in the notation of §11, Table 2) $|\mathbf{a}| = 3.646$Å, $|\mathbf{b}| = 2.196$Å, $|\mathbf{c}| = 2.437$Å. In the crystal hydroxylamine, we can take the origin Q such that in the coordinate system $(Q, \mathbf{a}, \mathbf{b}, \mathbf{c})$, the points of the axis ℓ have coordinates $(x, 0, 0)$, those of the axis m $(0, y, 1/2)$, and the coordinates of the atoms N and O can be obtained from two of them

$$O(0.120, -0.624, -0.046) \text{ and } N(0.242, -0.012, 0.126)$$

by the action of the symmetry group, which in these coordinates obviously acts as follows:

$$R_{\ell,\pi}{}^{\mathbf{a}}\colon (x, y, z) \mapsto (x+1, -y, -z),$$

$$R_{m,\pi}{}^{\mathbf{b}}\colon (x, y, z) \mapsto (-x, y+1, -z).$$

Find the distance between the closest atoms N and O in the crystal hydroxylamine. Try to imagine the packing of the molecules (ignoring the atoms H). Represent its projection to the (x, y)-plane.

2. Prove that the patterns of type II.1, II.2 in Figure 12.24 and III.1, III.2 in Figure 12.23 have uniformly discontinuous groups of symmetry of types II.a, II.b and III.a, III.b respectively.

12.3. Crystallographic groups and geometries: discrete groups. Before proceeding to this classification, we invite the reader to stop and consider the curious situation in which we find ourselves in connection with the appearance in this section of a new class of groups of motion, the crystallographic groups. This kind of situation often occurs in mathematics, when different ideas lead to the appearance of similar mathematical objects.

The problem of describing locally Euclidean geometries led us in §§3–11 to the uniformly discontinuous groups corresponding to these geometries. A quite different motivation has led us in this section to a new class of groups, the crystallographic groups, arising in crystallography as the symmetry groups of crystals. Neither of these two classes of groups contains the other, but the uniformly discontinuous crystallographic groups are common to both classes. It might be expected that there exists another natural class of groups, which contains at the same time all of the uniformly discontinuous groups, and all the crystallographic groups; and that, as with the uniformly discontinuous groups, each group of this class defines some kind of geometry, which although not necessarily locally Euclidean may deserve attention, with the whole class of groups describing some class of geometries wider than the locally Euclidean class, but which can be defined without reference to groups. If this geometry originates from the crystallographic group of an actual crystal, then it is endowed with physical properties, so is a real physical world, describing the physical properties of the substance. We devote the remainder of this section to these interesting questions.

To realise this program, notice that both the uniformly discontinuous groups and the crystallographic groups satisfy condition (1) of the definition of crystallographic groups. It is natural to take the class of groups satisfying this property as the class of groups we are looking for.

Definition. A group Γ of motions of the plane (or of 3-space) is *discrete* if for every point A of the plane (or 3-space) there exists a constant $d(A) > 0$, depending on A, such that for every motion F in Γ with $F(A) \neq A$, it follows that $|AF(A)| \geq d(A)$.

The next step of our program consists of the construction of a geometry from a discrete group Γ. For this, we must simply repeat word-for-word the construction of a geometry from a uniformly discontinuous group: two points A and A′ are equivalent if one of them can be obtained from the other by a motion of Γ; a point a of the geometry corresponding to Γ is specified by a set **A** of equivalent points; the distance $|ab|$ between two points of the geometry is defined as the shortest of the distances $|AB|$ where A runs through the set **A** of equivalent points specifying a, and B runs through the set **B** specifying b. The only difficulty with this definition is the proof that the distance $|ab|$ we have defined exists. For this, as in the case of uniformly discontinuous groups, we must first of all observe that, in determining the distance, we can fix the point A of **A** and look for the minimum of the distances $|AB|$ where A is the fixed point, and B runs through **B**. Secondly, each set **B** of equivalent point is discrete (hence the term discrete group); that is, the constant d(B) is the same for all points of **B** , and only depends on **B**; this follows at once from the fact that Γ is a group.

We note that, just as easily as for the uniformly discontinuous groups, it can be checked that the crystallographic groups are exactly the discrete groups corresponding to bounded geometries. It is natural to expect, although not so simple to prove, that the uniformly discontinuous groups are exactly the discrete groups corresponding to locally Euclidean geometries.

Which class of geometries do discrete groups give rise to? In answering this question we will for simplicity restrict ourselves to discrete groups in the plane, although all our arguments and assertions can be carried over to 3-space.

In the case of uniformly discontinuous groups, as we know, the answer can be formulated as follows:

(1) there is one standard geometry, the plane, which corresponds to the trivial uniformly discontinuous group (consisting of the identity only);

(2) all the remaining geometries (including the plane) corresponding to uniformly discontinuous groups are precisely the geometries which are locally Euclidean, that is, are identical to the plane in sufficiently small regions;

(3) hence by classifying all uniformly discontinuous groups in the plane, we get a description of all locally Euclidean geometries.

In the case of discrete groups, we will see that the answer is similar, although there are many more standard geometries.

Theorem 1. (1) There is a certain collection of standard geometries; these correspond to the discrete groups for which all motions of the group have a common fixed point; we will describe all of these, and denote them by C_n and D_n for $n = 1, 2, \ldots$

(2) All the remaining geometries (including C_n and D_n) corresponding to discrete groups are precisely the geometries which are locally C_n or D_n, that is, which are identical with the standard geometries C_n or D_n in sufficiently small regions (a precise definition of what this means will be given below).

(3) Hence by giving a classification of all discrete groups in the plane, we get a description of all locally C_n or D_n geometries.

Thus first of all, we consider those discrete groups Γ which have a fixed point O; by this we mean that every motion of Γ takes O to itself. If Γ were uniformly discontinuous, it would follow from §8, Proposition 1, that Γ consists of just one element, the identity motion E. In the case of discrete groups, we will see that the answer is not much more complicated: we will prove that Γ is finite, that is, it has only a finite number of motions. Notice that the converse is also true: every finite group of motions is discrete and has a fixed point (prove this!).

Proposition 1. A discrete group Γ having a fixed point O is finite.

Proof. Consider a point A distinct from O, and the set **A** of points equivalent to A under Γ. Let's prove that **A** is a finite set. Indeed, any point equivalent to A is of the form F(A) for F in Γ, so that

$$|OA| = |F(O)F(A)| = |OF(A)|$$

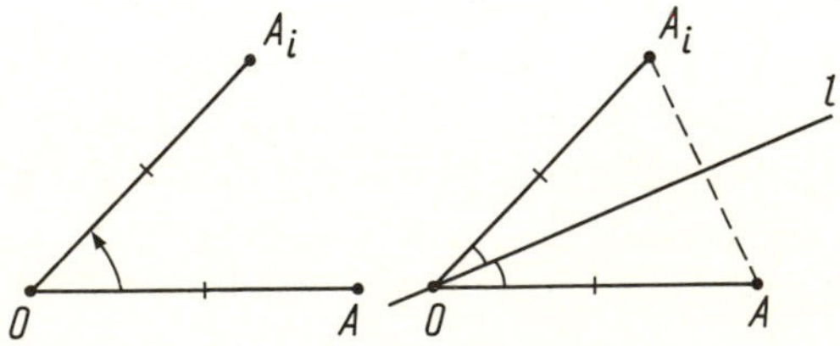

Figure 12.7

from which it follows that all the points of **A** lie on the circle of radius $|OA|$ centred at O, and in particular, inside the disc $D(O, 2|OA|)$. Thus the discrete set **A** is bounded and therefore finite by §7, Lemma. Suppose that **A** consists of points $\{A_1, ..., A_n\}$. Every motion F of Γ has the property that $F(O) = O$ and $F(A) = A_i$ for some $i = 1, ..., n$. It then follows from Chasles' theorem that F is either a rotation about O through the angle AOA_i^, or a reflection in the bisector ℓ of AOA_i^ (Figure 12.7). It follows that the number of motions of Γ is at most 2n (the same thing also follows from §8, Lemma 3). This proves the proposition.

Using this finiteness, it is easy to classify all discrete groups Γ with a fixed

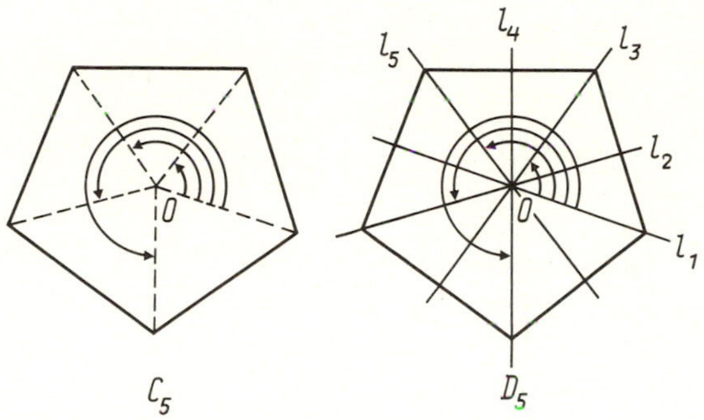

Figure 12.8

point O. We give the answer; the reader can prove it himself (see Exercise 1). The finite groups of motions of the plane with fixed point O are the groups C_n and D_n for $n = 1, 2, ...$ as follows: C_n consists of all the rotations of a regular n-gon, and D_n of all the symmetries of a regular n-gon with centre at O (see Figure 12.8). Hence

$$C_n = \langle R_O^{2\pi/n} \rangle = \{E, R_O^{\varphi}, R_O^{2\varphi}, ..., R_O^{(n-1)\varphi}\},$$

(where $\varphi = 2\pi/n$), is the group generated by the rotation $R_O^{2\pi/n}$, consisting of the n rotations around O through multiple of $2\pi/n$; and

$$D_n = \langle S_{\ell_1}, S_{\ell_2} \rangle = \{E, R_O^{\varphi}, R_O^{2\varphi}, ..., R_O^{(n-1)\varphi}, S_{\ell_1}, S_{\ell_2}, ..., S_{\ell_n}\},$$

(where $\varphi = 2\pi/n$), is the group generated by the two reflections S_{ℓ_1} and S_{ℓ_2} in axes ℓ_1 and ℓ_2 through O making an angle of π/n; it consists of the same rotations as C_n and the reflections in axes $\ell_1, \ell_2, ..., \ell_n$ through O and having angles with one another which are multiples of π/n (see Figure 12.8).

A fundamental domain for C_n is a wedge with vertex O and angle $2\pi/n$; and the equivalent points of this fundamental domain are points on the sides of the wedge, obtained from one another by a rotation through $2\pi/n$ (Figure 12.9). Glueing together the sides of this wedge along equivalent points, we get a cone with vertex at O. Thus the geometry corresponding to C_n, which we will continue to denote C_n, is the geometry on the cone. Consider in 3-space a sphere centred at O, and a conical surface C, swept out by rays OM emanating from O, as M traces out on the sphere a closed curve f without self-intersection, called a *directrix* of the cone. By analogy with the geometry on a cylindrical surface in §3, it can be shown that if the length of the generator f is $2\pi R/n$, where R is the radius of the sphere and n is a natural number, then the geometry on the conical surface C is identical with C_n. We should mention that by no means every conical surface C gives a geometry coinciding with any C_n for a natural number n.

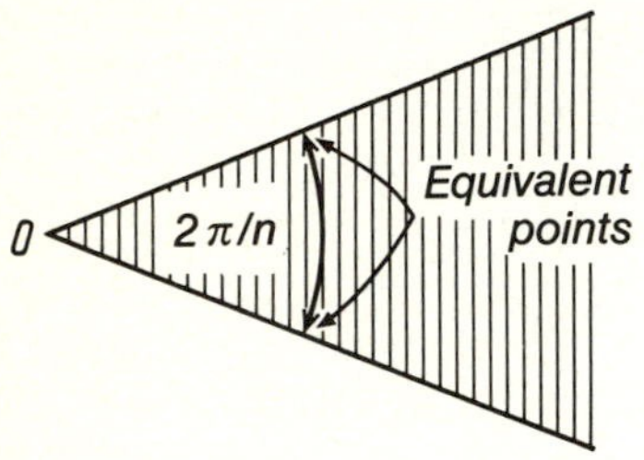

Figure 12.9

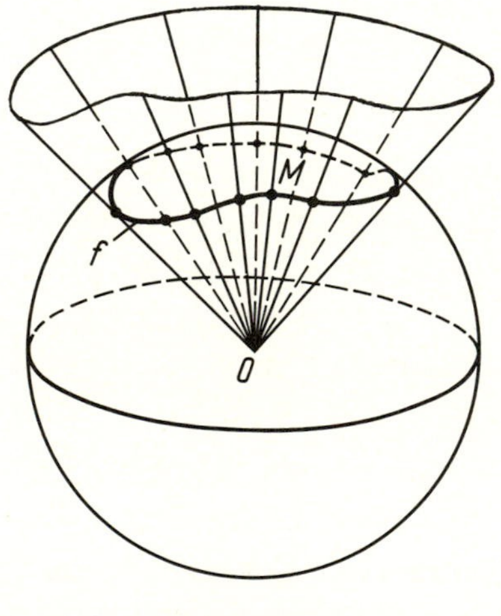

Figure 12.10

The fundamental domain for the group D_n is also a wedge, with vertex O and angle of π/n, but in contrast to C_n, no two points of the wedge are equivalent to one another (Figure 12.11). This means that the geometry D_n corresponding to D_n is just this wedge, considered as part of the plane, with distances measured as in the plane. This is easy to explain using the terminology of the jet service of §§3-5. The distance between two points A and B of the wedge in the geometry

D_n is equal to the shortest time required to travel from A to B moving about the wedge with unit speed and making use of the instantaneous jet service joining equivalent points; but since inside the wedge there are no distinct equivalent points, there is no jet service at all, and the distance between A and B in D_n is equal to their distance in the plane.

Notice that there is an important phenomenon appearing here: there are two kinds of geometries: with and without a jet service. The first kind is obtained from the corresponding fundamental domain by glueing. For the second kind, no glueing

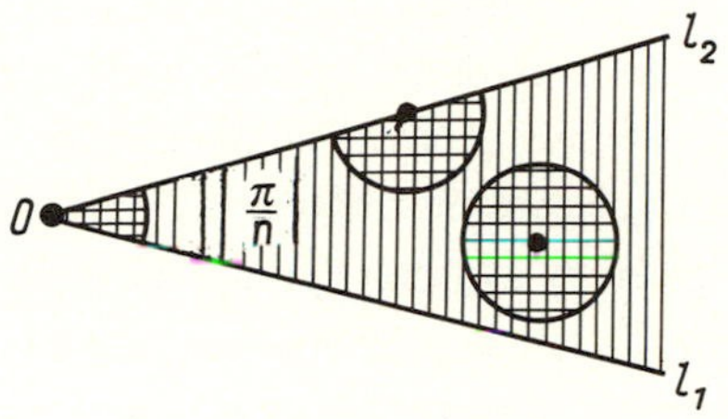

Figure 12.11

is required: the geometry just coincides with the fundamental domain, that is, it is just a plane figure.

The geometries C_n and D_n demonstrate the basic difference between geometries corresponding to discrete groups, as opposed to those coming from uniformly discontinuous groups. Geometries corresponding to uniformly discontinuous groups have the same structure in all sufficiently small regions: every point has a spherical neighbourhood of sufficiently small radius which is a disc in the plane. By contrast, in the geometry of D_n, there are three kinds of spherical neighbourhoods of points: a point not on the sides of the wedge, a point on the sides of the wedge, and the vertex of the wedge have as their respective spherical neighbourhoods a disc, a half-disc, and a wedge of angle π/n (see Figure 12.11). In the geometry C_n, the vertex of the cone is a singular point, for which an arbitrarily small spherical neighbourhood is not a disc; in D_n, the singular points are the sides and vertex of the wedge.

In the remainder of this section, we will say that a geometry Σ is locally C_n or D_n if Σ is identical with the standard geometries C_n or D_n in sufficiently small regions; to be more precise, this means the following: there exists $r > 0$ such that for every point a of Σ, we can find a standard geometry C_n or D_n, and a point A of this standard geometry, such that the spherical neighbouhood D(a, r) of a in Σ can be superposed on the spherical neighbourhood D(A, r) in the standard geometry C_n or D_n.

Exercises

1. Prove that the groups C_n and D_n give all the finite groups of motions of the plane. For which m and n do the groups C_m and D_m contain C_n or D_n? [Hint: follow the idea of proof of §8.2, Theorems 3 and 4; if the group only contains rotations, consider the rotation in Γ through the smallest angle. If Γ also contains reflections, consider the group Γ' consisting of rotations in Γ.]

2. We will say that lines in a geometry corresponding to a discrete group are curves which come from lines in the plane. What are the lines in C_n and D_n?

12.4. A typical example: the geometry of the rectangle. Let's now convince ourselves by examples of the truth of Theorem 1, verifying it experimentally. We will see that there are many purely geometrical methods of constructing geometries which are locally C_n or D_n. However, firstly, each time we will see that the newly constructed geometry is given by a discrete group; secondly, we will thus arrive at the complete list of locally C_n or D_n geometries, and a complete list of discrete groups; thirdly these constructions will show us that there exists a close connection between all such geometries (and obviously some connection between the corresponding discrete groups).

As a first example of a geometry which is locally C_n or D_n, but not one of them, consider a rectangle ABCD in the plane (Figure 12.12), with distance between points as in the plane. It is obviously a geometry: thus the hardest property

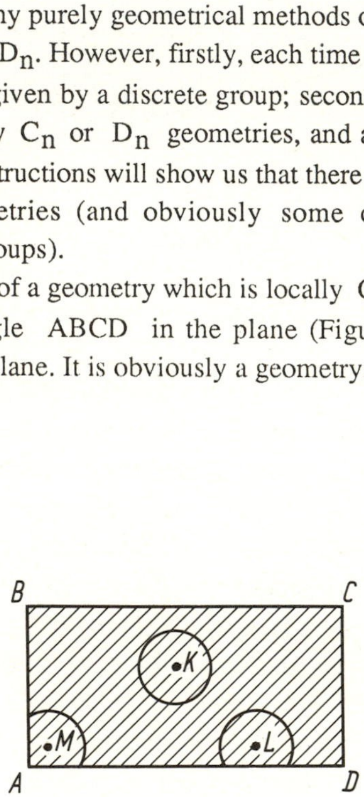

Figure 12.12. The geometry III.3

(d) in the definition of a geometry follows from the fact that ABCD is convex: the line segment joining any two of its points lies entirely in the rectangle. If we choose $r>0$ smaller than $|AB|/2$ and $|AD|/2$ then a spherical neighbourhood of radius r of any point of the rectangle ABCD is the same as the spherical neighbourhood of radius r in one of the geometries C_1 (the plane), D_1 (the half-plane) or D_2 (the right-angled wedge). For example, in Figure 12.12, the spherical neighbourhood of radius r of K is a disc with centre K in the plane; that of L is a disc of radius r with centre L in the half-space bounded by the line (AD); that of M is a disc of radius r with centre M in the quadrant BAD. This proves that the rectangle ABCD gives a locally C_n or D_n geometry (in fact, locally only C_1, D_1 or D_2). The fact that it is distinct from the geometries C_n or D_n is immediately apparent: firstly, the rectangle is bounded; secondly, its singular points are not as in C_n or D_n (it has 4 singular points of type D_2). Thus we really have some new geometry which is locally C_n or D_n.

On the other hand, one glance at the singular points of the rectangle ABCD is enough to suggest a candidate for the role of a discrete group of motions of the plane for which this rectangle is a fundamental domain and is the corresponding geometry. It is natural to take Γ to be the group

$$\Gamma = < S_{(AB)}, S_{(BC)}, S_{(CD)}, S_{(AD)} >$$

generated by reflections in the sides of ABCD.

Let's prove that it does have the properties we have indicated. The reflections $S_{(AB)}, S_{(BC)}, S_{(CD)}$ and $S_{(AD)}$ take the rectangle ABCD into the 4 adjacent rectangles, which together with ABCD itself make the figure Φ_1 of Figure 12.13, (b). By construction, the points of Φ_1 are equivalent to points of ABCD. The reflections $S_{(AB)}, S_{(BC)}, S_{(CD)}$ and $S_{(AD)}$ take the rectangles of Φ_1 into 8 new rectangles which, together with Φ_1, make up the figure Φ_2 of Figure 12.13, (c). By construction, points of Φ_2 are also equivalent to points of ABCD. Continuing this argument, we will get bigger and bigger figures Φ_n (for n = 1, 2, 3, ...), the points of which will be equivalent to ABCD, and which will fill out the whole plane. This proves that all points of the plane are equivalent to points of ABCD.

Now let's prove that, among points of ABCD, there are no equivalent points. One sees from the preceding argument that under composites of the generating motions $S_{(AB)}, S_{(BC)}, S_{(CD)}$ and $S_{(AD)}$ of Γ, that is, under motions of Γ, the images of ABCD fill out the plane, forming a lattice of congruent adjacent rectangles (Figure 12.13, (d)). This is obvious, since this lattice is taken to itself under each of the generators $S_{(AB)}, S_{(BC)}, S_{(CD)}$ and $S_{(AD)}$ of Γ, and therefore by any motion in Γ, since any such motion is a composite of the generators. Thus we can think of a motion in Γ as a symmetry of this lattice, and this is the idea of the proof of our assertion. However, for reasons which will become clear later, the lattice itself is not sufficient for our purpose. Let us in addition mark the rectangle ABCD by the letter L (Figure 12.13, (d)). Then arguing exactly as above, we get that the motions of Γ are the symmetries of the pattern of rectangles marked with L illustrated in Figure 12.13, (e). Now we can proceed to the proof of our assertion. Suppose that two points X and Y of the rectangle ABCD are equivalent, that is, Y = F(X), where F is a motion of Γ.

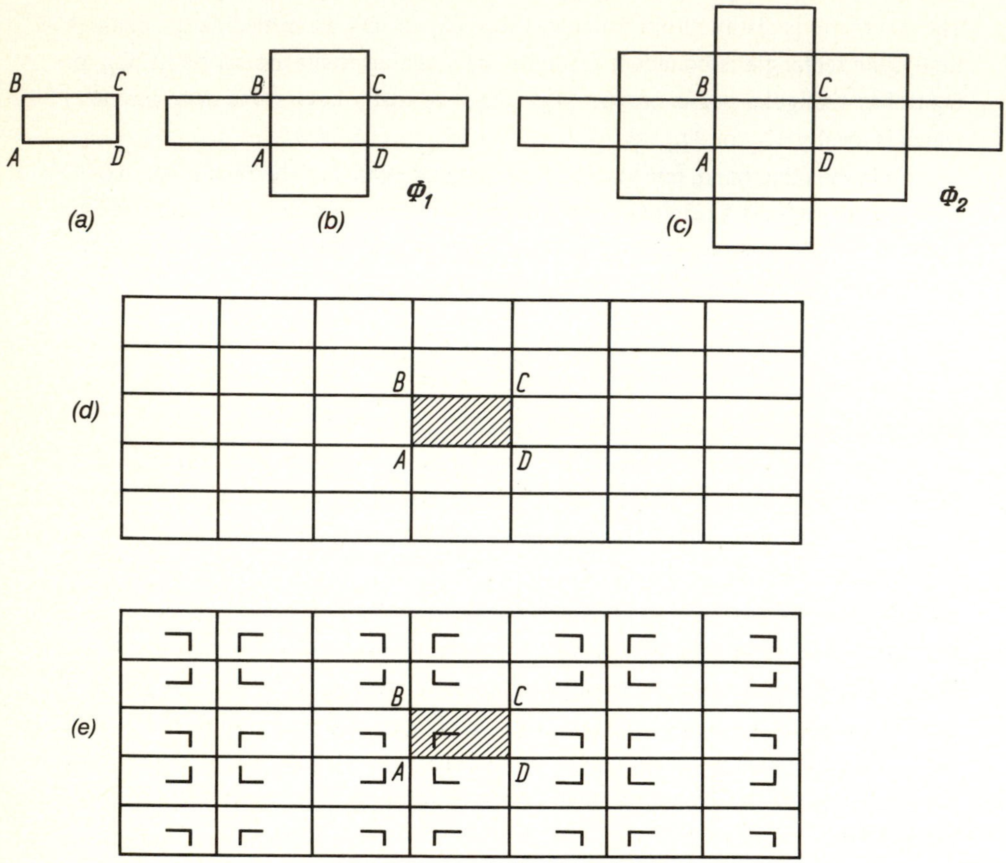

Figure 12.13. (a) ABCD; (b) Φ_1; (c) Φ_2;
(d) the lattice of rectangles; (e) the lattice of rectangles marked with L

The motion F takes the rectangle ABCD marked with L into some rectangle A′B′C′D′ of the pattern of Figure 12.13, (e), marked with its letter L, since motions of Γ are symmetries of this lattice. Now Y = F(X) is a point both of ABCD and of A′B′C′D′, so that these rectangles have points in common. It follows that A′B′C′D′ is one of the nine rectangles of the pattern illustrated in Figure 12.14.

But for each of these 9 rectangles marked with its letter L, there is exactly one motion of the plane which takes ABCD marked with its L into it (see §7, Lemma 3); it was precisely in order to get this uniqueness that we marked the rectangles by the letter L. As a result we get 9 possibilities for F, and it remains to deal with each of these. For example, suppose that F takes ABCD into the rectangle AMNB (see Figure 12.14); then $F = S_{(AB)}$ is the reflection in (AB). It is obvious that the points X of ABCD for which Y = F(X) also belongs to ABCD are exactly the points of the line segment [AB]; but for these points, Y = F(X) = X, so that this possibility for F does not give distinct equivalent points. The remaining 8 possibilities for F are dealt with similarly.

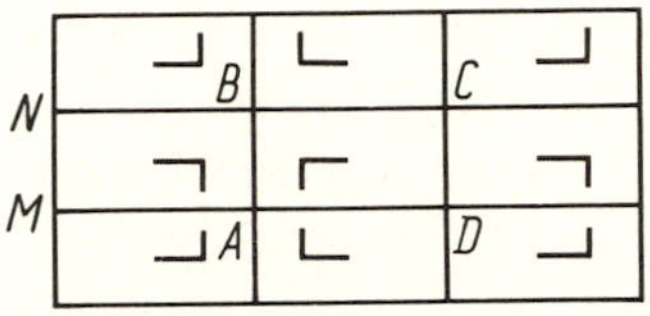

Figure 12.14

Notice that the same arguments prove that the symmetry group of the pattern of Figure 12.13, (e) equals Γ. Indeed, just as we proved above, each symmetry G of this pattern is uniquely determined by the rectangle G(ABCD) marked with L into which G takes ABCD marked with its L. If G does not belong to Γ then internal points of G(ABCD) would not be equivalent under Γ to points of ABCD, and this contradicts the fact that ABCD is a fundamental domain for Γ.

We prove finally that Γ is discrete. Notice that the constant d(X) (in the definition of discreteness) is the same for equivalent points, so that it is enough to consider only points X of ABCD. For this, we again refer to Figure 12.14, made up of the rectangles of the pattern surrounding ABCD. We use the fact that each of

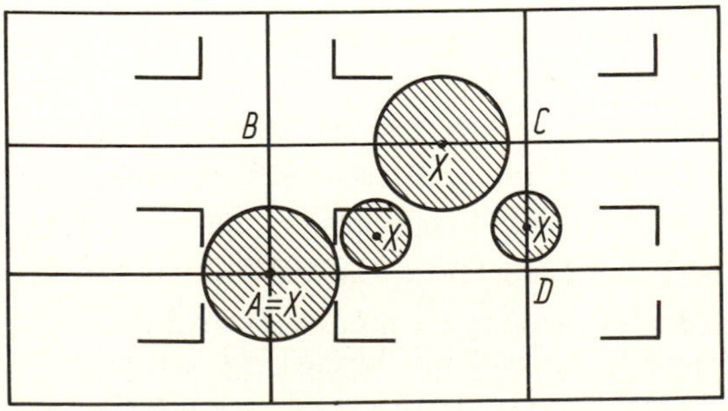

Figure 12.15

the rectangles of the pattern obviously has all the properties of ABCD, and in particular does not have distinct equivalent points. For each point X of ABCD, consider a disc D(X, d(X)) centred at X of radius d(X) chosen such that D(X, d(X)) does not intersect any sides of the rectangles of Figure 12.15 other than those passing through X.

We now claim that these numbers d(X) do what we require. For example, suppose that X = A is one of the vertices of ABCD. Then the disc D(A, d(A)) we have constructed is broken up by the 4 rectangles of the figure it meets into 4 sectors centred at A. In each of these sectors, there are no points equivalent to A (since this is true for the rectangle containing the sector), and hence there are no points equivalent to A in the whole disc D(A, d(A)); therefore d(A) is the required number for A. The other positions for X in ABCD are dealt with in an entirely similar way. This proves that the group Γ is discrete.

Thus we have proved that the locally C_n or D_n geometry we started from, the rectangle in the plane, turns out to be the geometry corresponding to a discrete group Γ of motions of the plane. We have constructed Γ, showing that it can be taken to be the group generated by reflections in the sides of the rectangle, and that the rectangle is then a fundamental domain for Γ; and we have proved that Γ is discrete. In this geometry, as in the geometries D_n, the rectangle is at the same time both a fundamental domain and the geometry for Γ. We denote this geometry and the corresponding group by III.3, leaving the numbers III.1 and III.2 for the uniformly discontinuous groups of Type III.a and III.b.

We recall that the proof we gave above consisted of constructing a pattern in the plane which has the rectangle as one of its elements, and has Γ as symmetry group.

12.5. Classification of all locally C_n or D_n geometries. We have deliberately treated the example of the geometry of the rectangle in such detail because in an entirely similar way one can verify that all the other geometries of Type II.3-7 and III.4-17, which we construct in what follows, correspond to the discrete groups, and have the fundamental domains and the equivalent boundary points indicated below. We will not repeat the proof, leaving it to the reader, and restrict ourselves to indicating the patterns which we need to use (see Figures 12.23-24 below), numbering them in the same way as the geometries and groups which correspond to them, so that the symmetry group of each of these patterns will be the discrete group with the same numbering. In our notation, roman II will denote geometries which are unbounded, but distinct from C_n and D_n; the corresponding groups are not crystallographic, although they are infinite. The numbers II.1 and II.2 remain for the familiar geometries and groups of Types II.a and II.b; roman III will denote the bounded geometries and the corresponding crystallographic groups.

Without any difficulty, we can think of other figures in the plane whose geometry is locally C_n or D_n, but is distinct from C_n and D_n: a strip, a half-strip, a rectangle, and three triangles with angles given by π/k for natural

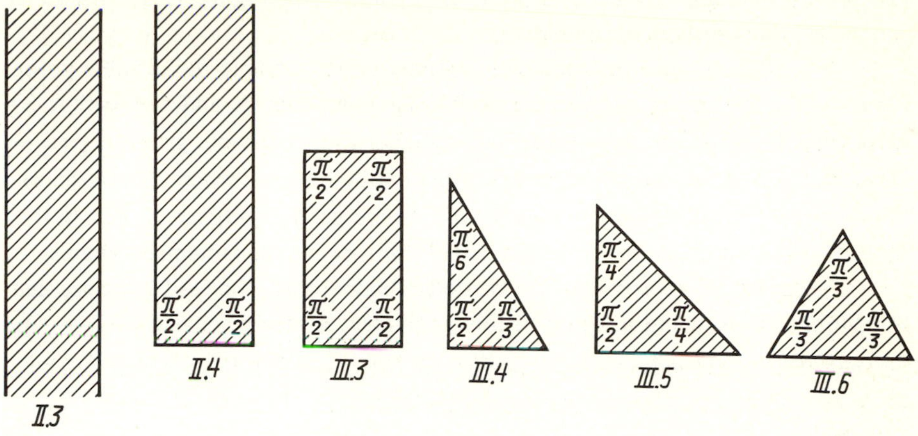

Figure 12.16. Geometries of Types II.3, II.4, III.3-6

(the groups are generated by reflections in the sides)

numbers k (Figure 2.16). As a non-trivial exercise, the reader can try to prove that there are no plane figures, other than the wedges D_n themselves and those of Figure 12.16, which give locally C_n or D_n geometries. In the same way as for the rectangle, the reader can see that these figures are geometries corresponding to the

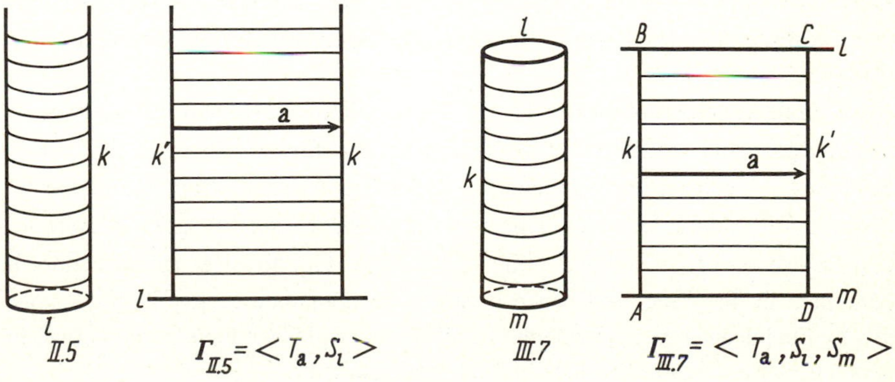

Figure 12.17. The geometries and groups of Types II.5 and III.7

discrete groups generated by reflections in their sides; the figures themselves are then fundamental domains.

Replacing the plane by the cylinder, we can also consider figures on the cylinder which are locally C_n or D_n. These give two new examples of geometries: the half-cylinder and the finite cylinder (see Figure 12.17, where ℓ and m are directrixes of the cylinder).

It is perfectly clear that the half-cylinder corresponds to the discrete group $\Gamma_{II.5} = \langle T_{\mathbf{a}}, S_\ell \rangle$, generated by a translation $T_{\mathbf{a}}$ in a vector **a** and a reflection S_ℓ in an axis ℓ parallel to **a**; and the finite cylinder to the discrete group $\Gamma_{III.7} = \langle T_{\mathbf{a}}, S_\ell, S_m \rangle$, generated by a translation and two reflections, where the translation vector **a** and the two reflection axes ℓ and m are parallel (Figure 12.17). In this figure, the half-strip and the rectangle ABCD will be the fundamental domains.

Recalling the geometry on the twisted cylinder of §4, the reader can see that the Möbius strip, being a figure of this geometry, is also a locally C_n or D_n geometry! It corresponds to the discrete group $\Gamma_{III.8} = \langle S_\ell^{\mathbf{a}}, S_m \rangle$, generated by a glide reflection and a reflection having parallel axes ℓ and m, and having a rectangle ABCD as fundamental domain (Figure 12.18). The reflection $S_{m'}$ (see

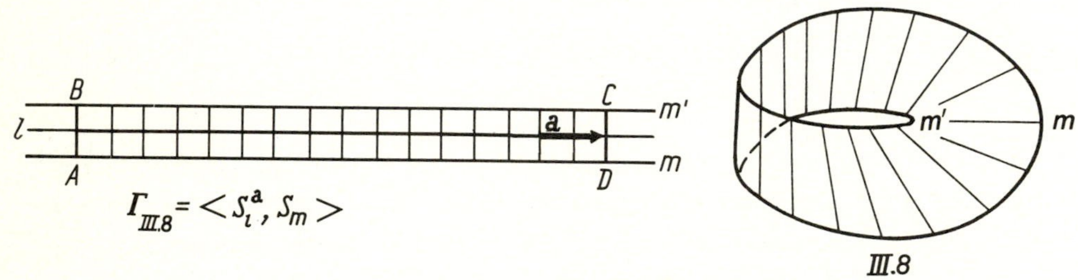

Figure 12.18. The geometry and group of Type III.8

Figure 12.18) is not one of the given generators of $\Gamma_{III.8}$, since it is a composite $S_{m'} = S_\ell^{\mathbf{a}} S_m (S_\ell^{\mathbf{a}})^{-1}$ of the generators $S_\ell^{\mathbf{a}}$ and S_m. Although the Möbius strip is like the finite cylinder, its singular points (the boundary) forms one curve rather than two.

In the examples of locally C_n or D_n geometries given so far, we have not had any geometry with singular points of type C_n for $n \geq 2$; indeed, in all the examples, the geometries were identical in sufficiently small regions to either C_1 (that is, the plane), or D_n. This is natural, since for $n \geq 2$, a spherical neighbourhood of the singular point of the cone C_n, however small, cannot be contained in the plane, and hence it cannot be contained in any 2-dimensional locally Euclidean geometry; whereas in all the examples we have given above, the geometries were figures in a locally Euclidean geometry.

Example of such geometries can also be given if we consider certain figures on certain cones C_n for $n \geq 2$. As we know, the cone C_2 is obtained by rolling

$$\Gamma_{\mathrm{II}.6} = <S_l\,,\,R_0^{\pi}>$$

II.6

$$\Gamma_{\mathrm{III}.9} = <S_{(AD)}\,,\,S_{(DC)}\,,\,R_0^{\pi}>$$

III.9

$$\Gamma_{\mathrm{III}.10} = <S_{(AB)}\,,\,R_0^{\frac{2\pi}{3}}>$$

III.10

$$\Gamma_{\mathrm{III}.11} = <S_{(AB)}\,,\,R_0^{\pi/2}>$$

III.11

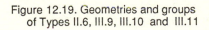

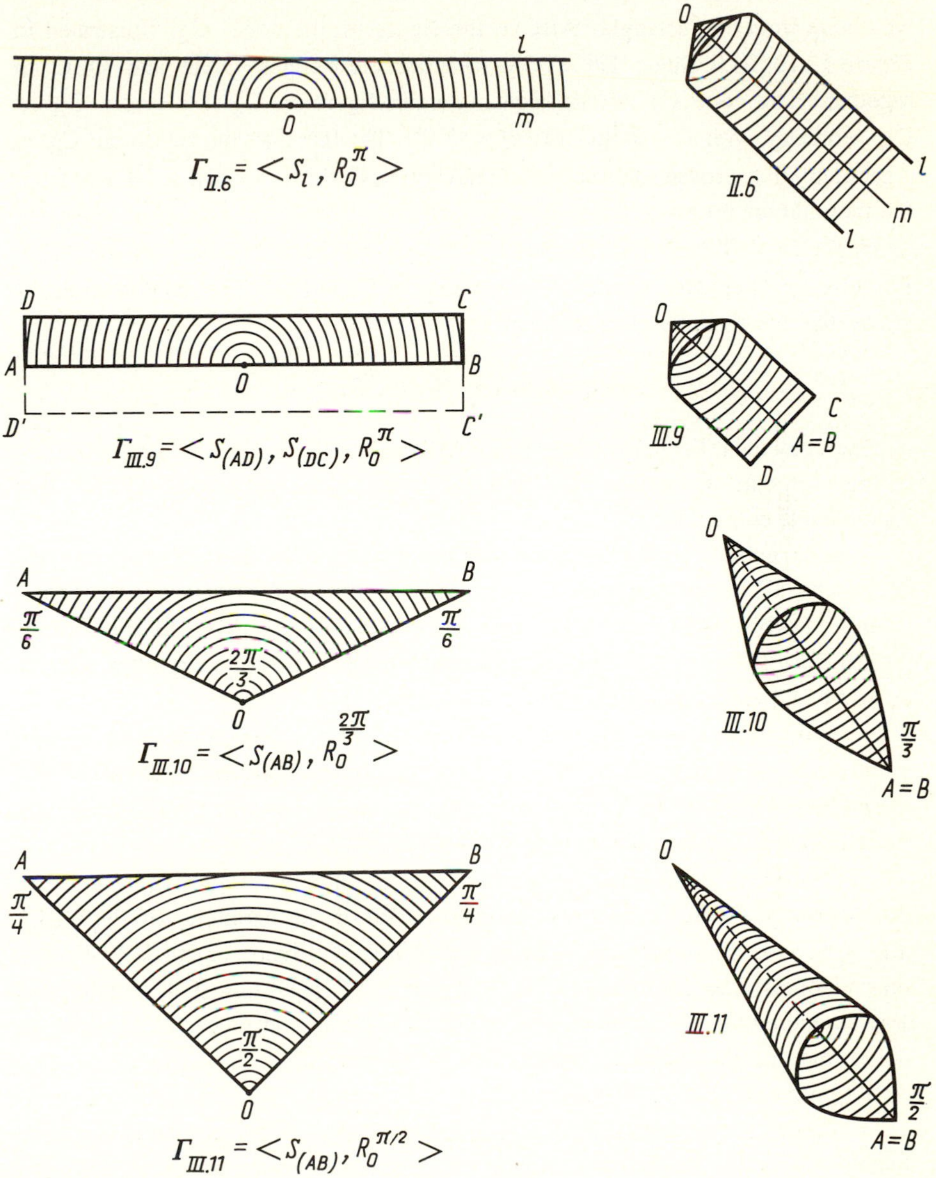

Figure 12.19. Geometries and groups
of Types II.6, III.9, III.10 and III.11

up a 'wedge' with vertex O and angle π. In this wedge, consider a rectangle ABCD, whose side AB is formed by the sides of the wedge, with O as its midpoint (Figure 12.19, $\Gamma_{III.9}$). Rolling up the angle of π at O into the cone C_2, we obtain from the rectangle ABCD the figure on the cone C_2 illustrated in Figure 12.19, III.9. Since $\hat{D} = \hat{C} = \pi/2$, and the wedges OAD and OBC fit together on the cone C_2 into an open angle, this figure on C_2 is a locally C_n or D_n geometry, with O a singularity of type C_2. The fact that this figure on C_2 is a geometry may arouse some suspicion; nevertheless, this is so, and it follows from the fact that the points of the rectangle ABCD equivalent under C_2 form the rectangle DCC'D' (dotted in Figure 12.19, $\Gamma_{III.9}$), which is a convex set (see §7.2, Exercise 6). Thus we obtain a new geometry, of Type III.9. The reader can easily check that it is given by the discrete group

$$\Gamma_{III.9} = \langle S_{(AD)}, S_{(DC)}, R_O{}^{\pi} \rangle,$$

generated by the reflections in the sides (AD) and (DC), and the central reflection in the midpoint O of the side AB of ABCD; the rectangle ABCD is a fundamental domain for Γ.

In an entirely similar way, one can construct another three geometries II.6, III.10 and III.11 on the cones C_2, C_3 and C_4, obtained from a strip or from a triangle OAB (see Figure 12.19), and check that they are given by the discrete groups $\Gamma_{II.6}, \Gamma_{III.10}$ and $\Gamma_{III.11}$, generated by a reflection and a rotation through π, $2\pi/3$ and $\pi/2$ respectively.

With this, we have exhausted the possibilities of this method: making new geometries as figures in already known locally Euclidean or locally C_n geometries. The point is, by this means we can never construct geometries with more than one singular point of type C_n with $n \geq 2$.

These can be obtained by another method, namely by *glueing geometries*. As an example, let's consider two identical copies of the geometry of Type III.3, that is, two identical rectangles ABCD and A'B'C'D' (Figure 12.20, (a)). We glue their identical sides [AB] and [A'B'], [BC] and [B'C'], [CD] and [C'D'], in the same way as one might sew a pillow-case out of two rectangular pieces of material. By doing this, we get a surface which is like the surface of a pillow-case if we fit it over a pillow (Figure 12.20, (a) and (b)).

Since a rectangle has right angles, the geometry of this surface is obviously locally C_n or D_n, having two singular points of type C_2, the points B = B' and C = C' of Figure 12.20, (b). If we adjoin the rectangle A'B'C'D' to ABCD as shown in Figure 12.20, (c), it is easy to see that this geometry is given by the discrete group generated by the two reflections $S_{(AD)}$ and $S_{(A'D')}$ and two central reflections $R_B{}^{\pi}$ and $R_C{}^{\pi}$, and the rectangle AA'D'D is a fundamental region for this group (Figure 12.20, (c)). If we notice that

$$S_{(A'D')} = R_B{}^{\pi} S_{(AD)} R_B{}^{\pi},$$

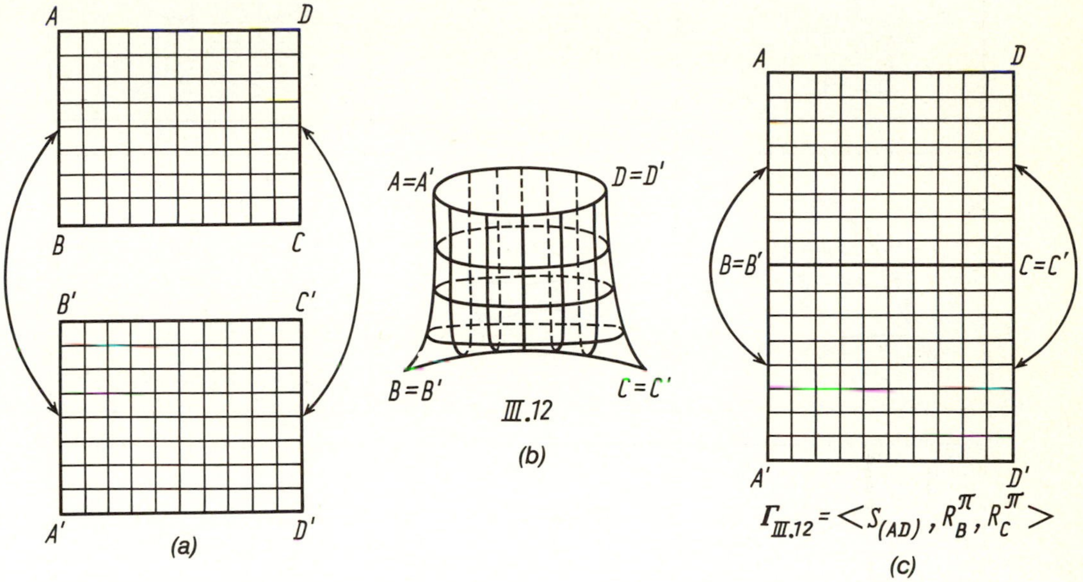

Figure 12.20. The geometry and group of Type III.12

then we see that we can make do with one fewer generator:

$$\Gamma_{III.12} = < S_{(AD)}, R_B^{\pi}, R_C^{\pi} >.$$

Note that in the previous example, we could also glue together the remaining free sides [AD] and [A′D′] of the rectangles, and get the geometry of Type III.13, which has 4 points of type C_2 as its singularities (Figure 12.21, (b)). We can see this geometry on the surface of a pillow. It is also given by the discrete group

$$\Gamma_{III.13} = < T_{2\mathbf{a}}, R_B^{\pi}, R_C^{\pi} >, \text{ where } \mathbf{a} = \overrightarrow{BA}$$

which is generated by two central reflections and a translation (Figure 12.21, (b), right). We just note that this is not the most general discrete group (or geometry) of this type: instead of the rectangle ABCD we could have taken a parallelogram (see Figure 12.21, (b), left). Now taking instead of the rectangle ABCD a half-strip and triangles, which are geometries of Types II.4, III.4, III.5 and III.6 (see Figure 12.16), we get by the same method another four new geometries of Types II.7, III.14, III.15 and III.16, given by the discrete groups

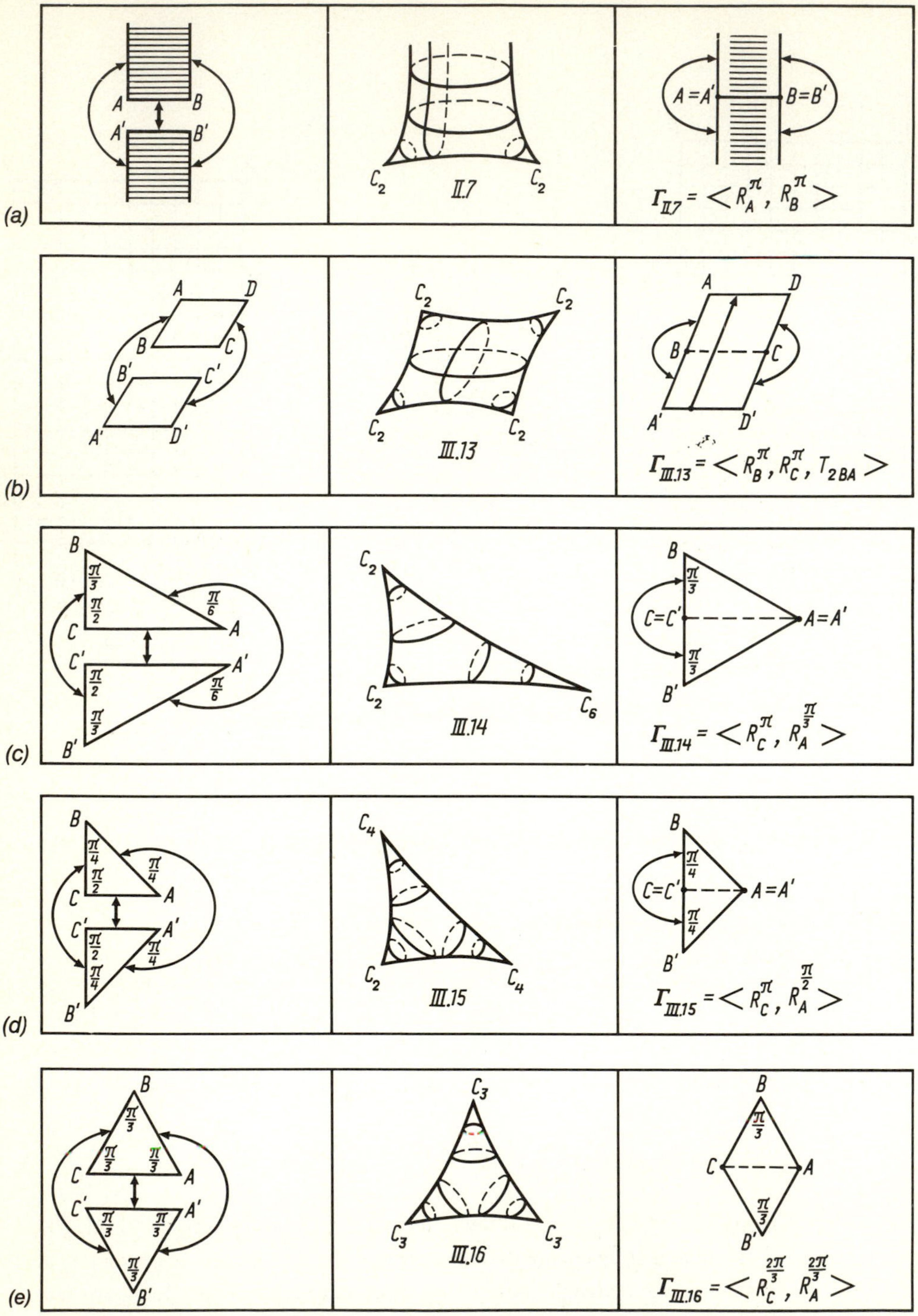

Figure 12.21. Geometries and groups of Types II.7 and III.13-16

$$\Gamma_{II.7} = <R_A{}^{\pi}, R_B{}^{\pi}>, \quad \Gamma_{III.14} = <R_C{}^{\pi}, R_A{}^{\pi/3}>,$$

$$\Gamma_{III.15} = <R_C{}^{\pi}, R_A{}^{\pi/2}>, \quad \Gamma_{III.16} = <R_C{}^{2\pi/3}, R_A{}^{2\pi/3}>$$

(see Figure 12.21, (a), (c), (d), (e), where fundamental domains for these groups are also shown). These geometries have as singular points respectively two singularities of type C_2; three singularities of types C_2, C_3, C_6; one singular point of type C_2 and two of type C_4; and three of type C_3.

Finally, in order to construct the final and most non-trivial locally C_n or D_n geometry, return to the pillow-case geometry III.12 constructed above. Its boundary f, made up of singular points of type D_1, is a circle. Let's say that two points of the boundary circle f are opposite if these points divide f up into two arcs of equal length (see Figure 12.22, (a), where opposite points E and E′, F and F′, G and G′, H and H′ of f are shown). Obviously, glueing opposite points of the boundary circle f of this geometry, we get a surface whose geometry is locally C_n or D_n, and has as its singularities the two points B and C of type C_2. Just as for the Klein bottle, this surface can only be realised in 3-space if we allow self-intersections (see Figure 12.22, (a - d), where we glue first E and E′, then F and F′, then the curves EGF and EG′F; finally, allowing the surface to pass through itself, we glue the curves EHF and EH′F). In mathematics this surface is called the projective plane (see below). The reader can easily see that the geometry we have just constructed is given by the discrete group

$$\Gamma_{III.17} = <S_{(EE')}{}^{EE'}, R_B{}^{\pi}, R_C{}^{\pi}>;$$

here, in the rectangle $FF_1F'_1F'$ of Figure 12.22, (f), the points E and E′ are the midpoints of the sides FF′ and $F_1F'_1$, and B and C the midpoints of FF_1 and $F'F'_1$; and $S_{(EE')}{}^{EE'}$ is the glide reflection with vector $\overrightarrow{EE'}$ and axis (EE′). This rectangle is a fundamental domain of Γ. Note that the generator $R_C{}^{\pi}$ of this group can be expressed in terms of the others, so that the same group can be given by generators

$$\Gamma_{III.17} = <S_{(EE')}{}^{EE'}, R_B{}^{\pi}>.$$

Ignoring the two singularities, the pillow-case (Figure 12.22, (a)) is a hemisphere: if we made the pillow-case out of rubber, we could smooth out the singularities without tearing the pillow-case. The identification of opposite points on the boundary circle of a hemisphere which we have here is the same as that which gives rise to the Riemann geometry of §7.2, Exercise 4 and §9, Exercise 6. Thus our geometry and the Riemann geometry are realised on the same surface. The lines of Riemann geometry satisfy the incidence axioms of the ordinary plane, but with the parallel axiom replaced by the projective plane axiom, that any two distinct

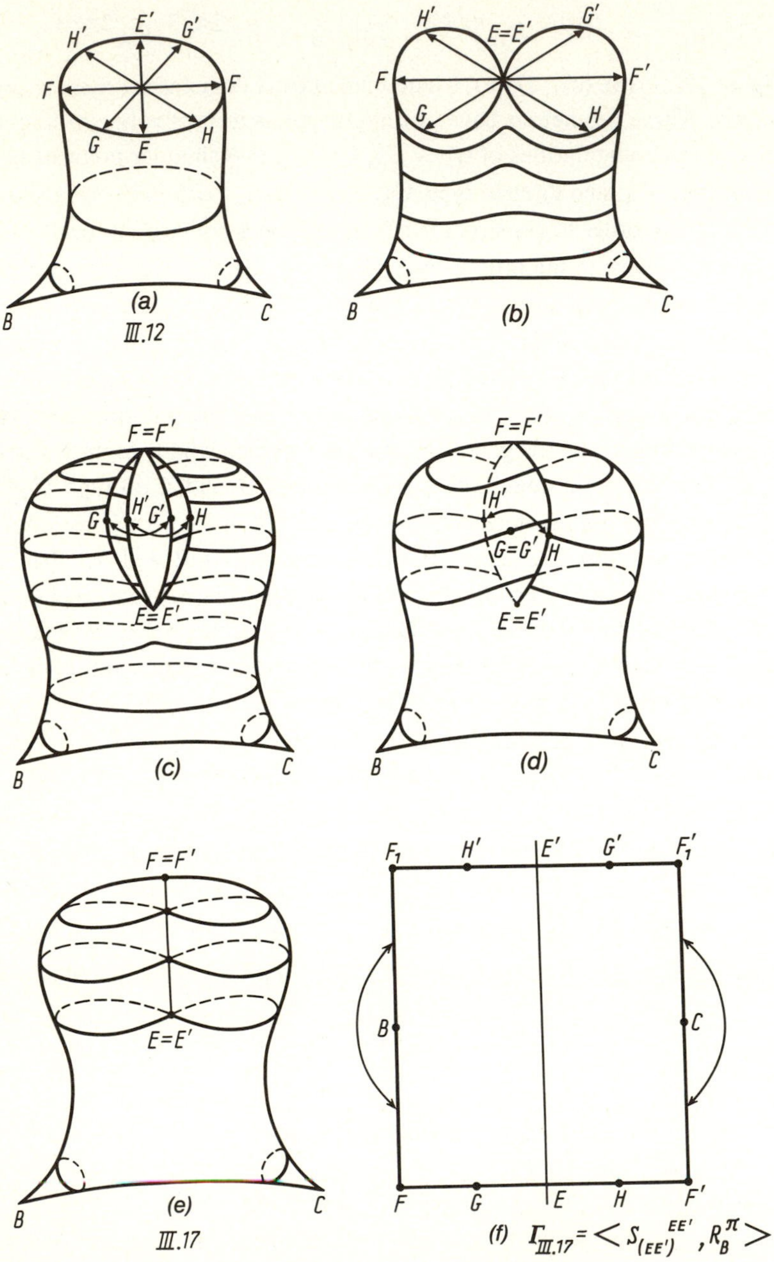

Figure 12.22. The geometry and group of Type III.17

lines meet (see §7.2, Exercise 4). This gives rise to the name projective plane for this surface.

In view of the many and varied methods for constructing locally C_n or D_n geometries which have appeared, it might appear that the process of constructing geometries will never terminate, and that from the geometries we have constructed we can always construct more, and so on indefinitely. But this is not the case.

Theorem 2. The geometries C_n and D_n themselves, and those of Type II.1-7 and III.1-17 exhaust all the locally C_n or D_n geometries. The corresponding groups

$$C_n, D_n, \Gamma_{II.1} - \Gamma_{II.7}, \Gamma_{III.1} - \Gamma_{III.17}$$

exhaust all the discrete groups of motions of the plane; for the description of these groups, see Figures 12.16-22. These are the symmetry groups of the patterns illustrated in Figures 12.23-24. In particular, the 17 groups $\Gamma_{III.1}$ - $\Gamma_{III.17}$ exhaust all plane crystallographic groups.

We will not prove this theorem, but the reader can convince himself that the methods of constructing geometries we have considered (or any others) starting from the geometries listed in Theorem 2 will not give any new locally C_n or D_n geometries. For example, the method of constructing the geometry III.17 from III.12 can be applied to geometry III.8 on the Möbius strip, for which the singular points of type D_1 form a circle. But we just get the geometry on the Klein bottle which we already know. The reader will be able to think of lots more similar examples.

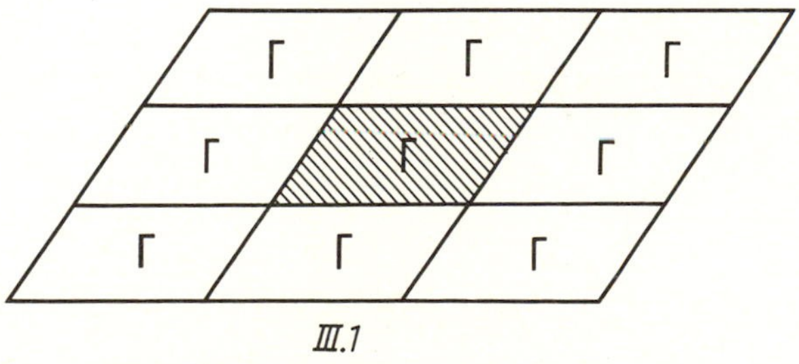

Figure 12.23. A pattern (ornament) for the group of Type III.1 (= III.a)

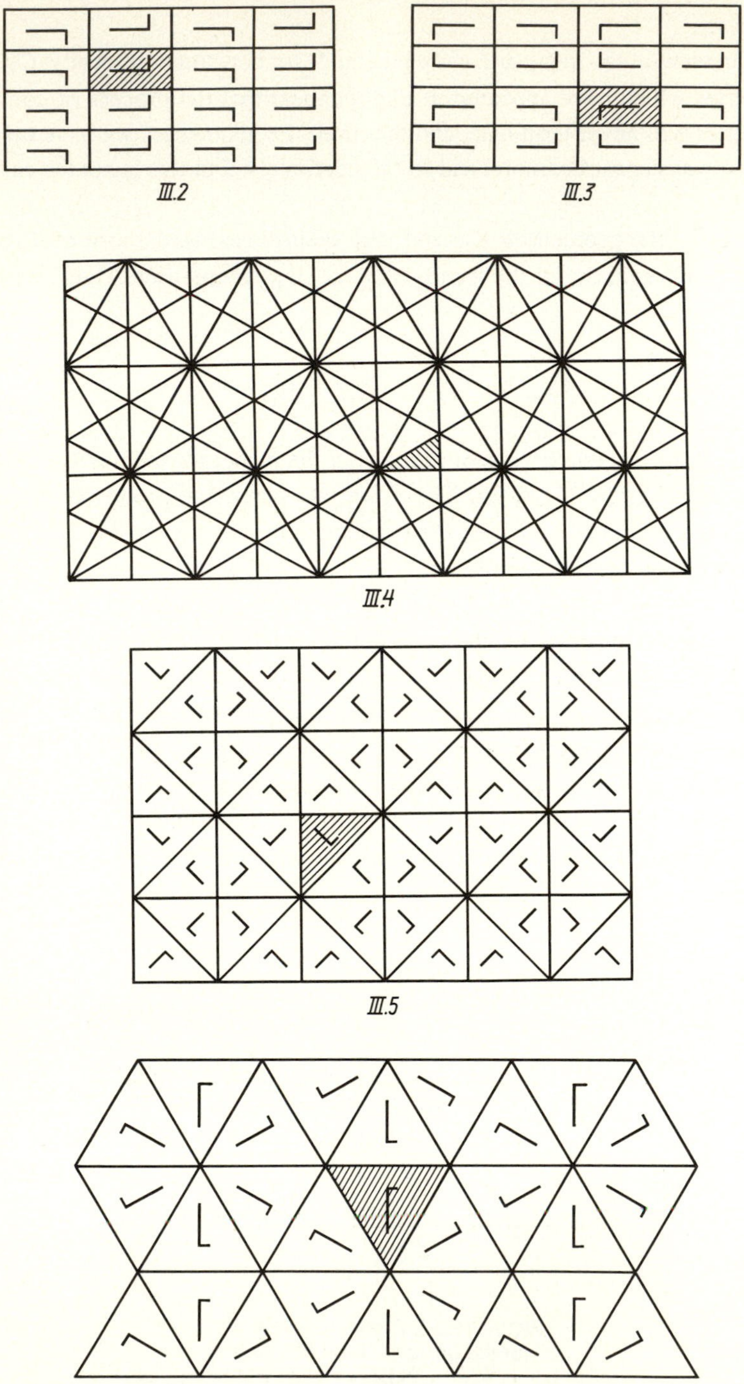

Figure 12.23. Patterns for the groups of Type III.2 (= III.b) - III.6

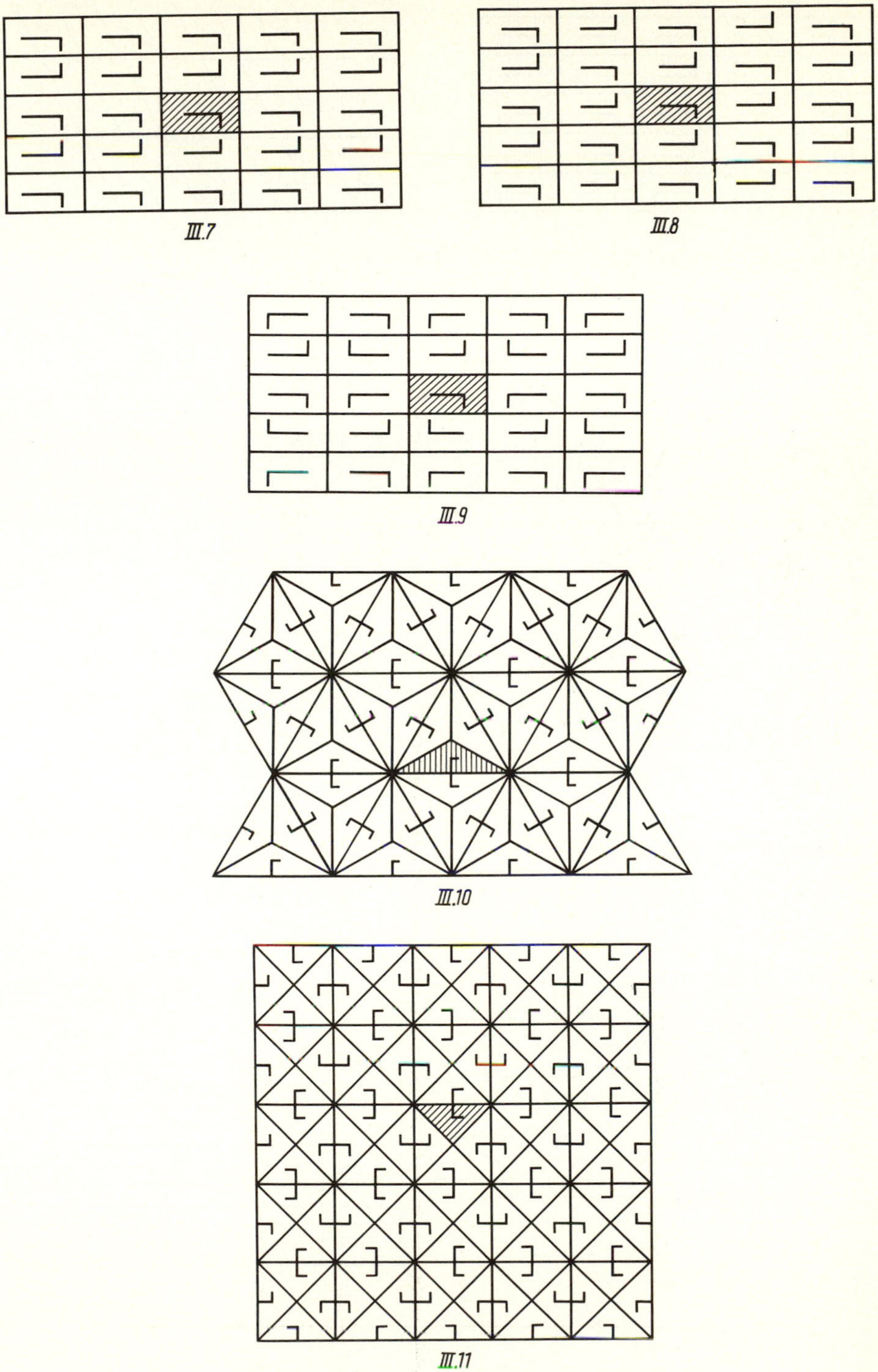

Figure 12.23. Pattern for the groups of Type III.7-11

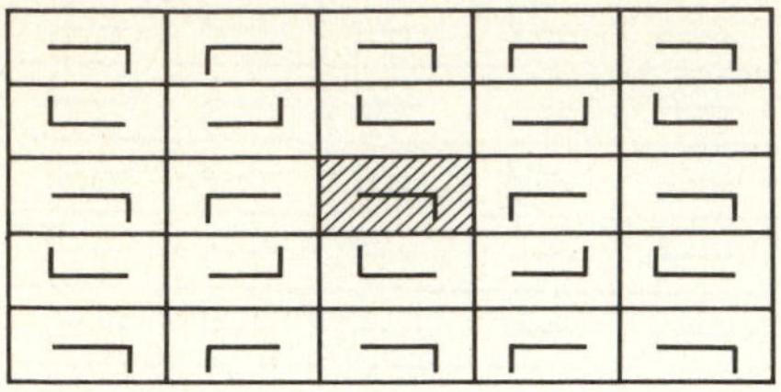

III.12

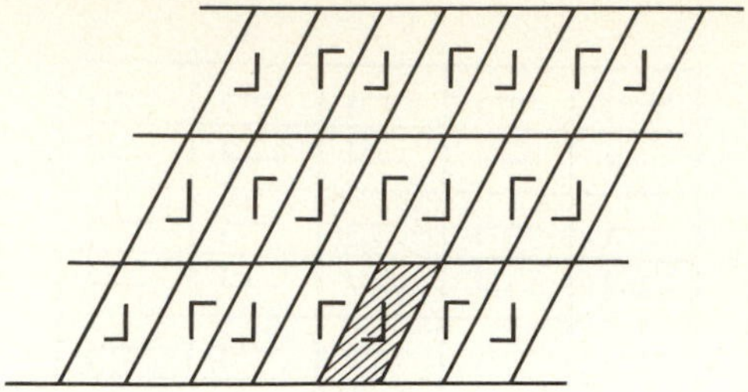

III.13

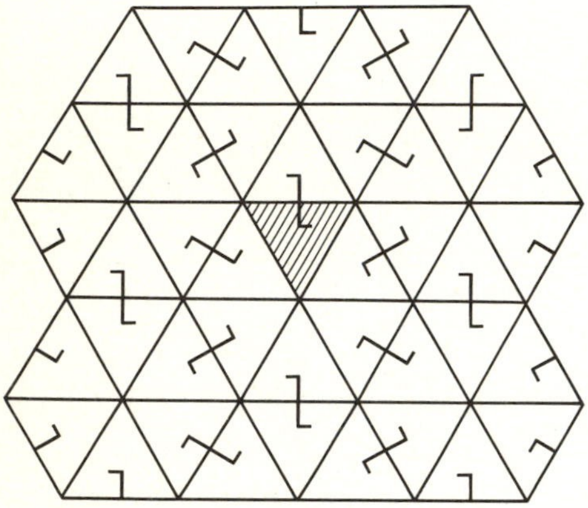

III.14

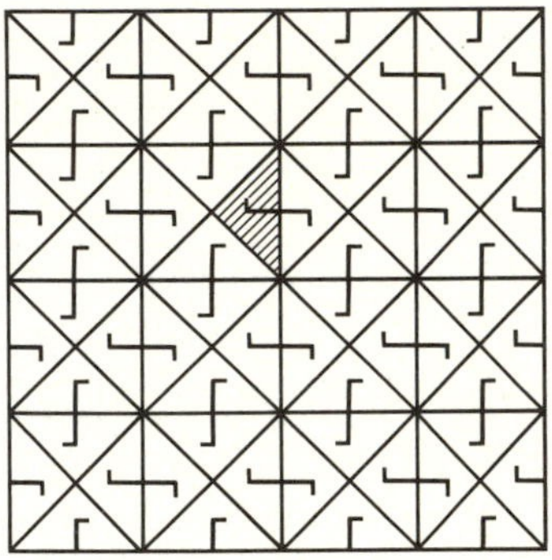

III.15

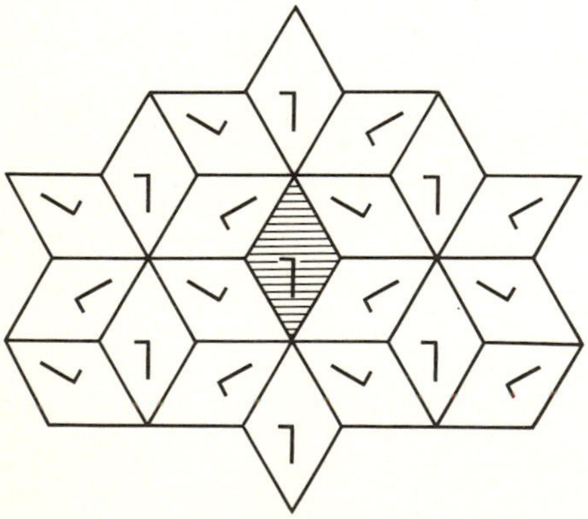

III.16

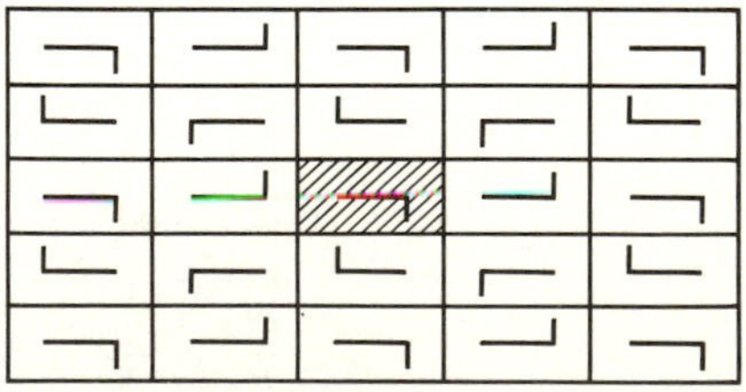

III.17

Figure 12.23. Patterns for the groups of Type III.12 - 17

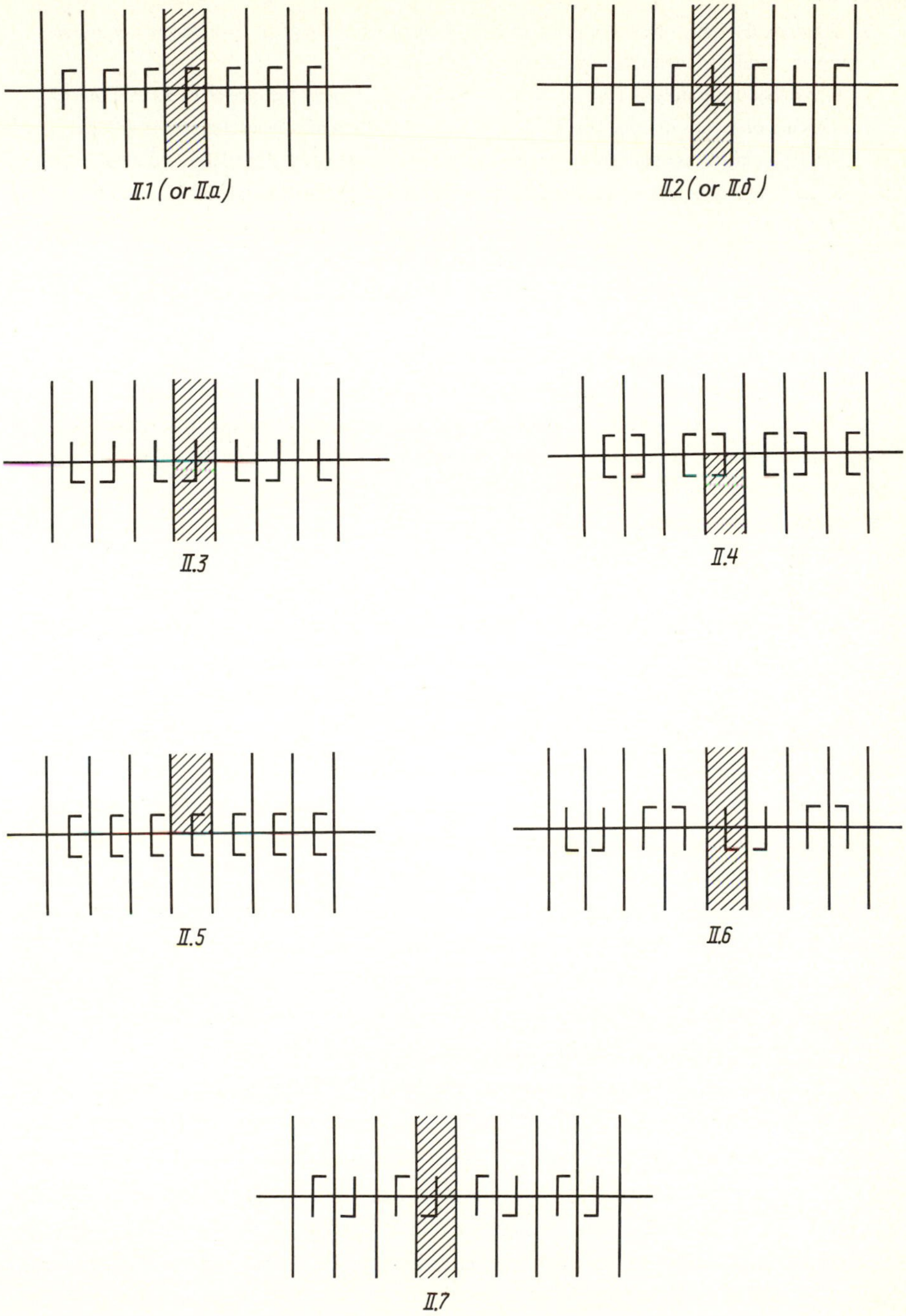

Figure 12.24. Strip (frieze) patterns
for the groups of Type II.1 - 7

Exercises

1. What are the types of the symmetry groups of the tiling of the plane by adjacent congruent rectangles, regular triangles, hexagons, squares?

2. Prove that the groups of Type II.7, III.13-16 are respectively the groups of all motions of the first kind of groups of Type II.4, III.3-6 (recall the method of constructing these groups).

3. Which geometry is obtained by glueing together a geometry of Type III.8 and another of Type III.12, having boundary circles of the same length, along their boundaries?

4. In billiards on a rectangular table, it is required to hit ball B with ball A.

(a) Find all ways of doing this (see §12.3, Exercise 2);

(b) Find all ways in which A is reflected n times in vertical cushions and m times in horizontal cushions;

(c) find all periodic trajectories of the motion of balls in this billiards;

(d) Consider the analogous problem for billiards played on an equilateral triangle.

5. For each of the regular n-gonal prisms and pyramids, the tetrahedron and the cube, consider the group Γ of symmetries of the first kind, and of the first and second kind. What is a fundamental domain of Γ? Which points on the boundary are equivalent? Imagine the corresponding geometries.

12.6. On the proof of Theorems 1 and 2. The proof of the fact that every locally C_n or D_n geometry is constructed from some discrete group differs very little from the proof we gave in §10 for the case of locally Euclidean (that is, locally C_1) geometries. For it, we just need to change the notion of charts and coverings to enable them to deal with the general case of locally C_n or D_n geometries (we invite the reader to think this notion through). But it turns out to be unexpectedly difficult (although still much simpler than the arguments of §10) to prove the direct assertion of Theorem 1: that every discrete group gives rise to a locally C_n or D_n geometry. This requires a construction of suitable fundamental domains for discrete groups (in all of our examples, we guessed the fundamental domains for specific groups, using the explicit way in which they were given), and some study of their properties. Incidentally, this difficulty can be got round by proving Theorem 2 first, then using the explicit form of discrete groups which it provides.

The proof of Theorem 2 is based on ideas which are already familiar to us. Discrete groups break up into three types, according to the translation group which they contain: for groups of Type I this consists of just the identity; for groups of Type II, translations in vectors whose ends when laid out from any point form a series O' of equidistant points (lattice) on a line; finally for groups of Type III, the ends of the translation vectors form a plane lattice O''. The groups of Type I are just C_n and D_n. For groups of Type II or Type III, entirely similarly to the situation considered in §11, all the motions in the group must take the lattice O' or O'' into itself. From this, all groups of Type II are easily determined. To find the groups of Type III, we must make use of the tabulation of all symmetries of plane lattices given in §11.3. From this it is easy to reconstruct all the groups of this type. (This explains in particular why groups of Type III only contain rotations through angles π, $\pi/2$, $\pi/3$ and $2\pi/3$.) We will not give the detailed proof here. For the case of crystallographic groups in the plane not containing any glide reflections, the reader can find it in D. Hilbert and S. Cohen-Vossen, Geometry and the imagination (Chelsea, New York, 1952), Chapter II, §12 .

12.7. Crystals and their molecules. We now return to crystals and their crystallographic groups. What is the significance for a crystal of the geometry corresponding to its crystallographic group Γ? Since by the very definition of Γ all the points of a set **A** of equivalent points have identical physical properties, assigning these properties to the point a corresponding to **A** we provide the points of the geometry with physical properties. Under this, obviously, the physical properties of each point of space will be represented among the physical properties of the points of our geometry, and different points of the geometry will have, by definition of Γ, different physical properties. In other words, the geometry, provided in this way with physical properties, is the elementary cell from which the whole crystal unfolds; and the correspondence between geometries and discrete groups can be interpreted by saying that the structure and physical properties of this elementary cell completely determine the structure and physical properties of the entire crystal, as it develops from this elementary cell. For example, what is a molecule of salt? Looking at Figure 12.5, which illustrates crystalline salt, it is quite unclear how to join atoms of sodium and chlorine into a molecule (even worse difficulties arise if the crystal consists of more than two elements, which do not join together into clearly expressed groups of atoms).

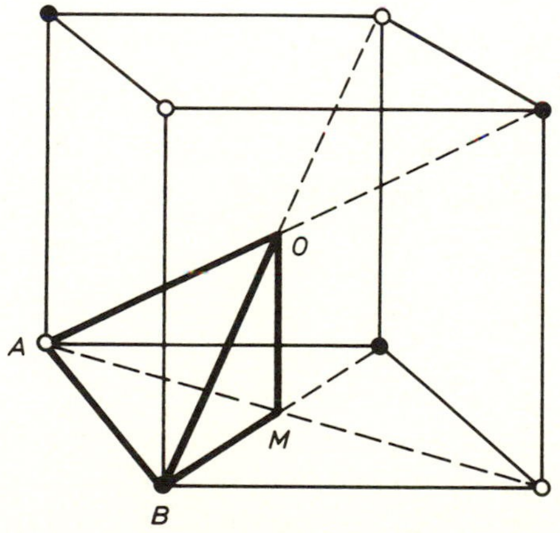

Figure 12.25

As a first approximation to the answer, it is natural to take the molecule to be a fundamental domain for the crystallographic group of salt; this would not be quite right, since a fundamental domain cannot be chosen uniquely. On the other hand, the geometry corresponding to this group is constructed similarly to the 2-dimensional geometries $\Gamma_{III.3}$ - $\Gamma_{III.6}$ (see Figure 12.16) and is a tetrahedron ABMO, with the atoms of sodium and chlorine at two of the vertices A and B (see Figure 12.25). This tetrahedron with atoms of sodium and chlorine at two of its vertices is the salt molecule; and the structure of this molecule entirely determines the structure of the crystal salt, since it is composed of these molecules. Defining in this way the molecule of a crystal to be the geometry corresponding to the crystallogrphic groups, provided with physical properties, we can say that the structure and the physical properties of a molecule of a crystal contains complete information on the structure of the whole crystal! The reader may object to this, since judging by the examples of geometries corresponding to plane crystallographic groups, for example geometry III.2 on the Klein bottle, and III.17 on the projective plane, molecules defined in this way may have very complicated structure. But then, none of us believes that the world we live in is simple, do we?

Chapter IV
Geometries on the torus, complex numbers and Lobachevsky geometry

§13. Similarity of geometries

13.1. When are two geometries defined by uniformly discontinuous groups the same? The description of locally Euclidean geometries, which made up Chapter II, contained one gap, and we now fill this. It was proved that every such geometry can be constructed as a geometry Σ_Γ for a certain uniformly discontinuous group Γ of motions of the plane. It would seem that the classification of all such groups given in Chapter II, §8 then solves the problem. However, this is not quite the case: what we have done is to present a list of geometries such that every geometry of the type we are interested in is identical to one in the list; however, we have not determined when two geometries from our list are identical (that is, can be superposed on one another), and when they are different. Hence the question as to how many distinct geometries there are in existence has not yet received a definitive answer. We can make some assertions in this direction at once; for example, geometries belonging to the different Types I, II.a, II.b, III.a and III.b are different, since they are distinguished by properties such as the existence of closed curves, boundedness, and whether right and left are distinguishable. But it remains unclear whether the geometries within each type are distinct or are the

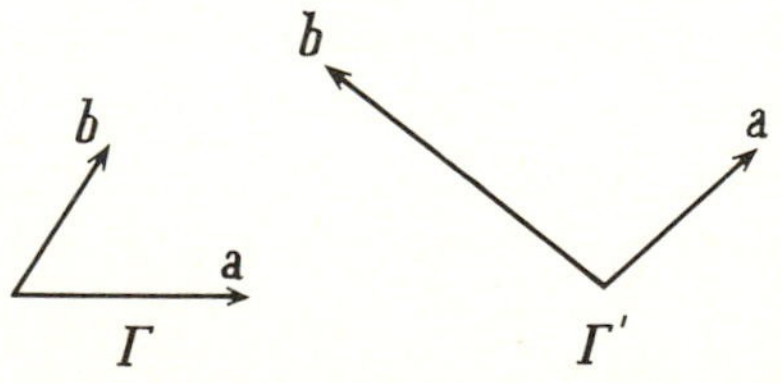

Figure 13.1

same: for example, are the two geometries defined by groups Γ and Γ' of Type II.a, when Γ and Γ' are generated by the pairs of vectors illustrated in Figure 13.1, the same or distinct?

To prepare the answer to this question, we now consider a very natural method which allows us to get from one group to another. We will show afterwards that this is just the way in which the groups defining identical geometries are related.

Suppose that G is a motion of the plane Π, and F another motion; it will be convenient to view F as a superposition of Π on a plane Π', possibly different from Π. In this case, we can of course use the superposition F to transfer the motion G to the plane Π': for any point X′ of Π' we must take the point X of Π which corresponds to X′ under the superposition F, then act on X by G, and finally take the point of Π' which corresponds to G(X) under F (Figure 13.2).

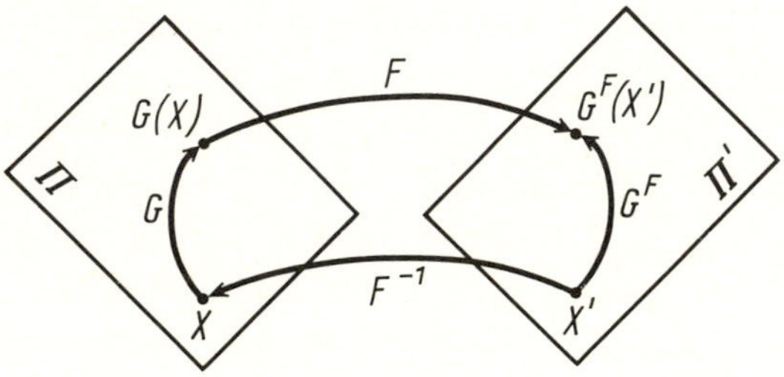

Figure 13.2

In terms of composite maps, we get the point

$$Y' = FGF^{-1}(X'). \qquad (1)$$

This assigns to X′ the point Y′, and gives a transformation of Π', which we denote by G′; that is, $G' = FGF^{-1}$. Since each of the three transformations F, F^{-1} and G appearing in the formula for G′ do not change distances, the same is true of G′, so that G′ is a motion of the plane Π'. This motion is denoted by G^F, and is called the *conjugate* of G by F. Thus, by definition we have

$$G^F = FGF^{-1}. \qquad (2)$$

Properties of conjugation. From (2) we get at once

(a) $(G_1G_2)^F = G_1{}^F G_2{}^F$,

(b) $(G^{-1})^F = (G^F)^{-1}$.

It follows from this that if G runs through a group Γ of motions of Π, and F is a given superposition of Π onto a plane Π', then the conjugate motions G^F of G by F form a group of motions of Π'; this group is denoted by Γ^F, and is called the *conjugate group of* Γ *by* F.

It is also easy to check the following properties, describing the effect of conjugating a translation, rotation or glide reflection:

$$\text{(c)} \qquad (T_{\mathbf{a}})^F = T_{F(\mathbf{a})},$$

$$\text{(d)} \qquad (R_O{}^{\alpha})^F = R_{F(O)}{}^{F(\alpha)},$$

$$\text{(e)} \qquad (S_{\ell}{}^{\mathbf{a}})^F = S_{F(\ell)}{}^{F(\mathbf{a})};$$

here in (c) and (e), $F(\mathbf{a})$ is the vector obtained by letting F act on $\mathbf{a}$, so that as in §11.2, if $\mathbf{a} = \overrightarrow{AB}$ then $F(\mathbf{a}) = \overrightarrow{F(A)F(B)}$; in (d), $F(\alpha)$ is the angle obtained by letting F act on α, which means that if α is an angle $\widehat{AOB}$ then $F(\alpha)$ is the angle $\widehat{F(A)F(O)F(B)}$ (of course, $F(\alpha) = \pm\alpha$, depending on whether F is of the first or second kind).

It follows from this that if Γ is (say) a group of Type II.a, generated by a pair of vectors $\mathbf{a}$ and $\mathbf{b}$, then Γ^F is also of Type II.a, and is generated by the pair of vectors $F(\mathbf{a})$ and $F(\mathbf{b})$. More generally, it follows from properties (a - e) that if a group Γ is given by one of the diagrams listed in §8.3, Theorem 7, then the conjugate group Γ^F is given by the diagram obtained by applying the superposition F to it.

Now we can answer the question we asked at the beginning of the section.

Theorem 1. Two geometries Σ_Γ and $\Sigma_{\Gamma'}$ corresponding to uniformly discontinuous groups Γ and Γ' can be superposed on one another if and only if the groups Γ and Γ' are conjugate by means of some motion of the plane.

Proof. Let Γ be a group of motions of a plane Π and Γ' a group of motions of a plane Π'. The proof of one implication of the theorem is very simple.

Suppose that F is some superposition of Π on Π' such that $\Gamma' = \Gamma^F$. If we assign to each point X of Π the point $F(X)$ of Π', then points X and Y equivalent under Γ go into points $F(X)$ and $F(Y)$ which are equivalent under Γ'. Thus we get a map of Σ_Γ to $\Sigma_{\Gamma'}$, taking a set $\mathbf{X}$ of points equivalent under Γ (that is, a point of Σ_Γ) into $F(\mathbf{X})$, which is a set of points equivalent under Γ' (that is, a point of $\Sigma_{\Gamma'}$); this map is a superposition of Σ_Γ onto $\Sigma_{\Gamma'}$. These assertions all follow at once from the definition (in (1) and (2) above), and the interested reader can easily carry out the verification for himself.

The other direction of the theorem is harder to prove. Suppose that we are given that Σ_Γ can be superposed on $\Sigma_{\Gamma'}$, and let f be this superposition. As in the first part of the proof, we will assume that Γ consists of motions of Π and Γ' of motions of Π'. We need to construct a superposition F of Π onto Π' and prove that $\Gamma' = \Gamma^F$.

Recall that there is a map φ of Π onto the geometry Σ_Γ, which takes a point X of Π into the point x of Σ_Γ determined by X; and φ has an important property (in §10.2, such maps were called coverings): a disc D(X, r) in the plane, centred at X and of sufficiently small radius r, is superposed onto the disc $D(\varphi(X), r)$ in Σ_Γ centred at $\varphi(X)$ and of radius r. Denote by φ' the corresponding map of Π' onto the geometry $\Sigma_{\Gamma'}$. Let P be some point of Π, and $p = \varphi(P)$ its image in Σ_Γ; let $p' = f(p)$ be the image of p in $\Sigma_{\Gamma'}$ under the superposition f, and finally, let P′ be some point of Π' which defines p′, that is, $p' = \varphi'(P')$. We know that the maps φ, φ' and f are superpositions on discs of radius r (Figure 13.3).

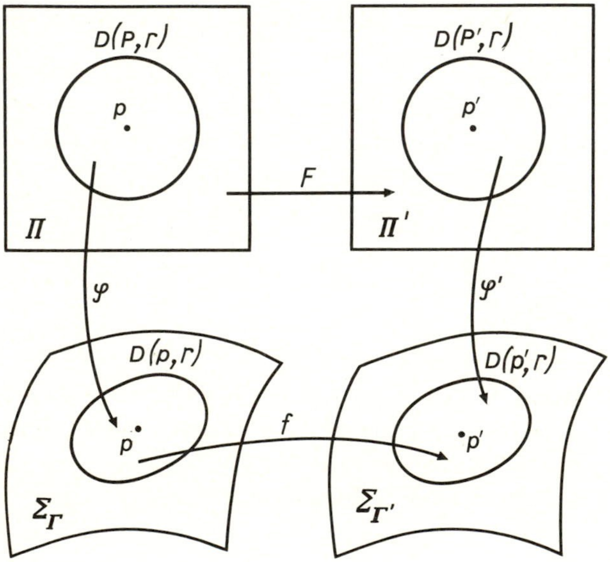

Figure 13.3

Consider the map $(\varphi')^{-1} f\varphi$ on the disc D(P, r); the expression we have written makes sense, since φ' is a superposition of D(P′, r) onto D(p′, r), and hence the inverse map $(\varphi')^{-1}$ exists. Obviously, in this way we get some superposition of the disc D(P, r) in Π onto the disc D(P′, r) in Π'. By §10.1, Lemma 1, this can always be realised by some superposition of the plane Π onto Π': there is a superposition F of Π onto Π' such that for points X of D(P, r), we have $F(X) = (\varphi')^{-1} f\varphi(X)$, that is

$$\varphi' F(X) = f\varphi(X). \tag{3}$$

Since this relation holds for all points of $D(P, r)$, it follows from §10.4, Lemma 4 (coverings agreeing on a disc are equal) that it holds for all points of Π.

From (3), which is now established for all points of Π, it follows easily that $\Gamma' = \Gamma^F$. Let G be any motion in Γ; we prove that FGF^{-1} belongs to Γ'. According to §7, Theorem 1, Γ' consists exactly of the motions G' of Π' which take equivalent points into equivalent points, that is, for which $\varphi' G'(X') = \varphi'(X')$ for any point X' of Π'. Therefore we need only check that $\varphi' FGF^{-1}(X') = \varphi'(X')$. If we set $F^{-1}(X') = X$, that is $X' = F(X)$, then our relation can be written in the form $\varphi' F(G(X)) = \varphi'(F(X))$. From (3) we get $\varphi' F(G(X)) = f\varphi(G(X))$, and $\varphi'(F(X)) = f(\varphi(X))$, and our relation takes the form $f\varphi(G(X)) = f(\varphi(X))$. This is obviously true, since X and $G(X)$ are equivalent under Γ, and hence $\varphi G(X) = \varphi(X)$. Thus Γ^F is contained in Γ'. If G' is any motion in Γ', then similar arguments show that $G = F^{-1}G'F$ belongs to Γ. It follows from this that $G^F = FGF^{-1} = G'$, that is, G' belongs to Γ^F, and hence also Γ^F contains Γ'. Thus $\Gamma^F = \Gamma'$, and the theorem is proved.

Exercises

1. Prove that if Σ_Γ is a geometry corresponding to a uniformly discontinuous group Γ, then any motion F of the plane for which $\Gamma^F = \Gamma$ gives rise to a motion of Σ_Γ. Prove that every motion of Σ_Γ arises in this way. (For the definition of motion of a geometry, see §7.2, Exercise 2.)
2. When will two motions F and F' of the plane with the above property give rise to the same motion of Σ_Γ?
3. Describe all motions of geometry on a cylinder, a twisted cylinder, and a Klein bottle.
4. Let $(\Sigma_{\Gamma'}, \Sigma_\Gamma, \varphi)$ be the covering considered in §10.2, Exercise 3. Prove that this covering satisfies the assertion of §10, Theorem 1 (c) if and only if $\Gamma'^F = \Gamma$ for any motion F of Γ. Consider examples of groups Γ', Γ of different types.

13.2. Similarity of geometries. The same question - when are two geometries defined by uniformly discontinuous groups the same? - can be considered in another, rather wider, setting. To explain this, recall the arguments which we used in discussing the notion that geometries be the same, or superposable. We imagined that the two geometries Σ and Σ' have inhabitants, who can communicate with one another by radio, and as a result of discussions, can convince themselves that measurements carried out in their geometries lead to the same results. Mathematically, we expressed this as a map F of Σ onto the whole of Σ' which does not change distances:

$$|F(A)F(B)| = |AB|. \tag{4}$$

But what does this equality mean physically? The point is that measurement of distances only leads to numbers (and (4) is a relation between actual numbers) after introducing a definite unit of length, and it is not clear in what way the inhabitants of different geometries could standardise their units of length. Thus physically, it makes much more sense to propose that all measurements (made in any units of length) made by the inhabitants of Σ and Σ' should lead to proportional results. Thus we arrive at a new notion relating two geometries.

Suppose that there exists a map F of Σ onto the whole of Σ', and a constant $\lambda > 0$ such that

$$|F(A)F(B)| = \lambda|AB|$$

for any two points A and B of Σ; then the geometries Σ and Σ' are said to be *similar* and the map F is a *similarity*. The coefficient λ is called the *ratio of similarity* of F; it is always a positive number.

It is similarity which most naturally reflects the fact that all the properties of two geometries are the same. This corresponds to the idea that if all the objects of our world, including ourselves and all our means of measurement, were simultaneously increased by the same factor, then we would not be able to notice the increase. The notion of equality or superposition of geometries was mathematically rather simpler, which is why we based our arguments on it to start with.

However, the notion of similarity often reduces without difficulty to that of superposition of geometries, and this enables us to carry over at once to the case of similarities the result of Theorem 1. For this we need first of all to consider a similarity from the plane Π to itself, that is, the case $\Sigma = \Sigma' = \Pi$; we will call this simply a similarity of the plane Π. One example will already be familiar to the

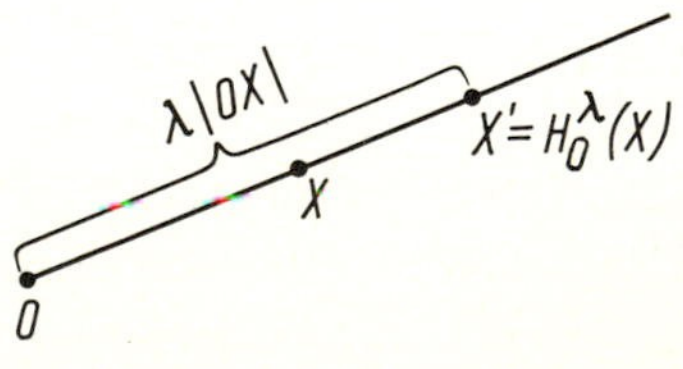

Figure 13.4

reader, the dilation with given centre O and ratio $\lambda > 0$; this is the transformation which takes a point X into the point X' lying on the same ray OX and at a distance $\lambda|OX|$ from O (Figure 13.4). If $\lambda > 1$ then the whole plane expands out from O, and if $\lambda < 1$ then it shrinks towards O in this ratio. This dilation will be denoted by H_O^{λ}.

Lemma 1. Suppose that O is an arbitrary fixed point of the plane. Any similarity F of the plane is obtained as the composite of some motion with the dilation H_O^{λ}, where λ is the ratio of similarity of F.

Indeed, let F be a similarity of the plane with ratio λ. Consider the map $G = FH_O^{1/\lambda}$. It follows at once from the definition that G is a motion. In the equality

$$G(X) = F(H_O^{1/\lambda}(X)),$$

substitute $H_O^{\lambda}(Y)$ for X. Obviously, $H_O^{1/\lambda}(H_O^{\lambda}(Y)) = Y$, so that

$$G(H_O^{\lambda}(Y)) = F(Y)$$

for all Y, that is $F = GH_O^{\lambda}$. Since G is a motion, this proves the lemma.

In the same way as for motions, we define the conjugate of G by a similarity F to be the transformation

$$G^F = FGF^{-1}.$$

If F has ratio λ then F^{-1} multiplies all distances by λ^{-1}, G leaves them unchanged, and F multiplies them by λ. Thus we see that G^F is again a motion. All of the properties (a - e) of conjugacy are preserved, together with their proofs. In particular, the conjugate group Γ^F of a group of motions Γ by a similarity F is defined.

Theorem 2. Two geometries Σ_Γ and $\Sigma_{\Gamma'}$ corresponding to uniformly discontinuous groups Γ and Γ' are similar if and only if the groups Γ and Γ' are conjugate by some similarity of the plane.

Proof. First of all, quite similarly to the proof of the first implication of Theorem 1, it can be checked that a similarity H of the plane defines a similarity h of Σ_Γ onto the geometry $\Sigma_{\Gamma'}$, where $\Gamma' = \Gamma^H$, and that the ratios of similarity of H and h are equal.

The proof of the converse follows from Theorem 1. Let f be a similarity of Σ_Γ onto $\Sigma_{\Gamma'}$, and let λ be its ratio. Consider the dilation $H = H_O^{1/\lambda}$ in the plane. By what we have said above, H defines a similarity h of $\Sigma_{\Gamma'}$ onto the geometry $\Sigma_{\Gamma''}$, where $\Gamma'' = (\Gamma')^H$, and h has the same ratio of similarity as H, that is $1/\lambda$; then hf is a similarity with ratio 1, that is a superposition of $\Sigma_{\Gamma'}$ onto $\Sigma_{\Gamma''}$. By Theorem 1, there is now a motion F of the plane such that $\Gamma^F = \Gamma''$ $= (\Gamma')^H$, and hence $(\Gamma^F)^{H^{-1}} = \Gamma'$. As the reader will easily check, $(\Gamma^F)^{H^{-1}} =$

$\Gamma(H^{-1}F)$, and hence $\Gamma' = \Gamma(H^{-1}F)$, where $H^{-1}F$ is a similarity. This completes the proof.

In conclusion we consider two examples. We start with spherical geometries. The two geometries on spheres of radiuses R_1 and R_2 are obviously identical if $R_1 = R_2$; moreover, the converse also holds. The point is that the radius of a sphere can be characterised as a property of the geometry: $R = (1/2\pi)\ell$, where ℓ is the length of any line of the geometry. On the other hand, if two spheres of radiuses R_1 and R_2 are positioned such that their centres coincide at a point O, then we obviously get a similarity, with ratio R_2/R_1, by taking a point X of the first to the point of the second lying on the same ray OX. Thus, there are many different spherical geometries, as many as there are values of the radius R, but up to similarity there is just one unique spherical geometry.

We now consider the geometry on a cylinder. This is of the form Σ_Γ, where Γ is a group of Type II.a, determined by a vector **a**, and consisting of translations in multiples $n\mathbf{a}$ of **a**. Choosing a point O and laying off the vectors $n\mathbf{a}$ from this point, we can represent the group as a series of equidistant points on a line (Figure 13.5).

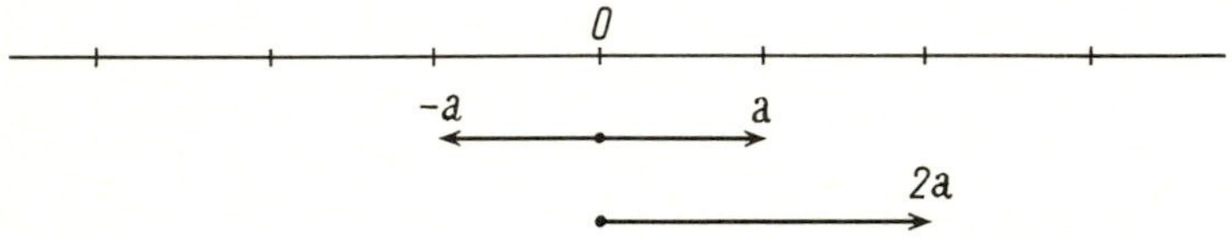

Figure 13.5

According to Theorem 1 two such geometries are identical if and only if the corresponding groups are conjugate under a motion F, that is, the series of points on a line representing them can be taken into each other by a motion F. For this to happen it is obviously necessary and sufficient that the distances between corresponding points (the mesh of the linear lattice) should be the same in both series. But this distance equals $|\mathbf{a}|$. The result we get is geometrically obvious: $|\mathbf{a}|$ is characterised as a property of the geometry Σ_Γ as the length of any closed line of Σ_Γ (see §3.3). On the other hand, it follows from Theorem 2 that two such geometries are similar if and only if the corresponding series of points on a line can be taken into one another by a similarity of the plane, and this can always be done. Thus all geometries corresponding to groups of Type II.a are similar.

One should not conclude from these two examples that in other cases too, the geometries corresponding to all groups of one type are similar. The question is much more interesting than that, and we now proceed to investigate it.

Exercises

1. Prove that if Γ is a uniformly discontinuous group then the conjugate group Γ^F of Γ by a similarity F is again uniformly discontinuous.

2. Prove that all geometries Σ_Γ corresponding to groups Γ of Type II.b (that is, geometries on the twisted cylinder) are similar.

3. (a) Let φ be a similarity of geometries of Σ and Σ', and suppose that Σ is locally Euclidean; prove that Σ' is also.

(b) In the conditions of (a) suppose that ℓ is a line of Σ. Prove that $\varphi(\ell)$ is a line of Σ'. If ℓ is closed, then how are the lengths of ℓ and $\varphi(\ell)$ related?

(c) For which locally Euclidean geometries Σ do there exist similarities of Σ into itself which are not motions?

§14. Geometries on the torus

14.1. Geometries on the torus and the modular figure. Let's consider geometries on the torus, and try to determine when they are similar. Of the types of groups corresponding to locally Euclidean geometries which have not yet been considered (that is, III.a and III.b), the groups of Type III.a, corresponding to geometries on the torus, are the most interesting. The groups of Type III.b are simpler, and we leave the consideration of these to the reader in §14.2, Exercise 2. Recall that every group Γ of Type III.a is determined by a pair of generating vectors $(\mathbf{e}_1, \mathbf{e}_2)$; for these we can take any two non-collinear vectors in the plane. Then Γ is the group of all translations in vectors of the form $m\mathbf{e}_1 + n\mathbf{e}_2$, where m and n run through the integers. If all of these vectors are laid off from some point O, then Γ is represented as a plane lattice (Figure 14.1). According to §13.2, Theorem 2,

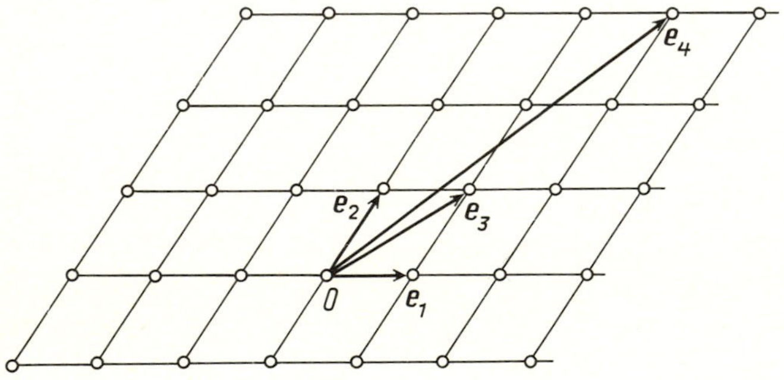

Figure 14.1

our problem reduces to the question of when two plane lattices are similar. We can fix the point O, and thus only consider similarities F which fix this point. By §13.1, Lemma 1, any such similarity is obtained as the composite of a dilation centred at O and some motion G which must also leave O fixed. From Chasles' theorem in §8.1, it follows that G can only be a rotation about O or a reflection in a line through O. From now on, we only consider this kind of similarities.

The problem facing us breaks up into two parts. The first part is to know when two given pairs of vectors $(\mathbf{e}_1, \mathbf{e}_2)$ and $(\mathbf{e}_1', \mathbf{e}_2')$ are similar; that is, when does there exist a similarity F such that $F(\mathbf{e}_1) = \mathbf{e}_1'$ and $F(\mathbf{e}_2) = \mathbf{e}_2'$? This question is very easy, and the whole difficulty of the problem lies in the second part, which asks when different pairs of generating vectors determine one and the same lattice. For example, the lattice of Figure 14.1 is generated by either of the pairs $(\mathbf{e}_1, \mathbf{e}_2)$ and $(\mathbf{e}_3, \mathbf{e}_4)$

We start with the simpler first question. Obviously, under a similarity the ratio of the lengths of vectors $|\mathbf{e}_1|/|\mathbf{e}_2|$ is preserved. It is also tempting to say that the angle between $\mathbf{e}_1$ and $\mathbf{e}_2$ is also preserved; here the angle means the angle of rotation from $\mathbf{e}_1$ to $\mathbf{e}_2$, where a positive angle corresponds to an anticlockwise rotation. However, this is not quite right: by §13.2, Lemma 1, the similarity F is of the form GH_O^λ, where H_O^λ is a dilation and G is a motion, and then the angle is preserved if G is a rotation (Figure 14.2, (a)), and is changed to its negative if G is a reflection (Figure 14.2, (b)).

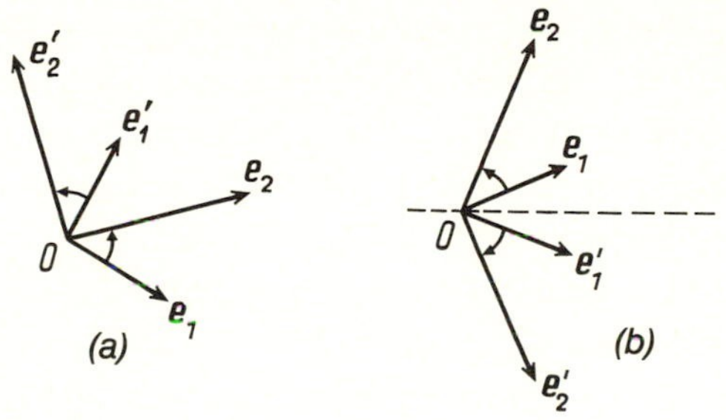

Figure 14.2

Thus the angle between $\mathbf{e}_1$ and $\mathbf{e}_2$ is preserved up to sign. We have obtained a necessary and sufficient condition for similarity of a pair of vectors: if $(\mathbf{e}_1, \mathbf{e}_2)$ and $(\mathbf{e}_1', \mathbf{e}_2')$ are pairs of vectors such that $|\mathbf{e}_1|/|\mathbf{e}_2| = |\mathbf{e}_1'|/|\mathbf{e}_2'|$, and $\theta = \pm\theta'$, where θ is the angle between $\mathbf{e}_1$ and $\mathbf{e}_2$, and θ' is that between $\mathbf{e}_1'$ and $\mathbf{e}_2'$, then these pairs are similar. Indeed, in view of the condition $\theta = \pm\theta'$, we can apply to $(\mathbf{e}_1, \mathbf{e}_2)$ a motion G to make the angle between $\mathbf{e}_1$ and $\mathbf{e}_2$ equal to that between $\mathbf{e}_1'$ and $\mathbf{e}_2'$. After this it remains to apply a dilation centred at O such that $G(\mathbf{e}_1) = \mathbf{e}_1'$; then by the condition on the lengths, $G(\mathbf{e}_2) = \mathbf{e}_2'$.

Using what we have just seen, given any pair of vectors, we can find a similar pair of a specially simple form. Namely, fix a vector $\mathbf{e}$ in the plane. Obviously any pair of non-collinear vectors $(\mathbf{e}_1, \mathbf{e}_2)$ is similar to a pair $(\mathbf{e}, \mathbf{f})$ for which the angle between $\mathbf{e}$ and $\mathbf{f}$ is less than π; and for every pair $(\mathbf{e}_1, \mathbf{e}_2)$, the pair $(\mathbf{e}, \mathbf{f})$ is unique (recall that the vector $\mathbf{e}$ is fixed once and for all). The vector $\mathbf{e}$ is usually drawn on some horizontal line ℓ, and then the condition on $\mathbf{f}$ is that its end should lie in the upper half-plane, situated above the line ℓ (Figure 14.3).

Thus the set of pairs of vectors up to similarity is in one-to-one correspondence with the points of the upper half-plane (see Figure 14.3).

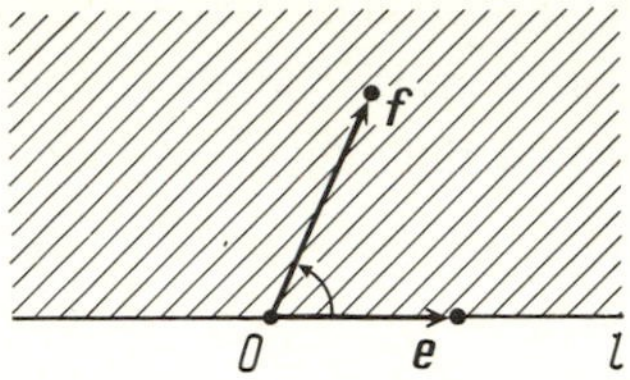

Figure 14.3

We now proceed to the second and harder question. Here there are two possible approaches. We can either specify in each lattice a definite, simplest possible, pair of generating vectors, so that the problem of similarity of lattices reduces to the question of similarity of such pairs of vectors, which we have already solved. Or we can determine when two pairs of vectors generate the same lattice, and use this to represent the set of all lattices considered up to similarity. Each of these approaches will turns out to be fruitful, but in combination they will be especially valuable. We now consider the first approach, and in §14.2, the second; §§15–17 will be devoted to the combination of the two approaches.

A way of choosing a simplest possible system of generating vectors in each lattice has actually already been indicated, in the proof of §8.3, Theorem 5 and in the remark after the proof of this theorem. It was shown there that we can take $\mathbf{e}_1$ to be the shortest vector of the lattice distinct from $\mathbf{0}$, and $\mathbf{e}_2$ to be a vector whose projection to the direction of $\mathbf{e}_1$ has length at most half the length of $\mathbf{e}_1$. To this we just add that, replacing $\mathbf{e}_2$ by $-\mathbf{e}_2$ if necessary, we can assume that the angle between $\mathbf{e}_1$ and $\mathbf{e}_2$ is not obtuse. We will say that a pair of vectors of the lattice satisfying these three conditions is a *reduced generating pair* of the lattice.

By definition, if $(\mathbf{e}_1, \mathbf{e}_2)$ is a reduced generating pair of some lattice, then it satisfies

(i) $|\mathbf{e}_2| \geq |\mathbf{e}_1|$;

(ii) the length of the projection of $\mathbf{e}_2$ to the direction of $\mathbf{e}_1$ is less than or equal to half the length of $\mathbf{e}_1$;

(iii) the angle between $\mathbf{e}_1$ and $\mathbf{e}_2$ is not obtuse.

Geometrically, these conditions mean that the endpoint of $\mathbf{e}_2$ belongs to the shaded area of Figure 14.4.

Conversely, if $(\mathbf{e}_1, \mathbf{e}_2)$ is a pair of non-collinear vectors satisfying conditions (i - iii), then $(\mathbf{e}_1, \mathbf{e}_2)$ will be a reduced generating pair for the lattice they generate. Moreover, every other reduced generating pair of vectors of this

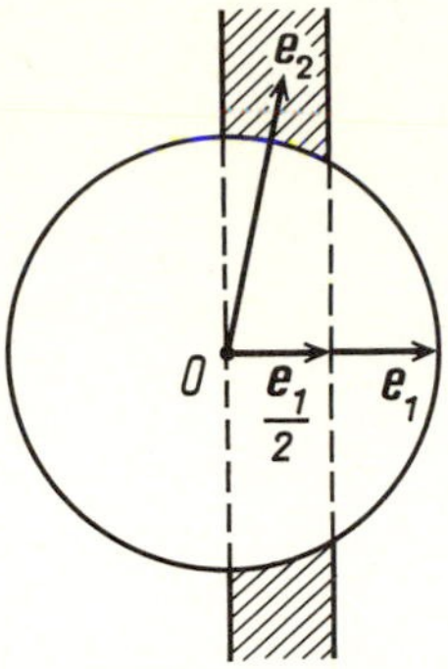

Figure 14.4

lattice is congruent to $(\mathbf{e}_1, \mathbf{e}_2)$. The reader will see this from Figure 14.5, (a) - (f), which illustrate all possible positions of a pair of vectors satisfying (i - iii), and list all the reduced generating pairs for the lattices they generate. It is interesting to note that because of this list, the existence of reduced generating pairs other than $(\mathbf{e}_1, \mathbf{e}_2)$ relates to the presence of symmetries of the lattice; these symmetries take $(\mathbf{e}_1, \mathbf{e}_2)$ precisely into the other reduced generating pairs of the lattice (see Exercise 1).

We can now present the solution of our problem on similarity of lattices. We know that every lattice has a reduced generating pair of vectors, and this pair is unique up to congruence; furthermore, similar lattices obviously have similar reduced generating pairs, and every pair of vectors satisfying conditions (i - iii) is a reduced generating pair for the lattice it generates. Therefore our problem reduces to describing the pairs of vectors satisfying (i - iii) up to similarity. For this, by the solution to our first question, we just need to describe the set of points of the upper half-plane which represent such pairs of vectors. Recall that in this representation, the pair $(\mathbf{e}_1, \mathbf{e}_2)$ is represented by $(\mathbf{e}, \mathbf{f})$, where $\mathbf{e}$ is the fixed vector, and $\mathbf{f}$ is a vector whose endpoint is in the upper half-space; this endpoint represents the pair $(\mathbf{e}_1, \mathbf{e}_2)$. The pair of vectors $(\mathbf{e}, \mathbf{f})$ again obviously satisfies conditions (i - iii) if $(\mathbf{e}_1, \mathbf{e}_2)$ did. Thus our problem reduces to describing the set of endpoints of vectors $\mathbf{f}$ for which the pair $(\mathbf{e}, \mathbf{f})$ satisfies conditions (i - iii). This description is obvious: the set of points in question forms the shaded region of the upper half-plane of Figure 14.6 (it is of course similar to one half of the shaded area of Figure 14.4). Thus all similar lattices are described by one point of the figure of the upper half-plane shaded in Figure 14.6. Different points of this figure represent non-similar lattices. This figure will be called the *modular figure* .

Thus we have obtained an answer to the question posed at the start of this section. However, the form of this answer is somewhat unusual. What is this figure we have obtained, whose boundary consists of curves of different types, two

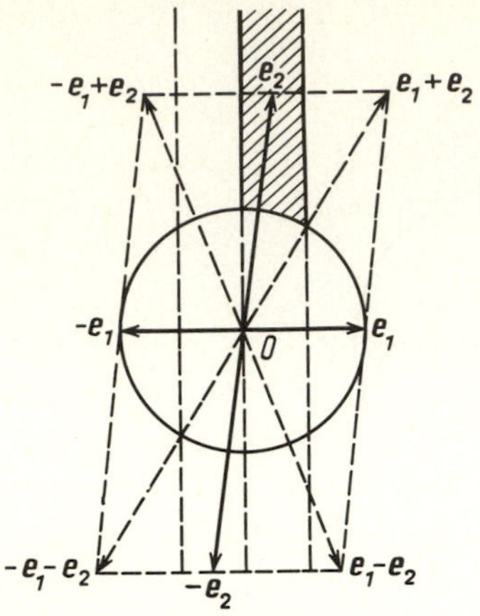

(a): (e_1, e_2) or $(-e_1, -e_2)$

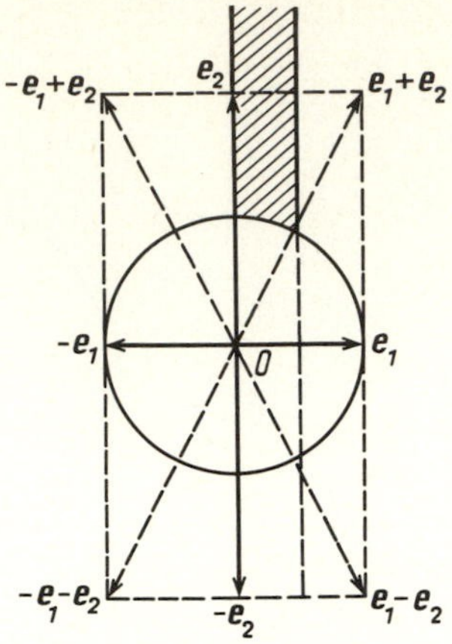

(b): $(\pm e_1, \pm e_2)$

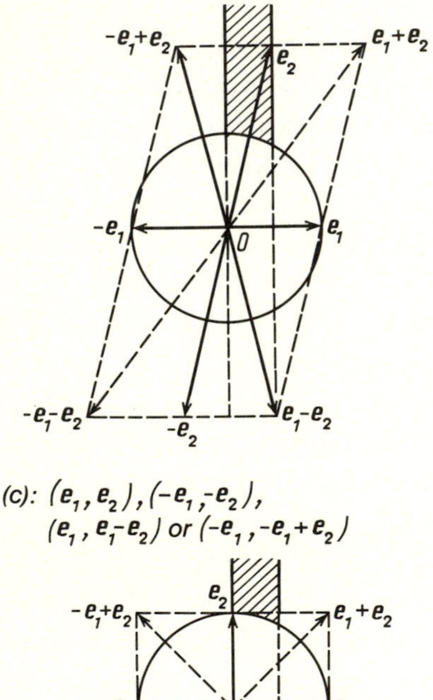

(c): $(e_1, e_2), (-e_1, -e_2),$
(e_1, e_1-e_2) or $(-e_1, -e_1+e_2)$

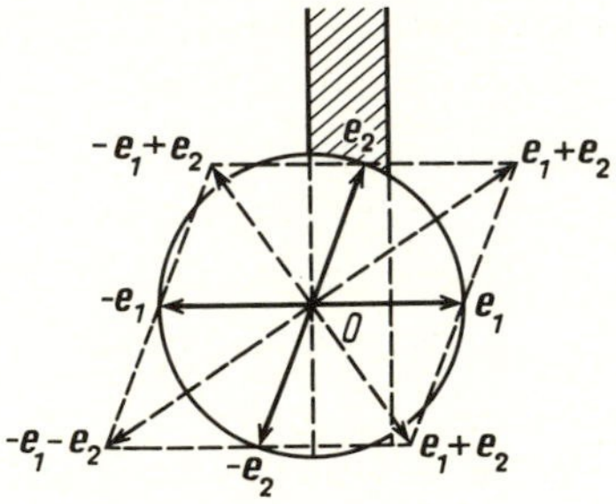

(d): $(e_1, e_2), (-e_1, -e_2),$
(e_2, e_1) or $(-e_2, -e_1)$

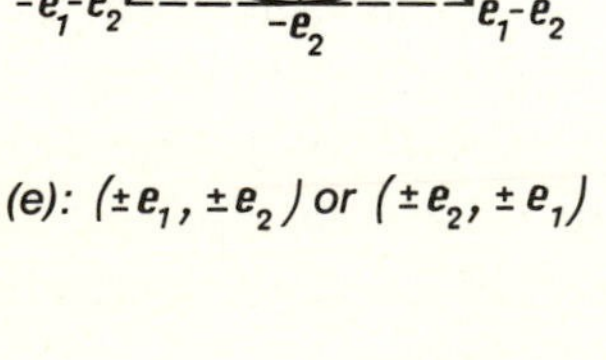

(e): $(\pm e_1, \pm e_2)$ or $(\pm e_2, \pm e_1)$

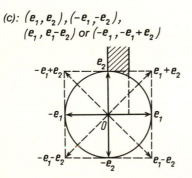

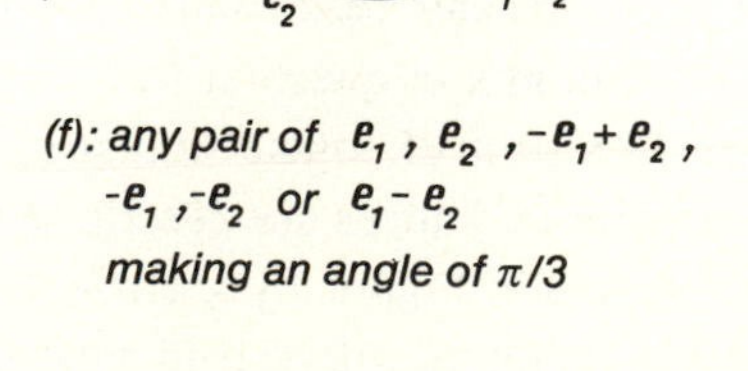

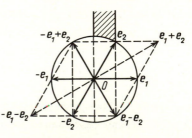

(f): any pair of e_1, e_2, $-e_1+e_2$,
$-e_1$, $-e_2$ or e_1-e_2
making an angle of $\pi/3$

Figure 14.5: Lattices with symmetries
and generating pairs of vectors

parallel rays and an arc of circle? In what follows we will attempt to get a better understanding of the answer we have obtained. This understanding will relate to the second approach to our problem, to which we proceed in §14.2.

Similar questions also arise in the theory of space lattices. These are of particular importance for crystallography, since as we saw in §12, every crystal has a related translation lattice, and for the description of the crystals we need to have a description of these lattices. This problem is solved by the same method as for plane lattices: by choosing in each lattice a certain simplest triple of generating vectors. However, here there are various different known ways of choosing the simplest triple: this is the so-called *reduction theory* of Seeber, Gauss, Dirichlet and Selling. In crystallography these questions go by the name of the theory of the *reduction cell* , or the conventional choice of the unit cell.

In addition, reduction theory, similarly to the 2-dimensional case, enables us to find all lattices in 3-space with symmetries; there are just 14 types of these. And this, similarly to the arguments of §§8, 11, 12 is a step towards the classification of all crystallographic and uniformly discontinuous groups in 3-space. The reader interested in these matters can refer to the books: J.J. Burckhardt, Die Bewegungsgruppen der Kristallographie, Birkhäuser, Basel, 1957 or M.J. Buerger, Introduction to crystal geometry, McGraw-Hill, New York, 1971.

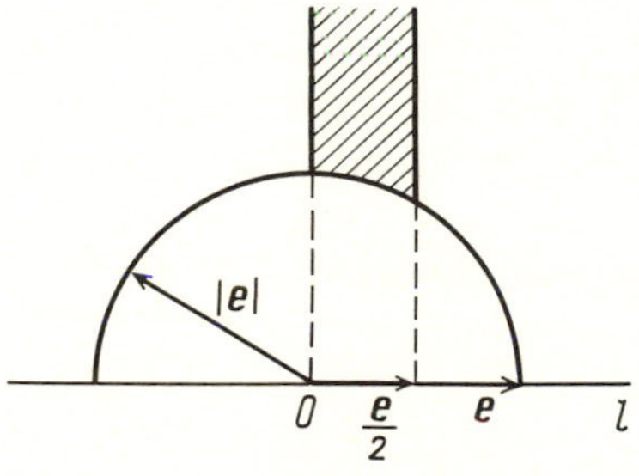

Figure 14.6

Exercises

1. Using all the possible cases for the positions of reduced generating pairs, as illustrated in Figure 14.5, (a) - (e), find all lattices with symmetries. Compare the answer with that of §11.3. Which points of the modular figure represent these lattices?

2. Prove that in a reduced generating pair (e_1, e_2), the vector e_2 is a next shortest vector of the lattice, after the integer multiples of e_1.

14.2. When do two pairs of vectors generate the same lattice? Let's determine the condition for two pairs of vectors (e_1, e_2) and (e_1', e_2') to generate the same lattice.

Certain necessary conditions spring to mind at once. First of all, we need that both of $\mathbf{e}_1'$ and $\mathbf{e}_2'$ should belong to the lattice generated by $\mathbf{e}_1$ and $\mathbf{e}_2$. This means that they should be expressible in the form

$$\begin{aligned} \mathbf{e}_1' &= m\mathbf{e}_1 + n\mathbf{e}_2, \\ \mathbf{e}_2' &= k\mathbf{e}_1 + \ell\mathbf{e}_2, \end{aligned} \tag{1}$$

where m, n, k and ℓ are integers. But conversely, $\mathbf{e}_1$ and $\mathbf{e}_2$ should also belong to the lattice generated by $\mathbf{e}_1'$ and $\mathbf{e}_2'$, so be expressible in the form

$$\begin{aligned} \mathbf{e}_1 &= m'\mathbf{e}_1' + n'\mathbf{e}_2', \\ \mathbf{e}_2 &= k'\mathbf{e}_1' + \ell'\mathbf{e}_2', \end{aligned} \tag{2}$$

where m', n', k' and ℓ' are integers. Moreover, relations (1) and (2) are obviously also sufficient. Indeed, it follows from (1) that the vector $p\mathbf{e}_1' + q\mathbf{e}_2'$, for any integers p, q, belongs to the lattice generated by $(\mathbf{e}_1, \mathbf{e}_2)$; just substitute for $\mathbf{e}_1'$ and $\mathbf{e}_2'$ as in (1), and multiply out the brackets. Thus the entire lattice generated by $(\mathbf{e}_1', \mathbf{e}_2')$ is contained in the lattice generated by $(\mathbf{e}_1, \mathbf{e}_2)$; using (2) similarly, we see that the second lattice is contained in the first, so that they are equal.

Let's try to get the answer in a more compact form. For this, multiply the first equation in (1) by ℓ and the second by n, and subtract; we get

$$\ell\mathbf{e}_1' - n\mathbf{e}_2' = (m\ell - nk)\mathbf{e}_1. \tag{3}$$

Similary, from (2) we get

$$-k\mathbf{e}_1' + m\mathbf{e}_2' = (m\ell - nk)\mathbf{e}_2. \tag{4}$$

Set $m\ell - nk = d$. This is a non-zero integer: indeed, if d were zero, then (3) and (4) would give $\ell\mathbf{e}_1' = n\mathbf{e}_2'$ and $k\mathbf{e}_1' = m\,\mathbf{e}_2'$, and it would follow from this that $\mathbf{e}_1'$ and $\mathbf{e}_2'$ are collinear. Dividing both sides of (3) and (4) by d, we get

$$\begin{aligned} \mathbf{e}_1 &= (\ell/d)\mathbf{e}_1' - (n/d)\mathbf{e}_2', \\ \mathbf{e}_2 &= -(k/d)\mathbf{e}_1' + (m/d)\mathbf{e}_2'. \end{aligned}$$

Now compare this with (2). Since the expression of $\mathbf{e}_1$ and $\mathbf{e}_2$ in terms of the non-parallel vectors $\mathbf{e}_1'$ and $\mathbf{e}_2'$ is unique, we must have

$$m' = \ell/d,\ n' = -n/d,\ k' = -k/d,\ \ell' = m/d, \tag{5}$$

and all of these numbers are integers. Transforming (2) in exactly the same way and comparing with (1), we arrive at the relations

$$m = \ell'/d', \quad n = -n'/d', \quad k = -k'/d', \quad \ell = m'/d', \tag{6}$$

where $d' = m'\ell' - n'k'$. It follows at once from these that $dd' = 1$, and since d and d' are both integers, this is only possible if $d = \pm 1$, that is $m\ell - nk = \pm 1$. Conversely, if $d = \pm 1$, then (3) and (4) imply (2) with the values of m', n', k' and ℓ' indicated in (5). Thus we have proved the following criterion.

Theorem 1. Suppose that m, n, k and ℓ are integers, and let $\mathbf{e}_1' = m\mathbf{e}_1 + n\mathbf{e}_2$ and $\mathbf{e}_2' = k\mathbf{e}_1 + \ell\mathbf{e}_2$; then the pair $(\mathbf{e}_1', \mathbf{e}_2')$ generates the same lattice as $(\mathbf{e}_1, \mathbf{e}_2)$ if and only if $m\ell - nk = \pm 1$.

This result, which we have obtained by computation, is clarified by the geometrical argument given below.

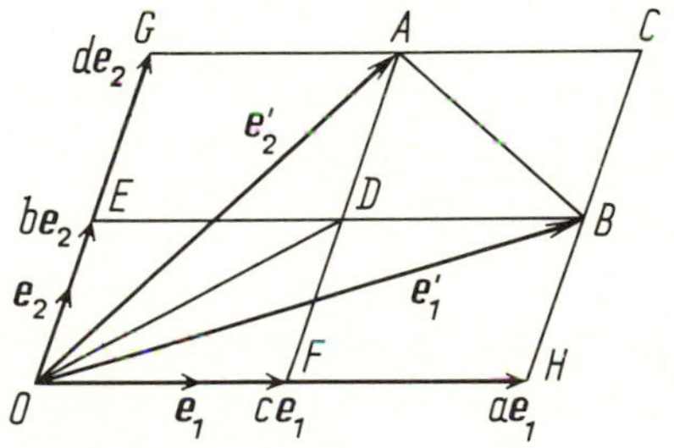

Figure 14.7

Lemma 1. If $\mathbf{e}_1' = a\mathbf{e}_1 + b\mathbf{e}_2$ and $\mathbf{e}_2' = c\mathbf{e}_1 + d\mathbf{e}_2$, then the area Δ' of the parallelogram constructed on $\mathbf{e}_1'$ and $\mathbf{e}_2'$ is equal to $|ad - bc| \cdot \Delta$, where Δ is the area of the parallelogram constructed on $\mathbf{e}_1$ and $\mathbf{e}_2$ (Figure 14.7).

We deal with the case in which the coefficients a, b, c and d are non-negative (this is the case illustrated in Figure 14.7), leaving the similar treatment of the other cases to the reader.

Denote by Δ and Δ' the areas of the parallelograms constructed respectively on $\mathbf{e}_1$ and $\mathbf{e}_2$, and on $\mathbf{e}_1'$ and $\mathbf{e}_2'$. Then $\Delta' = 2 \cdot \text{area}(OAB)$; on the other hand,

$$\text{area}(ADBC) = 2 \cdot \text{area}(ADB), \quad \text{area}(ADEG) = 2 \cdot \text{area}(ADO),$$

and

$$\text{area}(DBHF) = 2 \cdot \text{area}(DBO).$$

Therefore,

$$\Delta' = \text{area}(ADBC) + \text{area}(ADEG) + \text{area}(DBHF) = \text{area}(OGCH) - \text{area}(OEDF).$$

Finally,

$$\text{area}(OGCH) = ad\Delta, \quad \text{area}(OEDF) = bc\Delta,$$

whence

$$\Delta' = (ad - bc)\Delta,$$

which is the assertion of the lemma.

According to the lemma, the theorem reduces to saying that for all pairs of vectors $(\mathbf{e}, \mathbf{f})$ generating the same lattice, the area of the parallelogram constructed on them is the same. Indeed, we can give a characterisation of the area Δ of this parallelogram using only the lattice itself, independent of the pair of vectors $(\mathbf{e}, \mathbf{f})$. For this, consider the disc of radius R centred at a lattice point O, and let $n(R)$ denote the number of lattice points contained in this disc; it is intuitively clear that the area of this disc is very close to $n(R)\Delta$, and hence

$$\lim_{R \to \infty} \frac{\pi R^2}{n(R)} = \Delta.$$

However, the left-hand side depends only on the lattice. To justify this relation, associate with each lattice point P the parallelogram with vertices P, $T_{\mathbf{e}}(P)$, $T_{\mathbf{f}}(P)$ and $T_{\mathbf{e}+\mathbf{f}}$. The area of this parallelogram is of course equal to Δ, and the sum of the areas taken over all points P lying in the disc of radius R equals $n(R)\Delta$. On the other hand, if we denote by d the length of the longest diagonal of a parallelogram of this form, then all of the parallelograms under consideration are contained in a disc of radius $R + d$, and they entirely cover the disc of radius $R - d$ (Figure 14.8). Therefore

$$\pi(R - d)^2 \leq n(R)\Delta \leq \pi(R + d)^2,$$

and the required relation follows from this.

Now the problem considered in this section can be reformulated as follows: define pairs of vectors $(\mathbf{e}_1, \mathbf{e}_2)$ and $(\mathbf{e}_1', \mathbf{e}_2')$ to be equivalent if they generate the same lattice. By the very definition, this notion of equivalence between pairs of vectors satisfies the conditions $(\alpha - \gamma)$ considered in §7.1. Algebraically, it can be expressed as follows: there should exist integers m, n, k, ℓ such that

$$\mathbf{e}_1' = m\mathbf{e}_1 + n\mathbf{e}_2, \quad \mathbf{e}_2' = k\mathbf{e}_1 + \ell\mathbf{e}_2, \quad \text{and} \quad m\ell - nk = \pm 1.$$

We now define a new notion of equivalence by saying that pairs $(\mathbf{e}_1, \mathbf{e}_2)$ and $(\mathbf{e}_1', \mathbf{e}_2')$ are equivalent if they generate similar lattices. As we have seen, every

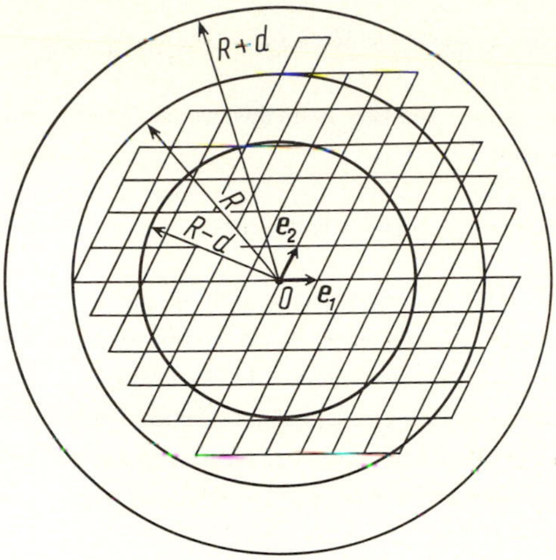

Figure 14.8

pair $(\mathbf{e}_1, \mathbf{e}_2)$ is equivalent to a pair $(\mathbf{e}, \mathbf{f})$, where $\mathbf{e}$ is a fixed vector and $\mathbf{f}$ lies in the upper half-plane, so that we can restrict ourselves to considering equivalence of pairs $(\mathbf{e}, \mathbf{f})$ and $(\mathbf{e}, \mathbf{f}')$. How is our equivalence expressed in this case? We note that there should exist integers m, n, k and ℓ satisfying $m\ell - nk = \pm 1$, and such that the pair of vectors $(m\mathbf{e} + n\mathbf{f}, k\mathbf{e} + \ell\mathbf{f})$ constructed from them satisfies the following condition: if F is the similarity for which $F(m\mathbf{e} + n\mathbf{f}) = \mathbf{e}$, and the endpoint of the vector $F(k\mathbf{e} + \ell\mathbf{f})$ lies in the upper half-plane, then $F(k\mathbf{e} + \ell\mathbf{f}) = \mathbf{f}'$.

This definition of equivalence is at present far from transparent, and we have to do some work to present it in a more comprehensible form.

Exercises

1. Given a pair of non-collinear vectors (e_1, e_2), the pair (e_1', e_2') is expressed in terms of (e_1, e_2) by $e_1' = me_1 + ne_2$, $e_2' = ke_1 + \ell e_2$, where m, n, k and ℓ are integers such that $m\ell - nk = \pm 1$. According to Theorem 1, (e_1, e_2) must be similarly expressible in terms of (e_1', e_2'); find the expression.

2. Prove that two geometries on the Klein bottle are similar if and only if, in the notation of §8, they have the same ratio $|a|/|b|$.

14.3. Application to number theory. The properties of lattices treated above relate to certain questions of number theory. Of special interest in number theory are the lattices for which the square of the distance from some fixed origin to any lattice point is an integer. For example, the lattice of points with integer coordinates is of this kind, since the square of the distance from the origin to the point (m, n) is the integer $m^2 + n^2$.

In the general case, if (e_1, e_2) generate the lattice, then the square of the distance from any of its points to a fixed point is $|me_1 + ne_2|^2$, where m and n are any integers. By the cosine theorem,

$$|me_1 + ne_2|^2 = |me_1|^2 + |ne_2|^2 + 2|me_1|\cdot|ne_2|\cdot\cos\psi,$$

where ψ is the angle between me_1 and ne_2. From this, working out the various cases for the signs of m and n, we get the formula

$$|me_1 + ne_2|^2 = am^2 + bmn + cn^2, \tag{7}$$

where $a = |e_1|^2$, $c = |e_2|^2$, and $b = 2|e_1|\cdot|e_2|\cdot\cos\varphi$, with φ the angle between e_1 and e_2. The expression (7) will be an integer for all integers n and m, provided that a, b and c are integers. From now on we assume this. Such lattices will be said to be *integral* .

We denote by Δ the area of the parallelogram constructed on the vectors e_1 and e_2 (thus, as we know, Δ is the area of the parallelogram constructed on any pair of vectors generating the same lattice). By a well-known theorem, $\Delta = |e_1|\cdot|e_2|\cdot\sin\varphi$, so that

$$\Delta^2 = |e_1|^2\cdot|e_2|^2\cdot\sin^2\varphi = |e_1|^2\cdot|e_2|^2(1 - \cos^2\varphi) = ac - (1/4)b^2.$$

It follows from this that the number

$$D = 4\Delta^2 = 4ac - b^2$$

is an integer and is positive. As we have already said, it does not depend on which pair of vectors (e_1, e_2) we choose as generators of the lattice. D is called the *discriminant* of the lattice.

Conversely, if we have three arbitrary integers a, b and c for which $a > 0, c > 0$ and $4ac - b^2 > 0$, then they correspond to some integral lattice in which (7) holds. Indeed, by assumption $b^2/4ac < 1$, and hence there exists an angle φ for which $\cos\varphi = b/(2\sqrt{(ac)})$. If we choose vectors e_1 and e_2 such that $|e_1| = \sqrt{a}$ and $|e_2| = \sqrt{c}$, and the angle between them is φ, then (7) will obviously be satisfied.

If (e_1, e_2) is a reduced generating pair of vectors, then by definition

$$|e_2| \geq |e_1| \text{ and } |e_2|\cdot|\cos\varphi| \leq |e_1|/2$$

that is, in our new notation,

$$c \geq a \geq b \geq 0. \tag{8}$$

It follows from this that $D \geq 4b^2 - b^2 = 3b^2$, that is

$$0 \leq b \leq \sqrt{D/3}. \tag{9}$$

Since b is an integer, for fixed D there are only finitely many possibilities for b, and for fixed b we have $4ac = D + b^2$, so that the number of possibilities for a and c is also finite. Finally, the numbers a, b and c determine the pair of vectors (e_1, e_2) uniquely up to a motion. Since every lattice has a reduced generating pair, as a result we get that for a given discriminant, there are only a finite number of distinct integral lattice (up to motions). Using the arguments given above, they can all be found concretely.

Example. The lattice of points with integer coordinates is generated by two orthogonal vectors e_1

and e_2 of length 1, so that for it we have $a = c = 1, b = 0$, and $D = 4$. Let's find all integer lattices of discriminant 4: in any such lattice there is a reduced generating pair (e_1, e_2). Since $D = 4$, b must be even. According to the inequality (9), we have $0 \le b \le 2/\sqrt{3} < 2$, and therefore $b = 0$; hence $ac = 1$, that is, $a = c = 1$. Our lattice is thus the same as the lattice of points with integer coordinates, that is, there exists a unique integral lattice of discriminant 4.

Application. Let's determine which positive integers can be written as a sum of two squares of integers:

$$a = m^2 + n^2. \tag{10}$$

First of all, we can assume that m and n are coprime; for if they had a common factor $d > 1$, then $m = dm_1$ and $n = dn_1$, so that

$$a = d^2 a_1,\ a_1 = m_1{}^2 + n_1{}^2, \tag{11}$$

and the problem reduces to the same problem for a smaller number a_1.

As we have seen, relation (10) just means that in the lattice of points with integer coordinates, there exists a vector $f = me_1 + ne_2$ for which $|f|^2 = a$. We can assume that m and n are coprime; this means that f cannot be of the form $f = df_1$, where f_1 is a vector of the same lattice. In §8.3 we saw that, given a vector f of this kind, we can always find a vector f' such that f and f' together generate the whole lattice of points with integer coordinates. If $a = |f|^2$, $c = |f'|^2$ and we write φ for the angle between f and f' and set $b = 2|f|{\cdot}|f'|{\cdot}\cos\varphi$, then a, b, and c are integers, and $4ac - b^2 = 4$. Conversely, to any integers a, b and c satisfying these conditions there exists a corresponding integral lattice with discriminant 4, and since there is only one such lattice, this will be the lattice of points with integer coordinates. Thus given a, we conclude that (10) is satisfied for some coprime m and n if and only if there are integers b and c with $c > 0$ such that $4ac - b^2 = 4$. In this case b must be even; suppose $b = 2t$, then $ac - t^2 = 1$. Since c is arbitrary, our condition simply means that there exists an integer t such that

$$a \mid (t^2 + 1), \tag{12}$$

that is, $t^2 + 1$ is divisible by a.

This reduction of the question of the solvability of (10) to the divisibility question (12) is an application of the properties of lattices. The question as to when (12) can be solved for given a is answered using completely different ideas.

We go through this simple, but very beautiful, arithmetical argument. Consider any prime number p dividing a. Then a fortiori, $p \mid (t^2 + 1)$. We determine first of all when this holds, that is, according to what we have said above, when a prime number p can be written as a sum of two squares of integers. Since for $p = 2$ the answer is clear ($2 = 1^2 + 1^2$), we will from now on assume that p is odd.

Any integer n gives a certain residue r on dividing by p, that is $n = pk + r$, where r is one of the numbers $0, 1, \ldots, p-1$. If n is not divisible by p then r can only be one of the $p-1$ numbers $1, 2, \ldots, p-1$. A number r with $1 \le r \le p-1$ which is the residue of a perfect square is called a *quadratic residue*. For example, for $p = 11$, among the residues 1, 2, 3, 4, 5, 6, 7, 8, 9, 10, the quadratic residues are 1, 3, 4, 5 and 9; for $p = 13$ the quadratic residues are 1, 3, 4, 9, 10, 12 (the reader should verify both of these assertions!). If n is a number such that $n + 1$ is divisible by p, that is $n + 1 = pm$, then $n = p-1 + p(m-1)$, and its residue is $p-1$. It follows from this that $p \mid (t^2 + 1)$ holds for some t if and only if $p - 1$ is a quadratic residue for p. (For example, this holds for $p = 13$, but not for $p = 11$.)

To find all quadratic residues, it is obviously sufficient to take the squares of each of the numbers $1, ..., p-1$, and take the residues of these; however, this is not efficient: since $(p-k)^2 = k^2 + p(p-2k)$, it follows that $(p-k)^2$ and k^2 have the same residue on dividing by p, so that it is enough just to take the squares of $1, 2, ..., (p-1)/2$. Hence we get $(p-1)/2$ residues. Let's prove that they are all distinct. If n and m are two numbers such that $1 \le n, m \le (p-1)/2$ and n^2 and m^2 have the same residue on dividing by p, then $p \mid (n^2 - m^2)$, that is $p \mid (n-m)(n+m)$, and so $p \mid (n-m)$ or $p \mid (n+m)$. Here we are using the fact that if a product of two numbers is divisible by a prime number, then one of the two must be divisible by p; this follows easily from the fact that any integer has a unique expression as a product of primes. By the assumption on n and m, we have $|n-m| \le (p-1)/2$, $n+m \le p-1$, so that the only possibility is $n = m$. Thus we have proved that there exist exactly $(p-1)/2$ quadratic residues.

We now take any residue k, and multiply it successively by $1, 2, ..., p-1$, and write out the resulting residues:

$$k \cdot 1 = r_1 + pn_1 \quad (\text{here } r_1 = k \text{ and } n_1 = 0)$$

$$k \cdot 2 = r_2 + pn_2,$$

$$\dots$$

$$k \cdot (p-1) = r_{p-1} + pn_{p-1}.$$

Firstly, none of the integers $k \cdot r$ for $1 \le r \le p-1$ is divisible by p, and hence the residues r_i are all contained among $1, 2, ..., p-1$. Furthermore, they are all distinct: indeed, if $r_i = r_j$ for $i > j$, this means that $k \cdot i$ and $k \cdot j$ would have the same residue on dividing by p, so that p would divide $k \cdot i - k \cdot j = k \cdot (i-j)$; and this is impossible, since p does not divide either k or $i-j$. So the numbers $r_1, r_2, ..., r_{p-1}$ we get in this way are $p-1$ distinct residues, and therefore they are all of the possible residues $1, 2, ..., p-1$. In particular, what is important for us is that one of the residues, say r_ℓ, equals 1; this proves that for any residue k with $1 \le k \le p-1$ there exists exactly one residue ℓ with $1 \le \ell \le p-1$ such that $k \cdot \ell$ has residue 1 on dividing by p. When this happens we say that k and ℓ are *reciprocal*. It could happen that some residue k is its own reciprocal; this means that k^2 has residue 1, and therefore $p \mid (k^2 - 1)$, that is $p \mid (k-1)(k+1)$. It follows that $p \mid (k-1)$ or $p \mid (k+1)$; the first case happens for $k = 1$, and the second for $k = p-1$. Thus all the residues except for 1 and $p-1$ break up into pairs of reciprocal residues. For example, for $p = 11$, the reciprocal residues are 2 and 6, 3 and 4, 5 and 9, 7 and 8; and for $p = 13$, they are 2 and 7, 3 and 9, 4 and 10, 5 and 8, 6 and 11.

Now let's see how this breaking up into reciprocal pairs relates to quadratic residues. If k is a quadratic residue, that is, the residue of a number m^2 on dividing by p, and n is the reciprocal of m (so mn has residue 1), then the residue ℓ of n^2 is the reciprocal of k, since the residue of $k\ell$ is the same as that of $m^2n^2 = (mn)^2 = 1$. Thus the reciprocal of a quadratic residue is again a quadratic residue. It remains to see which quadratic residues are reciprocal to themselves; we know that there are just two such residues, namely 1 and $p-1$, and 1 is of course a quadratic residue. As far as $p-1$ is concerned, this is a quadratic residue if and only if $p \mid (t^2 + 1)$ for some t. For such p, the set of $(p-1)/2$ quadratic residues breaks up in 1, $p-1$, and a further number, say N, of reciprocal pairs of quadratic residues. Therefore $(p-1)/2 = 2 + 2N$, and thus p is of the form $p = 4n + 1$, with $n = N + 1$; in other words, p has residue 1 on dividing by 4. This is what happens for $p = 13$: the quadratic residues break up into 1, 12, and two reciprocal pairs 3 and 9, 4 and 10. However, if $p - 1$ is not a quadratic residue, that is if the relation we're interested in $p \mid (t^2 + 1)$ does not hold for any t, then the set of $(p-1)/2$ quadratic residues breaks up into 1 and n reciprocal pairs. Therefore $(p-1)/2 = 2n + 1$, so $p = 4n + 3$; in other words, p has residue 3 on dividing by 4. This is what happens for $p = 11$: the quadratic residues break up into 1 and two reciprocal pairs 3 and 4, 5 and 9. Hence we have proved that an odd prime number p can be written as the sum of two squares if and only if it is of the form $4n + 1$.

Turning now to the representation of any positive integer a as a sum of two squares, we see that if p is a prime factor of a of the form $4n + 3$, then (10) is only possible if n and m are both divisible by p, that is if a is divisible by p^2, and therefore in (11), p goes into d_1 and cannot go into a_1. In other words, the representation (10) is only possible for those a of the form $a = d^2a_1$ where a_1 is a product of 2 and prime numbers of the form $4n + 1$. We now check that any such number can actually be written as the sum of two squares. Let $a = d^2b$, with $b = p_1 \dots p_m$ and each of the p_i is either 2 or a prime of the form $4n_i + 1$. We already know that any of the p_i can be written as a sum of two squares. Thus it is enough to check that if two integers r and s can be written as the sum of two squares, then so can their product rs. Let

$$r = x^2 + y^2 \quad \text{and} \quad s = u^2 + v^2;$$

then

$$rs = (x^2 + y^2)(u^2 + v^2) = (xu - yv)^2 + (xv + yu)^2.$$

The final identity is easy to check; its meaning will become clearer a little later in §15.1.

Thus we have finally proved the theorem: a positive integer a can be written as a sum of two squares if and only if $a = d^2b$, where b is only divisible by 2 and prime numbers of the form $4n + 1$.

The reader wishing to get to know this area of number theory can for example refer to the following books:

C.F. Gauss, Disquitiones Arithmeticae, Werke, Band I, or Yale Univ. Press, New Haven - London 1966, or Chelsea, New York, 1965;

P.G. Dirichlet, Vorlesungen über Zahlentheorie, Chelsea, New York, 1965;

J-P. Serre, A course in arithmetic, Springer, Berlin - New York, 1973;

H. Davenport, The higher arithmetic, Cambridge Univ. Press, Cambridge - New York, 1982.

Exercises

1. Prove that in every plane lattice there is a non-zero vector $\mathbf{v}$ for which

$$|v| \le \frac{\sqrt{2\Delta}}{\sqrt[4]{3}},$$

where Δ is the area of the parallelogram constructed on a pair of generating vectors. For which lattice does this inequality turn into equality (you will see that there is only one, up to similarity)?

Find an upper bound on the smallest non-zero value taken by the expression $ax^2 + bxy + cy^2$ for integers x and y, where $a > 0, c > 0$ and $4ac - b^2 > 0$.

2. Find all integral lattices of discriminant $D \le 15$. Which points of the modular figure correspond to these lattices? Prove that there exist two distinct (non-similar) lattices of discriminant $D = 15$.

3. Prove that the expression $19x^2 + 7y^2 + 23xy$ for coprime integers x and y takes the integer value a if and only if a is odd and (-3) is a quadratic residue on dividing by a.

§15. The algebra of similarities: complex numbers

15.1. The geometrical definition of complex numbers. According to §13.1, Lemma 1, any similarity F can be written in the form $F = GH_O^{\lambda}$ where G is a motion, and H_O^{λ} is a dilation centred at some fixed point O. At present we consider the case that G is an (anticlockwise) rotation through an angle φ around the same point O, that is $G = R_O^{\varphi}$. In what follows, the point O will be fixed throughout, so that a similarity of this kind is uniquely determined by the ratio of similarity λ and the angle φ. We will denote this transformation by $F = F_{\lambda,\varphi}$, so that F rotates the plane through an angle φ around O and expands or shrinks it (according to whether $\lambda > 1$ or $\lambda < 1$) by the ratio λ. Obviously,

$$\begin{aligned} F_{\lambda,\varphi}F_{\lambda',\varphi'} &= F_{\lambda\lambda',\varphi+\varphi'}, \\ (F_{\lambda,\varphi})^{-1} &= F_{\lambda^{-1},-\varphi}. \end{aligned} \tag{1}$$

An angle φ will always be considered up to addition of $2k\pi$, with k an integer; hence we can write $2\pi - \varphi$ instead of $-\varphi$.

On the plane, we fix once and for all a vector $\mathbf{e} = \overrightarrow{OP}$ of length 1. Obviously the similarity F is uniquely determined by its effect on $\mathbf{e}$, that is, by the vector $\mathbf{x} = F_{\lambda,\varphi}(\mathbf{e})$. Clearly, $\mathbf{x} \neq \mathbf{0}$; we have $|\mathbf{x}| = \lambda$, and the angle (measured anticlockwise) between $\mathbf{e}$ and $\mathbf{x}$ is equal to φ. We sometimes write $F_{\mathbf{x}}$ instead of F. This connection between vectors of the plane and similarities $F_{\lambda,\varphi}$ suggests that we could transfer to vectors the simple composition rule (1) for similarities. The resulting operation on vectors will be called multiplication. Thus the product of vectors $\mathbf{x}$ and $\mathbf{y}$ is defined to be the vector $\mathbf{z}$ such that

$$F_{\mathbf{z}} = F_{\mathbf{x}}F_{\mathbf{y}}. \tag{2}$$

Comparing this with the rule (1) for multiplying similarities, we can say that $\mathbf{z}$ is the product of $\mathbf{x}$ and $\mathbf{y}$ if its length is the product of theirs (that is, $|\mathbf{z}| = |\mathbf{x}|\cdot|\mathbf{y}|$), and the angle which it makes with $\mathbf{e}$ is the sum of those made by $\mathbf{x}$ and $\mathbf{y}$ (Figure 15.1).

Now apply $F_{\mathbf{z}}$ to the vector $\mathbf{e}$. By (2) we get

$$F_{\mathbf{z}}(\mathbf{e}) = F_{\mathbf{x}}(F_{\mathbf{y}}(\mathbf{e})).$$

By definition of $F_{\mathbf{z}}$ and $F_{\mathbf{y}}$ we have $F_{\mathbf{z}}(\mathbf{e}) = \mathbf{z}$ and $F_{\mathbf{y}}(\mathbf{e}) = \mathbf{y}$. Therefore the definition of multiplication of vectors can also be written as follows:

$$\mathbf{z} = F_{\mathbf{x}}(\mathbf{y}). \tag{3}$$

According to (1), the composition of similarities $F_{\lambda,\varphi}$ reduces to multiplication of the ratios λ and addition of the angles φ. Since both these operations satisfy the commutative law, we have $F_{\lambda,\varphi}F_{\lambda',\varphi'} = F_{\lambda',\varphi'}F_{\lambda,\varphi}$, and therefore $\mathbf{xy} = \mathbf{yx}$. In exactly the same way, we can check that the multiplication of vectors we have defined satisfies the associative law $\mathbf{x}(\mathbf{yz}) = (\mathbf{xy})\mathbf{z}$. The similarity $F_{\mathbf{e}}$ is characterised by the fact that it takes $\mathbf{e}$ to itself, so that it is the identity

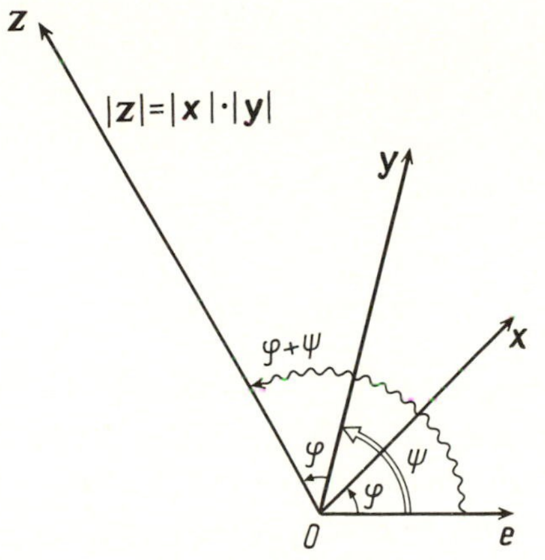

Figure 15.1

transformation. In view of this, $\mathbf{ex} = \mathbf{x}$; this can also be seen directly from Figure 15.1. Thus $\mathbf{e}$ plays the role of the number 1. Finally, we can see from (1) that for any non-zero vector $\mathbf{x}$ there exists a vector $\mathbf{x}^{-1}$ for which $\mathbf{xx}^{-1} = \mathbf{e}$; this vector has length $|\mathbf{x}^{-1}| = |\mathbf{x}|^{-1}$, and the angle it makes with $\mathbf{e}$ is minus that of $\mathbf{x}$ (Figure 15.2).

These properties of multiplication of vectors are very similar to those of multiplication of numbers. The reader knows that the properties of addition of vectors are also similar to those of addition of numbers: the commutative and associative laws hold, there exists a zero vector $\mathbf{0}$ and a negative $-\mathbf{x}$ of a vector. We therefore pay attention to the relation between addition and multiplication of vectors. Since a similarity takes a parallelogram into a parallelogram, we have (Figure 15.3)

$$F(\mathbf{x} + \mathbf{y}) = F(\mathbf{x}) + F(\mathbf{y}).$$

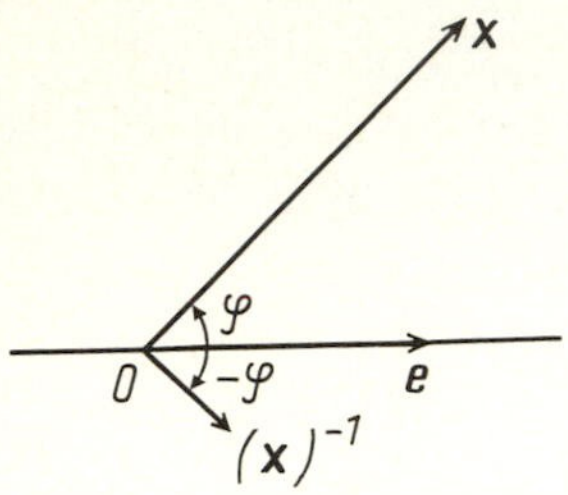

Figure 15.2

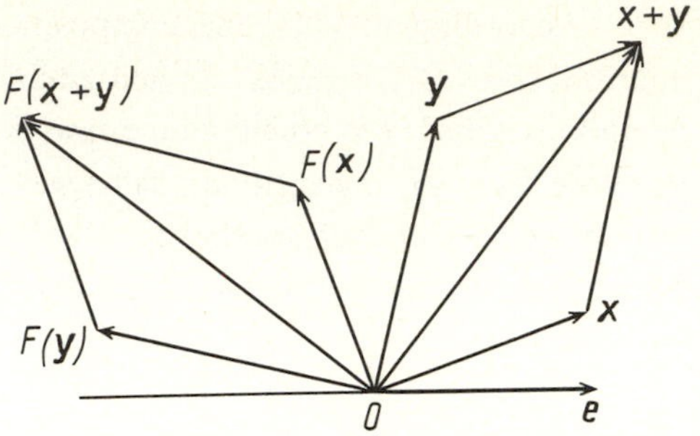

Figure 15.3

In particular, if $F = F_{\mathbf{z}}$, we get

$$\mathbf{z}(\mathbf{x} + \mathbf{y}) = \mathbf{z}\mathbf{x} + \mathbf{z}\mathbf{y},$$

that is, the distributive law.

In view of this complete analogy, vectors in the plane (with the above definition of the multiplication operation, and the addition operation which the reader already knows) will be also dignified with the name of numbers. The ordinary real numbers are included among the new numbers as a special case: they are the vectors $a\mathbf{e}$, where a is a real number. The reader will check easily that the definition of multiplication of vectors we have given agrees with multiplication of numbers: that is, $(a\mathbf{e})(b\mathbf{e}) = ab\mathbf{e}$; and furthermore, that it also agrees with the operation which he knows of multiplying a vector by a number: that is, for any vector $\mathbf{x}$ and number a, we have $(a\mathbf{e})\mathbf{x} = a\mathbf{x}$.

The numbers which correspond in this way to vectors will be called complex numbers. The reader will of course already have heard of them. We have constructed them here from first principles to show how naturally they appear in the flow of ideas we are interested in.

From now on we will not use vector notation ($\mathbf{x}$ and so on) for complex numbers, but simply denote them by letters, like real numbers. Furthermore, we will represent them as before in the plane, writing not the whole vector $\mathbf{x} = \overrightarrow{OX}$, but just its endpoint X. The complex number $\mathbf{e}$ will be denoted simply by 1, and the number $a\mathbf{e}$ will be written just as the corresponding real number a. In this connection the line through O on which the vector $\mathbf{e}$ lies will be called the *real line*; (it will be drawn in graphs as the x-axis Ox).

Consider the vector $\mathbf{f}$ of length 1 which makes an angle of $\pi/2$ (measured anticlockwise) with $\mathbf{e}$. The corresponding similarity is simply rotation through an angle $\pi/2$, that is $F_{\mathbf{f}} = R_O^{\pi/2}$. Since $R_O^{\pi/2} R_O^{\pi/2} = R_O^{\pi} = F_{-\mathbf{e}}$, we have $\mathbf{f}^2 = -\mathbf{e}$. We reserve the special notation i for **f**, considered as a complex number; as we have seen, $i^2 = -1$. Since every vector has a unique expression as a combination of **e** and **f**, any complex number can be written in a unique way in the form $z = a + bi$, where a and b are real numbers. a is called the *real part* of z, and denoted by Re(z); and b the *imaginary part*, b = Im(z). From the fact that $i^2 = -1$, and from the rules we know for operating with complex numbers, we deduce the following expression for the product of complex numbers written in the above form:

$$(a+bi)(c+di) = ac + bdi^2 + adi + bci = (ac-bd) + (ad+bc)i. \quad (4)$$

We introduce some additional notions which we require later. If z is a complex number, then its length as a vector is called the ***modulus*** of z, and denoted by $|z|$. If $z = a + bi$, then obviously $|z| = \sqrt{(a^2 + b^2)}$. From the multiplication rule for complex numbers it follows that $|z_1 z_2| = |z_1| \cdot |z_2|$ (see Figure 15.1). This implies of course that

$$|z_1 z_2|^2 = |z_1|^2 |z_2|^2.$$

If $z_1 = x + iy$ and $z_2 = u + iv$, then from this relation we get the identity

$$(xu - yv)^2 + (xv + yu)^2 = (x^2 + y^2)(u^2 + v^2),$$

which has already appeared in §14.3. It thus expresses the fact that

$$|z_1 z_2| = |z_1| \cdot |z_2|.$$

The angle φ which the vector of z forms with the real axis is called the *argument* of z, and denoted by arg z. We will assume that it is defined up to adding an integer multiple of 2π. From the multiplication rule for complex numbers, it follows that $\arg(z_1 z_2) = \arg z_1 + \arg z_2$. Finally, suppose that $z = a + bi$ is represented by the point P. Then the complex number $a - bi$ is represented by the point P′ which is the reflection of P in the real line; it is called the *complex conjugate* of z, and denoted by $\bar{z}$. The reader can check the following properties:

$$z\bar{z} = |z|^2, \quad \overline{z_1 + z_2} = \bar{z}_1 + \bar{z}_2,$$
$$\overline{z_1 z_2} = \bar{z}_1 \bar{z}_2 \quad \text{and} \quad \overline{(z_1/z_2)} = \bar{z}_1/\bar{z}_2. \quad (5)$$

Using complex numbers, we can write down in a very simple form any similarity of the plane which fixes the given point O. By §13.2, Lemma 1, such a similarity can be expressed as $F = GH_O^{\lambda}$, where H_O^{λ} is a dilation centred at O, and G is a motion, which in our case must also fix O. Such a motion is either a rotation, or a reflection in a line through O. In the first case, when $G = R_O^{\varphi}$, we have $F = F_{\lambda,\varphi}$, and to every such similarity we have assigned a complex number; if we denote this complex number by z_0, then $F_{\lambda,\varphi}(z) = z_0 z$, which is actually the definition of multiplication of complex numbers. Thus F takes a complex number z (considered as a point of the plane) into the number $z_0 z$.

We now consider the other case, when G is a reflection in an axis ℓ passing through O. If ℓ is just the real line Ox, then our reflection takes each complex number z into its complex conjugate $\bar{z}$. In the general case, we can use the fact that a reflection in any axis ℓ can be represented as the composite of a rotation and a reflection in a given axis (in our case it is convenient to take the real line): if the axis ℓ forms an angle of φ with the real line, then we must first perform the rotation $R_O^{-2\varphi}$ through the angle -2φ, followed by reflection in the real axis. The reader will easily convince himself of this from Figure 15.4. Since a rotation and a dilation reduce to multiplication by some complex number z_0, and a reflection in the real line Ox to complex conjugation, our transformation takes z

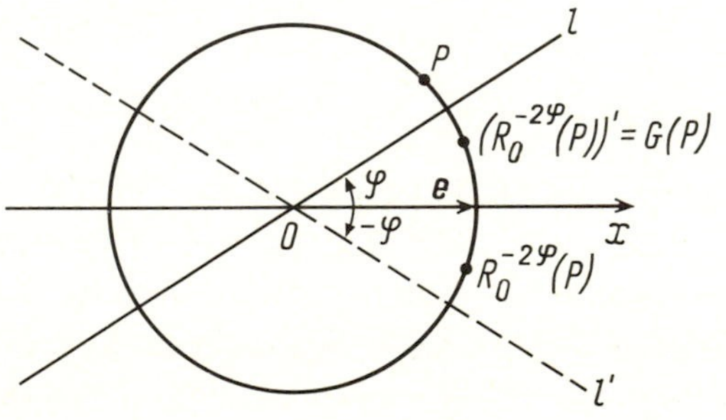

Figure 15.4

to the point $\overline{(z_0 z)}$. We have thus proved the following assertion.

Lemma 1. We consider the points of the plane as complex numbers. Any similarity which fixes O can be written either as the transformation taking z into $z_0 z$, or taking z into $\overline{(z_0 z)}$; here z_0 is some complex number which determines the transformation.

In what follows, when we consider points of the plane as complex numbers (as described above), we will speak of the plane as the *complex plane*. We recall that to have this representation of the points of the plane we must choose a point O of the plane (from which vectors are laid off), and a unit vector **e** (from which angles are measured anticlockwise).

The reader can get a more detailed introduction to complex numbers for example from the book: L.V. Ahlfors, Complex analysis, McGraw-Hill, New York 1979.

Exercises

1. Prove that for fixed complex numbers a and b, with $a \neq 0$, the map of the complex plane taking z into $az + b$ or $a\bar{z} + b$ is a similarity; prove that every similarity of the plane can be written in this way.
2. Prove that every similarity which is not a motion has exactly one fixed point.
3. Prove that all the similarities of the plane are exhausted by those which have already appeared: motions, composites $R_O{}^{\varphi}H_O{}^{\lambda}$ of a dilation and a rotation centred at the same point O, and composites $S_{\ell}H_O{}^{\lambda}$ of a dilation and a reflection in a line, with the centre of the dilation lying on the axis of the reflection.

15.2. Similarity of lattices and the modular group. Now we can return to the question discussed at the beginning of the previous section. We saw in §14.1 that a pair of non-collinear vectors considered up to similarity is specified by a point of the upper half-plane. We choose the vector **e** related to this half-plane as the unit vector, and considering the points of the half-plane as complex numbers, call it the *complex upper half-plane*. The condition that a complex number z belongs to the complex upper half-plane is obviously written as $\operatorname{Im} z > 0$. As a pair of vectors represented by z, we can take $(1, z)$, or any pair similar to it. Now we write down the condition that complex numbers z_1 and z_2 represent similar lattices. According to §14, Theorem 1, a pair of vectors generating the same lattice as $(1, z_1)$ can be written $(m+nz_1, k+\ell z_1)$. To find the point of the upper half-plane which represents the lattice generated by these vectors, we have to perform a similarity satisfying two conditions: (a) it takes the first vector to 1; (b) it takes the second vector to some vector of the upper half-plane (which will then give us the required number). According to Lemma 1, any similarity is determined by a complex number z_0, and two cases are possible: it takes any complex number z into z_0z or $\overline{(z_0z)}$. First of all let's try to satisfy condition (a), so that our similarity takes the first vector to 1. This means that $z_0(m + nz_1) = 1$ in the first case, or $\overline{(z_0(m + nz_1))} = 1$ in the second; since by good fortune $\bar{1} = 1$, these two conditions are the same. This determines the value of z_0 for us: $z_0 = (m + nz_1)^{-1}$.

Hence the similarity takes any complex number z into either $z/(m + nz_1)$ or $\overline{z/(m + nz_1)}$. Which of these two cases happens is determined by condition (b), which requires that the similarity takes the second vector $k + \ell z_1$ into the upper half-plane. Now $k + \ell z_1$ is taken to $(k+\ell z_1)/(m+nz_1)$ in the first case, or its conjugate $\overline{(k + \ell z_1)/(m + nz_1)}$ in the second; using the properties (5) of complex conjugation, we get

$$\overline{\left(\frac{k + \ell z_1}{m + nz_1}\right)} = \frac{k + \ell\overline{z_1}}{m + n\overline{z_1}}.$$

Since the condition for z to belong to the upper half-plane is $\operatorname{Im} z > 0$, we must study $\operatorname{Im} (k+\ell z_1)/(m+nz_1)$. For this, we get rid of the complex number in the denominator by multiplying top and bottom by the complex conjugate:

$$\frac{k + \ell z_1}{m + nz_1} = \frac{(k + \ell z_1)(m + n\overline{z_1})}{(m + nz_1)(m + n\overline{z_1})} = \frac{km + \ell n z_1\overline{z_1} + \ell m z_1 + kn\overline{z_1}}{|m + nz_1|^2}.$$

Since $z_1 = \operatorname{Re} z_1 + i \operatorname{Im} z_1$, we finally get:

$$\frac{k + \ell z_1}{m + nz_1} = \frac{km + \ell n z_1\overline{z_1} + \ell m z_1 + kn\overline{z_1}}{|m + nz_1|^2} + \left(\frac{(\ell m - kn) \operatorname{Im} z_1}{|m + nz_1|^2}\right)i.$$

Now z_1 was chosen in the upper half-plane, so that $\operatorname{Im} z_1 > 0$. Therefore, $\operatorname{Im} (k+\ell z_1)/(m+nz_1)$ has the same sign as $\ell m - kn$. We have seen that $\ell m - kn = \pm 1$. It follows that if $\ell m - kn = 1$, then $\operatorname{Im} (k+\ell z_1)/(m+nz_1) > 0$, so that $(k+\ell z_1)/(m+nz_1)$ is in the upper half-plane. Conversely, if $\ell m - kn = -1$, then the conjugate number

$$\overline{\left(\frac{k + \ell z_1}{m + nz_1}\right)} = \frac{k + \ell\overline{z_1}}{m + n\overline{z_1}}.$$

is in the upper half-plane. Thus we have obtained the rule formulated in the following theorem.

Theorem 1. The necessary and sufficient condition for numbers z_1 and z_2 lying in the complex upper half-plane to represent similar lattices is that z_2 is obtained from z_1 by one of the two methods:

$$z_2 = \frac{k + \ell z_1}{m + nz_1} \quad \text{with } \ell m - kn = 1, \tag{6}$$

or

$$z_2 = \frac{k + \ell \overline{z_1}}{m + n\overline{z_1}} \quad \text{with } \ell m - kn = -1, \tag{7}$$

where m, n, k and ℓ are integers.

We arrive at a situation which is reminiscent of the discussion of §7, when we defined the geometry Σ_Γ corresponding to a uniformly discontinuous group Γ. There we started by considering an arbitrary notion of equivalence of points in the plane, and then went on to consider the especially important case when this equivalence is defined in terms of motions. In the present situation, we are only considering points of the upper half-plane (or the complex numbers corresponding to them), and the following notion of equivalence: z_1 and z_2 are equivalent if they represent similar lattices. Theorem 1 establishes that this notion of equivalence can be defined by means of certain transformations of the upper half-plane, namely those given by (6) and (7). The analogy with §7 suggests that our set of transformations should be a group. In fact, a check shows that this is the case. Suppose, for example, that we have two transformations of the type of (6): one given by (6), and another by

$$z_3 = (k' + \ell' z_2)/(m' + n' z_2) \quad \text{with } \ell' m' - k' n' = 1.$$

Substituting (6) in this shows that $z_3 = (k'' + \ell'' z_1)/(m'' + n'' z_1)$, where $k'' = k'm + \ell' k$, $\ell'' = k'n + \ell'\ell$, $m'' = m'm + n'k$ and $n'' = m'n + n'\ell$. From this we deduce easily that

$$\begin{aligned} \ell'' m'' - k'' n'' &= knn'k' + \ell m\ell' m' - \ell mk'n' - kn\ell' m' = \\ &= (\ell m - kn)(\ell' m' - k'n') = 1. \end{aligned}$$

In exactly the same way it can be checked that the composite of transformations of types (6) and (7) (in either order) gives a transformation of type (7), and that a composite of two transformations of type (7) gives a transformation of type (6). Finally, the inverse of a transformation of type (6) or (7) will be a transformation of the same type with coefficients $\ell' = m$, $k' = -k$, $m' = \ell$, $n' = -n$. The group consisting of all transformations of type (6) and (7) is called the *modular group*.

Now we can give a much more substantial interpretation of the preceding theorem by observing that two points of the upper half-plane represent similar lattices if and only if they are equivalent under the modular group; and the modular figure turns out to be a fundamental domain for the modular group (in the same way as the strip is a fundamental domain for a group of motions of Type II.a and a parallelogram for a group of Type III.a, ...).

Furthermore, it follows from the main assertion of §14.1 that the modular figure does not contain distinct equivalent ponts (even on the boundary!).

The analogy with the situation which we have met many times in this book fails in two respects only: firstly, we are interested in the set of points of the upper half-plane, and it is on these that the modular group acts; but this set does not (as yet) form any kind of geometry; secondly, it is not clear whether the transformations of the modular group form a discrete group of motions. For instance, if we define distance in the upper half-plane in the usual way (as in the plane) then we get a geometry (this follows from the fact that a half-plane is convex), but the transformations of the modular group will not preserve this distance, as you can see by considering the element F of the modular group given by $F(z) = -1/z$.

Nevertheless, a geometry can be constructed, and this will occupy us in the final two sections.

It is interesting to give at once some idea of the extremely important consequences which this construction will have. First of all, since the modular group will be a discrete group of motions of this geometry, by the general construction of §7 and §12 of a geometry from a discrete group, it will correspond to a certain geometry Σ. This geometry Σ will give another description of the set of geometries on the torus (or of lattices in the plane) up to similarity, since the points of Σ are nothing but the sets of equivalent points.

In consequence we will have two descriptions of the set of geometries on the torus: the modular figure and the geometry Σ. This will be the key to understanding what the modular figure really is. As we will see, the modular figure will be the geometry Σ, in the same way that in §12, the geometries corresponding to the discrete groups of motions of the plane of Types II.3-4 and III.3-6 are figures in the plane (see Figure 12.16).

Another most important consequence will be the fact that the set of all geometries on the torus, considered up to similarity, will itself be an object of the same nature - the *geometry* Σ. The distance between points A and B of this geometry Σ will measure how close are the properties of the geometries on the torus represented by points A and B.

These corollaries justify the effort which we will have to spend on the construction of these geometries: on the upper half-plane, and on Σ.

Exercises

1. Consider the transformations F of the modular group which take the modular figure into its neighbouring figures, that is, such that its image under F intersects it in at least one point. Prove that there are 8 such transformations $F_1 - F_8$, given by the equations listed below, and that they take the modular figure I respectively into the figures I - VIII of the upper half-plane (see Figure 17.2, p. 239):

Equation of transformation	Fixed points
$F_1(z) = z,$	all points;
$F_2(z) = -\bar{z},$	points of the ray $\operatorname{Re} z = 0$;
$F_3(z) = 1/\bar{z},$	points of the semicircle $\|z\| = 1$;
$F_4(z) = -\bar{z} + 1,$	points of the ray $\operatorname{Re} z = 1/2$;
$F_5(z) = F_3(F_2(z)) = -1/z$	the point i;
$F_6(z) = F_3(F_4(z)) = 1/(1-z),$	the point $(1 + i\sqrt{3})/2$;
$F_7(z) = F_4(F_3(z)) = (z-1)/z,$	the point $(1 + i\sqrt{3})/2$;
$F_8(z) = F_4F_6(z)) = \bar{z}/\bar{z}-1),$	the point $(1 + i\sqrt{3})/2$.

2. Prove that the points of the modular figure which are fixed under one of the transformations $F_2 - F_8$, as listed above, correspond precisely to the lattices having symmetries other than central reflections; and that a transformation $F_2 - F_8$ which fixes one of these points z corresponds exactly to a symmetry of the lattices defined by z, considered up to composition with a central reflection (compare §11.3 and §14.1, Exercise 1).

§16. Lobachevsky geometry

The aim of this section is to construct a geometry on the complex upper half-plane for which the modular group will be a discrete group of motions.

As we know from §6, the basic notion in the definition of a geometry is that of distance, and all the remaining notions, such as lines, angles, group of motions and so on, are defined in terms of distance. In the present case however, we are given a certain group of transformations, the modular group, and we would like to construct a geometry, that is, define a distance, in such a way that transformations of this group are motions. In this situation it is most natural for us to start by constructing the group of motions of the future geometry. We define a certain group of transformations of the upper half-plane, which is similar in many of its properties to the group of motions of the Euclidean plane. Of course, so far we have not introduced any kind of distance between points of the upper half-plane, so that it is meaningless to ask whether the transformations belonging to the group are motions. But our aim is to define distance precisely so that this property holds. To emphasise that we have this in mind all the time, the group elements will be called 'motions', in quotation marks; once we have defined a distance and proved that it is preserved by the transformations belonging to our group, we will have won the right to remove the quotation marks. As a second stage, we define on the upper half-plane certain curves, related to our 'motions' by certain properties which are very similar to those relating motions and lines in the Euclidean plane; for this reason these curves will be called 'lines', also in quotation marks. Finally, we would like to define distance in such a way that it is related to 'motions' and 'lines' by the same kind of properties as those relating distance to motions and lines in Euclidean geometry. As we will see, this is possible, and in an essentially unique way. This is the plan of our investigations in this section.

16.1. 'Motions'. Notice that if in the equations (6) and (7) of §15.2, Theorem 1, which describe transformations belonging to the modular group, the coefficients k, ℓ, m, n are allowed to be any real numbers, rather than just integers, then we also get transformations of the upper half-plane, and these form a certain group $\mathcal{G}$; this is proved exactly as for the transformations belonging to the modular group. Thus we get a certain group $\mathcal{G}$ of transformations of the upper half-plane, which take z into

$$\frac{a + bz}{c + dz} \quad \text{if} \quad bc - ad > 0 \qquad (1)$$

or into

$$\frac{a + b\bar{z}}{c + d\bar{z}} \quad \text{if} \quad bc - ad < 0 \qquad (2)$$

for real number a, b, c and d. Obviously, $\mathcal{G}$ contains the modular group.

We will construct the new geometry in such a way that after we have defined distance, the elements of $\mathcal{G}$ will be its motions; this will certainly ensure that the transformations belonging to the modular groups will be motions. Moreover, as we will constantly be seeing, the group $\mathcal{G}$ is very similar in its properties to the group of motions of the Euclidean plane.

The transformations given by (1) and (2) will be called 'motions'; this is justified by the fact that we will subsequently introduce a distance which they will preserve.

We start from the fact that just as in the Euclidean plane, 'motions' divide into two kinds: motions of the first kind, given by (1), and of the second kind, given by (2). As in the case of the modular group, a composite of two 'motions' of the first kind is again of the first kind, whereas a composite of a 'motion' of the first kind with one of the second kind is of the second kind.

The fixed points of 'motions' in $\mathcal{G}$ look just like those of motions of the Euclidean plane. 'Motions' of the first kind either have no fixed points (these are the analogues of translations), or have exactly one fixed point (the analogue of rotations), or fix every point of the upper half-plane (the identity). We leave the verification of these properties to the reader; it reduces to considering the quadratic equation $(a + bz)/(c + dz) = z$. 'Motions' of the second kind either have no fixed points (these are the analogues of glide reflections), or have a whole curve of fixed points in the upper half-plane (the analogue of reflections in lines). We now check this property, at the same time determining what the fixed-point curves are.

Suppose that a 'motion' given by (2) fixes some point z of the upper half-plane. This means that

$$z = \frac{a + b\bar{z}}{c + d\bar{z}} \quad \text{with} \quad bc - ad < 0.$$

Multiplying both sides of the equation by $c + d\bar{z}$, and using the fact that $z\bar{z} = |z|^2$, we get

$$d|z|^2 + cz = a + b\bar{z}.$$

Let $z = x + iy$; then equating imaginary parts of both sides of this equation, we get $(b + c)y = 0$. Now z is in the upper half-plane, so that $y > 0$, and hence $b + c = 0$.

Thus $c = -b$. If $d \neq 0$, then divide the numerator and denominator of (2) by d, and write $\alpha = a/d$, $\beta = b/d$; dividing the inequality $bc - ad < 0$ by d^2 gives $\beta^2 + \alpha > 0$. Now the formula for our 'motion' takes the form $(\alpha + \beta\bar{z})/(-\beta + \bar{z})$. Adding and then subtracting β^2 to the numerator, we get the same formula in the form

$$\frac{\alpha + \beta^2 + \beta(\bar{z} - \beta)}{\bar{z} - \beta} = \beta + \frac{k}{\bar{z} - \beta}, \tag{3}$$

where $k = \alpha + \beta^2 > 0$. The condition that z remains fixed under the 'motion' takes the form $z - \beta = k/(\bar{z} - \beta)$, that is $|z - \beta|^2 = k$. Obviously this is the equation of a circle centred at the point β of the real line Ox and of radius $\sqrt{k}$. Since we are only interested in points of the upper half-plane, we should only consider the upper half of this circle (Figure 16.1), that is, a semicircle.

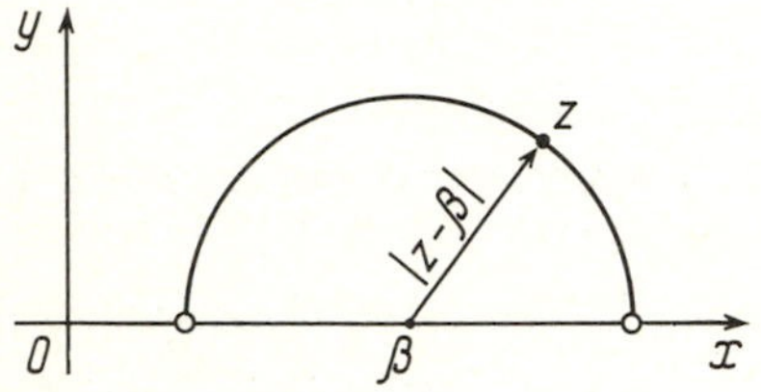

Figure 16.1

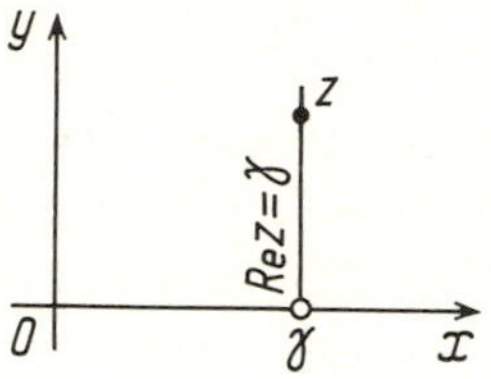

Figure 16.2

This semicircle is the required fixed-point set of the 'motion'. It is distinguished among all semicircles by the fact that it is orthogonal to the real line. (The angle between a line and a circle is defined to be the angle between this line and the tangent line to the circle at the point of intersection; orthogonal is defined accordingly. Note that a circle is orthogonal to the real line if and only if its centre is on the real line.)

If $d = 0$ then from $bc - ad < 0$ it follows that $b \neq 0$, and the 'motion' (2) is given by

$$2\gamma - \bar{z} = \gamma - (\bar{z} - \gamma), \tag{4}$$

where $\gamma = -a/(2b)$. In this case the required fixed-point set of the 'motion' (4) is given by the equation $2\gamma - \bar{z} = z$, that is $\gamma = \operatorname{Re} z$, and is the ray from γ orthogonal to the real line (Figure 16.2). Notice that geometrically this 'motion' is just a reflection in the ray.

Thus we have prove our assertion: if a 'motion' of the second kind has fixed points, then its fixed-point set is a curve in the upper half-plane; this curve will be called the fixed-point curve of the 'motion'. Furthermore, we have given a complete description of all such curves: the fixed-point curves of 'motions' of the second kind are exactly the semicircles and rays of the upper half-plane orthogonal to the real line (see Figures 16.1-2).

By analogy with the fact that a reflection of the ordinary plane is uniquely determined by its axis of symmetry, a 'motion' in $\mathcal{G}$ of the second kind is uniquely determined by its fixed-point curve. This follows at once from (3) and (4). Therefore such a 'motion ' will again be called a reflection in this curve. If the fixed-point curve is a ray, then as we saw above, the reflection is just the ordinary reflection of the plane in the line containing the ray, but considered only on the upper half-plane. If the fixed-point curve is a semicircle, then reflection in it also has a simple geometrical meaning; see Exercise 1.

Exercises

1. Prove that the reflection (3) in the semicircle $|z - \beta|^2 = k$ has the following geometrical meaning (inversion in a circle): a point z_1 and its image z_2 lie on the same ray emanating from the centre β of the semicircle, and $|z_1 - \beta|\cdot|z_2 - \beta| = k$. This reflection interchanges the inside and outside of the semicircle in the upper half-plane.

2. Prove that any 'motion' belonging to $\mathcal{G}$ is a composite of the following 4 'motions': (i) z goes to pz; (ii) z goes to $z + q$; (iii) z goes to r/z; (iv) z goes to $-\bar{z}$, where p, q and r are real numbers, and $p > 0$, $r < 0$. [Hint: if $d \neq 0$, write $(a + bz)/(cz + d)$ in the form $p + r/(z-q)$.]

3. Prove that every 'motion' is a composite of a number of reflections; a 'motion' of the first kind is a composite of evenly many reflections, and one of the second kind of oddly many. Deduce from this that a 'motion' of the first kind preserves the sense of going round a curve, whereas a 'motion' of the second kind reverses it.

16.2. 'Lines'. In introducing lines, we base ourselves on the following observation: if a motion of the Euclidean plane of the second kind has a fixed point (so is a reflection), then its fixed-point curve is a line, the axis of symmetry of the reflection. It would be natural if the same property also held in the new geometry, that is, if rays and semicircles orthogonal to the real line, which are just the fixed-point curves of such 'motions', were lines in it. The situation becomes clearer if we try to prove in Euclidean geometry, as far as possible without using concrete properties of the geometry, the fact that a fixed-point set of a motion which is a curve is a line.

Given a motion F of the plane, consider two distinct fixed points A and B of F. In Euclidean geometry, there is a line (AB) through A and B. Since F is a motion, it takes (AB) into a line passing through $F(A) = A$ and $F(B) = B$, that is, into itself, since in the Euclidean plane there is just one line through the two distinct points A and B. Moreover, F takes any point X of (AB) into itself: $F(X) = X$, since the position of X on (AB) is determined by its distances to A and B, and since F is a motion, $|F(X)A| = |XA|$ and $|F(X)B| = |XB|$. Hence the line (AB) is the fixed-point curve of F.

From the above proof one sees that the only property of the Euclidean plane we have used is the existence and uniqueness of the line through two distinct points. We now note that rays and semicircles of the upper half-plane orthogonal to the real line obviously satisfy this property: through any two distinct point A and B of the half-plane there is exactly one such curve (Figure 16.3).

After these observations, it becomes natural to construct the new geometry in such a way that it has one and only one 'line' through any two distinct points. As

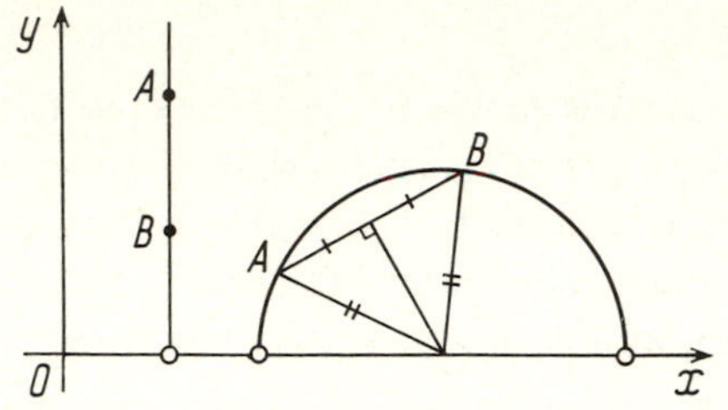

Figure 16.3

we have just seen, this implies that the 'lines' of the new geometry must be exactly the rays and semicircles of the upper half-plane orthogonal to the real line.

However, we must see that defining the 'lines' of the new geometry in this way does not contradict the group of 'motions' $\mathcal{G}$, which we have already defined; that is, we must check that 'motions' of $\mathcal{G}$ take 'lines' (rays and semicircles of the upper half-plane orthogonal to the real line) into 'lines', (that is, into these same rays and semicircles).

For this, recall that every such 'line' C is a fixed-point set of some 'motion'

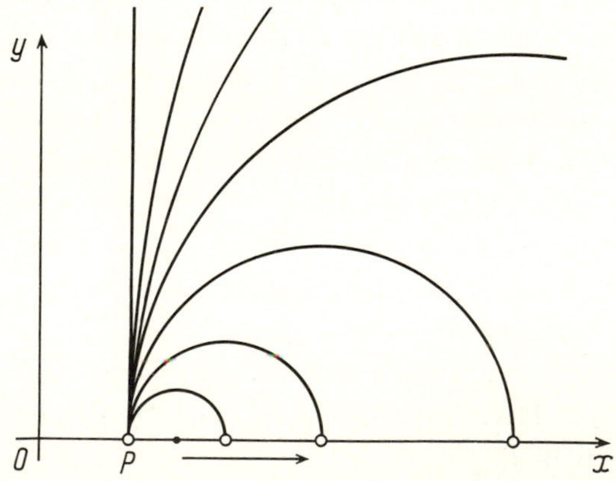

Figure 16.4

F of the second kind (in §16.1 we called F the reflection in C). Now suppose that G is any motion in $\mathcal{G}$. Notice that the image G(C) of the 'line' C under G is the fixed-point set of the 'motion' GFG^{-1} belonging to $\mathcal{G}$. Indeed, if A is a fixed point of GFG^{-1}, that is, $GFG^{-1}(A) = A$, then applying G^{-1} to either side, we get $F(G^{-1}(A)) = G^{-1}(A)$; this means that $G^{-1}(A)$ is a fixed point of the 'motion' F, hence a point of C, so that A is a point of G(C). Note also that GFG^{-1} is a 'motion' of the second kind in view of the rules already mentioned for composing motions in $\mathcal{G}$ of the first and second kinds. Thus we have verified that G(C) is the fixed-point curve of some 'motion' in $\mathcal{G}$ of the second kind. But as we know from §16.1, such curves are rays and semicircles of the upper half-plane orthogonal to the real line, that is, our 'lines'. This proved what was required.

After this necessary verification, we finally arrive at the definition underlined above of 'lines' of the geometry under construction. However, as with the 'motions' of $\mathcal{G}$, until we have a definition of distance, we will continue to refer to rays and semicircles of the upper half-plane orthogonal to the real line as 'lines' of the new geometry, in quotation marks.

At first sight it might seem that 'lines' belong to two families of curves of very different kinds, rays and semicircles. But there is a connection between the two: if we move the centre of a semicircle away to infinity (Figure 16.4, infinity is to the right) keeping one point P of intersection of the semicircle with the real line fixed, then the semicircles will become straighter and straighter, tending to the ray through P (see Figure 16.4). Thus rays are limiting positions of semicircles.

Exercise

1. The angle betwen two curves at a point of intersection is defined as the angle between their tangent lines at the point. Prove that the angle between two intersecting semicircles C_1 and C_2 orthogonal to the real line is not changed by a 'motion' in $\mathcal{G}$. [**Hint:** suppose that C_1 and C_2 meet the real line in α, β and γ, δ respectively; express the cosine of the angle of intersection of C_1 and C_2 in terms of $\alpha, \beta, \gamma, \delta$; then use §16.1, Exercise 2.]

Deduce from this that 'motions' in $\mathcal{G}$ do not change angles between arbitary intersecting curves. In view of this, we will define the angle between two curves of the new geometry at a point of intersection to be the angle (in the sense of Euclidean geometry) between their tangents.

16.3. Distance. Here we finish the construction of our geometry by defining the distance between its points. To avoid confusion with the ordinary distance between points in the plane, we will denote the new distance by $\|z_1, z_2\|$, where z_1, z_2 are points of the upper half-plane.

First of all we define the distance between points of a 'line' of the new geometry. The simplest such 'line' is the ray L in the upper half-plane given by $\operatorname{Re} z = 0$, consisting of points of the form it, with $t > 0$ a real number.

We write out the conditions which the definition of distance should satisfy. The first of these is the condition that L should be a 'line' of the new geometry:

(I) $$\|it_1, it_3\| = \|it_1, it_2\| + \|it_2, it_3\|,$$

whenever the point it_2 lies between it_1 and it_3 in L, that is, whenever $0 < t_1 \le t_2 \le t_3$ (Figure 16.5). The second is that the distance should not change under 'motions' belonging to $\mathcal{G}$, that is

(II) $$\|H(z_1), H(z_2)\| = \|z_1, z_2\|,$$

where H is a 'motion' in $\mathcal{G}$ such that all four points $z_1, z_2, H(z_1), H(z_2)$ belong to L. We can assume that $z_1 \neq z_2$, hence $H(z_1) \neq H(z_2)$, since otherwise we would have $z_1 = H^{-1}(H(z_1)) = H^{-1}(H(z_2)) = z_2$.

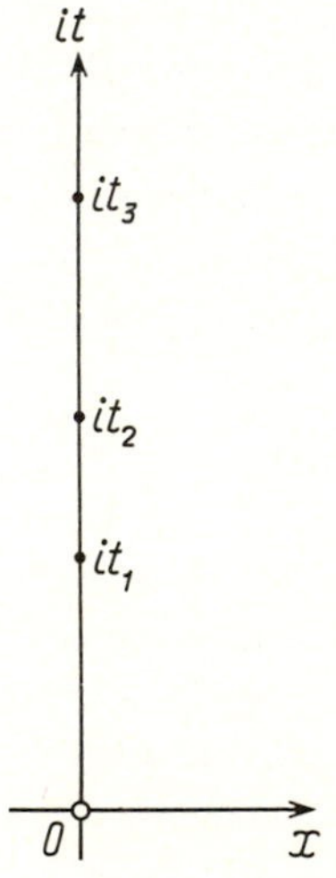

Figure 16.5

To write out condition (II), let's find explicitly all pairs of distinct points $z_1 \neq z_2$ and all 'motions' H such that z_1, z_2 and $H(z_1), H(z_2)$ all belong to L. For this, we note first that the 'motion' H takes the whole ray L into itself. Indeed, any 'motion' H takes the 'line' of the new geometry given by L into a 'line', which is then a ray or semicircle of the upper half-plane orthogonal to the real line Ox. This 'line' contains the points $H(z_1)$ and $H(z_2)$, since L contains z_1 and z_2; it follows from this that it is just L, since the two points $H(z_1)$ and $H(z_2)$ are distinct and lie on L, and there is exactly one 'line' through two distinct points.

It is easy to write out all these 'motions' H in $\mathcal{G}$ taking L, the ray Re z = 0 into itself. It turns out that four cases are possible:

(a) $H(z) = \beta z$, where $\beta > 0$;

(b) $H(z) = -\alpha/z$, where $\alpha > 0$;

(c) $H(z) = -\beta\bar{z}$, where $\beta > 0$;

(d) $H(z) = \alpha/\bar{z}$, where $\alpha > 0$.

Here the cases (a) and (b) correspond to H given by (1), whereas (c) and (d) correspond to (2). We restrict ourselves to the case that H is given by (1). From (1) we get at once

$$H(it) = \frac{a + ibt}{c + idt} = \frac{(a + ibt)(c + idt)}{c^2 + d^2t^2} = \frac{ac + bdt^2}{c^2 + d^2t^2} + \left(\frac{(bc - ad)t}{c^2 + d^2t^2}\right) i .$$

Note that since $\operatorname{Re} H(it) = 0$ for all $t > 0$, we must have $ac + bdt2 = 0$ for all $t > 0$, therefore $ac = bd = 0$. Using the fact that $bc - ad > 0$, we get two cases: either

$a = d = 0$ and $bc > 0$, so $H(z) = bz/c = \beta z$, where $\beta = b/c > 0$;

or

$b = c = 0$ and $ad < 0$, so $H(z) = a/(dz) = -\alpha/z$, where $\alpha = -a/d > 0$.

We now proceed to the analysis of condition (II). Writing the point z of the ray L given by Re z = 0 as z = it, with t > 0, and substituting in (II) all possible H given by formulas (a - d), we can rewrite condition (II) as two equations:

$$\|it_1, it_2\| = \|i\beta t_1, i\beta t_2\|$$

and

$$\|it_1, it_2\| = \|i(\alpha/t_1), i(\alpha/t_2)\|,$$

which must hold for all $\alpha, \beta > 0$ and all $t_1, t_2 > 0$. Now let us take logarithms of t_1, t_2, α and β, that is, introduce $u = \log_A t_1$, $u' = \log_A t_2$, $r = \log_A \alpha$, and $s = \log_A \beta$, where $A > 0, A \neq 1$; we write $\rho(u, u') = \|iA^u, iA^{u'}\| = \|it_1, it_2\|$. Then the previous relation takes on the familiar form

$$\begin{aligned} \rho(u, u') &= \rho(r + u, r + u'), \\ \rho(u, u') &= \rho(s - u, s - u'), \end{aligned} \tag{5}$$

where u, u′, r and s are any four real numbers. The relation (I) takes the form

$$\rho(u, u'') = \rho(u, u') + \rho(u', u''), \tag{6}$$

for any $u \le u' \le u''$.

Properties (5) and (6) are well known to be satisfied by distance on the real line, that is, the absolute value of the difference. Thus (I) and (II) will be satisfied if we set

$$\rho(u, u') = |u - u'|,$$

and this gives

$$\|it_1, it_2\| = \rho(\log_A t_1, \log_A t_2) = |\log_A t_1 - \log_A t_2| = |\log_A t_1/t_2|.$$

(Although we do not need this, the reader can verify that, up to multiplication by a constant, that is, up to the choice of the base of logs, this function is the unique possibility.) Observing that $t_1/t_2 = (it_1)/(it_2)$, we get the required distance

$$\|z_1, z_2\| = |\log_A z_1/z_2| \tag{7}$$

for points z_1 and z_2 of L, the ray $\operatorname{Re} z = 0$, where $A \neq 1$ is some fixed positive number.

We now consider how to define the distance between arbitrary points of the geometry under construction. Our aim is to define the distance $\|z_1, z_2\|$ between points of the upper half-plane to satisfy the following conditions:

(A) 'Motions' in $\mathcal{G}$ should actually be motions, that is, they should preserve the distance $\|z_1, z_2\|$.
(B) For points of L, the distance should be given by (7).

We show below that for any two points z_1, z_2 of the upper half-plane, there exists a 'motion' F in $\mathcal{G}$ taking z_1 and z_2 into points of L, for which the distance is given by (7). It obviously follows from this that under conditions (A) and (B) the distance $\|z_1, z_2\|$ between points z_1 and z_2 can be defined in one way only: we must choose a 'motion' F in $\mathcal{G}$ such that both $F(z_1)$ and $F(z_2)$ belong to L, and then set the required distance to be equal to

$$\|z_1, z_2\| = \|F(z_1), F(z_2)\| = |\log_A F(z_1)/F(z_2)|. \tag{8}$$

Defining distance in this way, we need only verify the following:

(i) for any two points z_1 and z_2 there exists a 'motion' F taking them both into L;

(ii) the definition of distance between z_1 and z_2 given in (8) does not depend on the choice of this 'motion' F;

(iii) 'motions' of $\mathcal{G}$ preserve this distance.

We leave the verification in (i) for the time being, and prove first properties (ii) and (iii), which have actually already been checked.

Let z_1 and z_2 be given points, and suppose that F and G are motions belonging to $\mathcal{G}$ such that the two pairs $F(z_1), F(z_2)$ and $G(z_1), G(z_2)$ both lie on L. To prove (ii) we need to show that the distances between points of L defined in (8) are equal, that is, that $\|F(z_1), F(z_2)\| = \|G(z_1), G(z_2)\|$. This however follows from property (II) of the distance between points of L, since $G(z_1) = HF(z_1)$, $G(z_2) = HF(z_2)$, where $H = GF^{-1}$ is a 'motion' in $\mathcal{G}$, and the points $F(z_1), F(z_2)$, $G(z_1), G(z_2)$ all lie on L. This proves (ii).

Now consider any two points z_1 and z_2 of the upper half-plane, and an arbitrary 'motion' P in $\mathcal{G}$. To prove (iii), we prove that

$$\|z_1, z_2\| = \|P(z_1), P(z_2)\|. \tag{9}$$

By definition (8),

$$\|P(z_1), P(z_2)\| = \|Q(P(z_1)), Q(P(z_2))\|, \tag{10}$$

where Q is some 'motion' in $\mathcal{G}$ for which both of $Q(P(z_1))$ and $Q(P(z_2))$ lie on L. But by the same definition (8), we then have

$$\|z_1, z_2\| = \|QP(z_1), QP(z_2)\| = \|Q(P(z_1)), Q(P(z_2))\|, \tag{11}$$

since both of $QP(z_1) = Q(P(z_1))$ and $QP(z_2) = Q(P(z_2))$ lie on L, and QP is a 'motion' in $\mathcal{G}$. The required equality (9) follows from (10) and (11), which proves (iii).

It remain for us to prove (i). We consider two cases.

Case 1. Suppose that both z_1 and z_2 lie on one ray $\operatorname{Re} z = a$, so that $\operatorname{Re} z_1 = \operatorname{Re} z_2 = a$; then we can obviously take

$$F(z) = z - a \tag{12}$$

as the required 'motion'.

Case 2. Suppose $\operatorname{Re} z_1 \neq \operatorname{Re} z_2$; then, as in Figure 16.3, draw a semicircle C through z_1 and z_2 orthogonal to the real line Ox, and let a and b be the endpoints of this semicircle, on the real line, with $a < b$ (Figure 16.6). We claim that we can take the required motion F to be

$$F(z) = (z - b)/(z - a). \tag{13}$$

For this it is enough to show that F takes the semicircle C into L. We already know from §16.2 that F(C) is either a ray or a semicircle orthogonal to the real line. Hence we only have to see that F(C) can only be the ray $\operatorname{Re} z = 0$, that is, L.

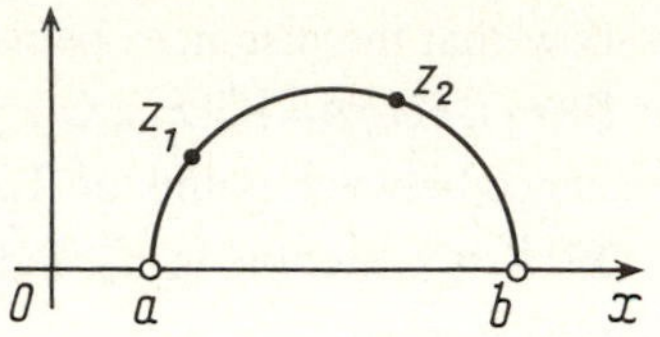

Figure 16.6

Let's show that F(C) can't be a semicircle. Indeed, if z is a variable point of the semicircle C, and tends towards the endpoint a, then the modulus |F(z)| grows without limit, since in (13) the numerator z - b tends to the non-zero number a - b, whereas the denominator z - a tends to 0. But this would be impossible if F(z) belonged to a semicircle.

Therefore F(C) is a ray (see Figure 16.7, (a)). To see that this ray is L, suppose that z is a variable point of the semicircle C, and tends towards its other endpoint b. In a similar way to the above, it follows by the formula (13) for F(z) that F(z) tends to 0 (see Figure 16.7, (b)). For any ray orthogonal to the real line,

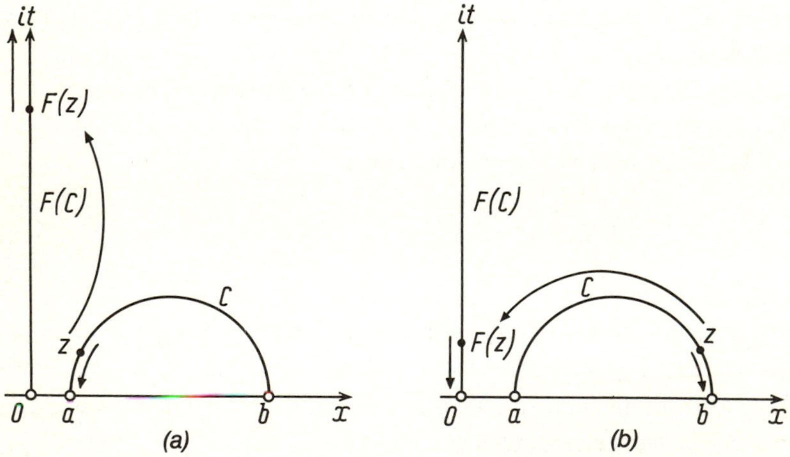

Figure 16.7

this is only possible if the endpoint is 0, that is, the ray is $\operatorname{Re} z = 0$. Thus $F(C)$ is L. This proves (i).

This completes the construction of distance: in the new geometry, distance is given by the definition (8) above, where for F we can take the motion given by (12) or (13), depending on the positions of z_1 and z_2 on the upper half-plane.

We would also like to have an explicit formula for the distance. This can be obtained without difficulty from (8), (12) and (13): the distance betwen points z_1 and z_2 of the geometry is given by one of the following formulas:

$$\|z_1, z_2\| = \left|\log_A \frac{z_2 - a}{z_1 - a}\right| \quad \text{if} \quad \operatorname{Re} z_1 = \operatorname{Re} z_2 = a, \tag{14}$$

or

$$\|z_1, z_2\| = \left|\log_A \frac{z_2 - b}{z_2 - a} : \frac{z_1 - b}{z_1 - a}\right| \quad \text{if} \quad \operatorname{Re} z_1 \neq \operatorname{Re} z_2, \tag{15}$$

where a and b are the endpoints of the semicircle through z_1 and z_2 orthogonal to the real line.

An interesting and useful remark is that in both of these formulas, by our constructions, the argument of log is a positive real number, as it should be. For (15) this is not obvious, since z_1 and z_2 are arbitrary complex numbers of the upper half-plane. The final remark is useful, because it allows us to rewrite (14) and (15) in another form which is often convenient. For this, note that in virtue of this positivity, using properties of modulus of complex numbers, the expressions in the argument of log can be rewritten

$$\frac{z_2 - a}{z_1 - a} = \frac{|z_2 - a|}{|z_1 - a|} \tag{16}$$

in one case, or

$$\frac{z_2 - b}{z_2 - a} : \frac{z_1 - b}{z_1 - a} = \frac{|z_2 - b|}{|z_2 - a|} : \frac{|z_1 - b|}{|z_1 - a|} \tag{17}$$

in the other. In turn, the modulus can be considered as the distance between the points correponding to complex numbers, for example, $|z_2 - a|$ is the distance from z_2 to a in the complex plane.

Exercises

1. Prove that the group of motions $\mathcal{G}$ has the following crucial property: if $z_1 \neq z_2$ and $w_1 \neq w_2$ are four given points such that $\|z_1, z_2\| = \|w_1, w_2\|$ then there exist motions belonging to $\mathcal{G}$ taking z_1 to w_1 and z_2 to w_2, and in fact exactly two such motions, one of the first kind and one of the second. [Hint: reduce to the case of points lying on a ray.]

2. Prove that the distance of (14) and (15) can also be given by the formula

$$\|z_1, z_2\| = \log_A\left(\frac{1+t}{1-t}\right), \qquad \text{where} \quad t = \left|\frac{z_1 - z_2}{z_1 - \overline{z}_2}\right|.$$

[Hint: prove this first for points of L, then prove that $|(z_1-z_2)/(z_1-\overline{z}_2)|$ does not change under motions in $\mathcal{G}$.]

3. Prove that the circles in the sense of the distance (14) and (15), that is, the set of points equidistant from a given point, are represented in the upper half-plane by ordinary circles (in the sense of the Euclidean distance) not intersecting the real line. [**Hint:** you can use the preceding exercise.] What is the effect of motions in $\mathcal{G}$ on ordinary circles of the upper half-plane?

4. Prove that four distinct points u_1, u_2, u_3 and u_4 of the complex plane lie on a circle or line if and only if the ratio

$$\frac{u_2 - u_4}{u_2 - u_3} : \frac{u_1 - u_4}{u_1 - u_3}$$

is a real number. What is the meaning of its sign?

16.4. Construction of the geometry concluded. It remains for us to check properties (a - d) of the definition of a geometry in §6.

First of all, notice that all of these properties involve two points z_1 and z_2 of the geometry. Corresponding to the definition of distance in the new geometry given in §16.3, checking these properties for z_1 and z_2 can be replaced by checking them for $F(z_1)$ and $F(z_2)$, where F is a motion in $\mathcal{G}$ for which $F(z_1)$ and $F(z_2)$ both lie on the ray L, given by $\operatorname{Re} z = 0$, since F does not alter distances between any two points. This proves properties (a), (b) and (d), since by construction of distance in §16.3, L is a line, for which all these properties are obviously satisfied.

Checking property (c) of a geometry reduces to proving the inequality

$$\|z_1, z\| + \|z, z_2\| \geq \|z_1, z_2\|, \tag{18}$$

where z_1 and z_2 are points of L. Note first of all that this inequality is obvious for $z_1 = z_2$, since then $\|z_1, z_2\| = 0$. Moreover, if z belongs to the interval $[z_1, z_2]$ of L, the inequality turns into the equality

$$\|z_1, z\| + \|z, z_2\| = \|z_1, z_2\|;$$

this means that the ray itself will be a line through z_1 and z_2.

Hence in checking (18), we will from now on assume that z_1 and z_2 are two distinct points of L, and that z does not belong to the interval $[z_1, z_2]$ of this ray. In this case we will even prove the strict inequality

$$\|z_1, z\| + \|z, z_2\| > \|z_1, z_2\|. \tag{19}$$

This final inequality proves in addition that there is a unique shortest curve joining the points z_1 and z_2 of the geometry, namely L. In this case, applying a motion F in $\mathcal{G}$, we deduce that this uniqueness property of the line will hold for any two distinct points of our geometry. We would also like to prove this, because it was assuming it that we arrived in §16.2 at the definition of lines of our geometry.

The idea of the proof of (19) is suggested by one of the methods of proof of the analogous inequality in the Euclidean plane. There the inequality $|AC| + |CB| > |AB|$ for C not belonging to the line segment [AB] can be deduced from the fact

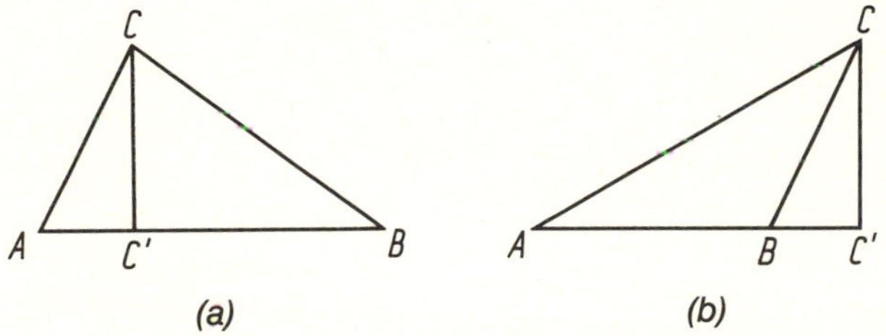

Figure 16.8

that the length of a segment decreases under orthogonal projection (Figure 16.8). Indeed, let C′ be the orthogonal projection of C to the line (AB). If C′ lies between A and B, then $C \neq C'$, and $|AC| > |AC'|$, $|BC| > |BC'|$, so that

$$|AC| + |BC| > |AC'| + |C'B| = |AB|$$

(Figure 16.8, (a)). If on the other hand, say, B lies between A and C′, then

$$|AC| + |BC| \geq |AC'| + |BC'| = |AB| + |BC'| + |BC'| = |AB| + 2|BC'| > |AB|$$

(Figure 16.8, (b)).

In our geometry, in an entirely similarly way, the proof of (19) reduces to the inequality

$$\|z, z_0\| > \|z', z_0\|, \tag{20}$$

where z_0 is a point of L, z is a point not on this ray, and z′ is the point of intersection of L with the semicircle of radius $|z|$ centred at O, which is perpendicular to L and to the real line (Figure 16.9).

Using (15) and (17), we get the following equality:

$$\|z, z_0\| = \left| \log_A \left(\frac{|z - b|}{|z - a|} : \frac{|z_0 - b|}{|z_0 - a|} \right) \right|$$

Since by a well-known property of circles in Euclidean geometry, bza and bz_0a are right-angled triangles,

$$\frac{|z - b|}{|z - a|} = \tan \varphi \quad \text{and} \quad \frac{|z_0 - b|}{|z_0 - a|} = \tan \varphi_0,$$

where φ and φ_0 are the angles of these triangles at the vertex a (Figure 16.9). From this we get

$$\|z, z_0\| = \left| \log_A \frac{\tan \varphi_0}{\tan \varphi} \right|.$$

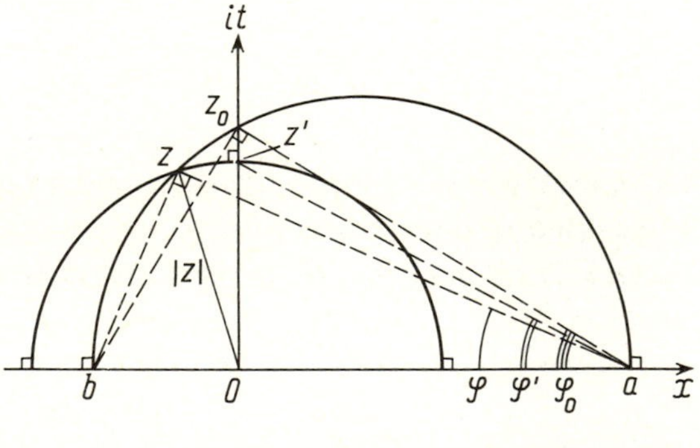

Figure 16.9

From (14) and (16) it follows that

$$\|z', z_0\| = \left| \log_A \frac{|z_0|}{|z'|} \right|.$$

From now on we consider the case $|z_0| \geq |z'|$ (Figure 16.9); the case $|z_0| \leq |z'|$ can be dealt with similarly. In this case it can be seen from the picture that

$\pi/2 > \varphi_0 > \varphi$, from which $\tan \varphi_0 > \tan \varphi$. Now the inequality (20) reduces to

$$\frac{\tan \varphi_0}{\tan \varphi} > \frac{|z_0|}{|z'|},$$

or equivalently,

$$\frac{\tan \varphi_0}{|z_0|} > \frac{\tan \varphi}{|z'|} \qquad (21)$$

Similarly, it can be seen from the picture that $\varphi < \varphi' < \pi/2$, and hence $\tan \varphi < \tan \varphi'$, where φ' is the angle of the triangle $Oz'a$ at the vertex a. Hence instead of (21), it is enough to prove that

$$\frac{\tan \varphi_0}{|z_0|} > \frac{\tan \varphi'}{|z'|}.$$

However, this is obvious, since $\tan \varphi_0/|z_0| = \tan \varphi'/|z'| = 1/|a|$. The proof is complete.

This concludes the construction of our geometry. By construction, $\mathcal{G}$ is a group of motions of the geometry, and as we have seen many times, it is very similar to the group of motions of the Euclidean plane. The lines of the geometry are exactly the rays and semicircles orthogonal to the real line. As on the ordinary plane, through any two distinct points there is exactly one line. This suggests that we should try to check whether or not the other properties of the Euclidean plane are satisfied, for instance, the axioms. We will not go through this verification: although very simple, it is quite tedious, since there are quite a large number of axioms; moreover, in different treatments of geometry the choice of axioms is different. We just state the final result: in our geometry all the axioms of Euclidean geometry are satisfied except for one alone: the fifth postulate of Euclid, or the parallel axiom. Instead of this axiom, our geometry satisfies the opposite assertion, Lobachevsky's axiom: given a line ℓ and a point A not lying on ℓ, there is more than one line through A parallel to ℓ (that is, not meeting ℓ). This can be seen at once from Figure 16.10: any line lying between the lines ℓ_1 and ℓ_2 of our geometry does not intersect ℓ.

Thus we are dealing with the celebrated Lobachevsky geometry, in which all the axioms of Euclidean geometry are satisfied, except for the parallel axiom, and instead of it the opposite, Lobachevsky's axiom, holds. Thus from now on we will call the geometry we have constructed *Lobachevsky geometry* (or the *Lobachevsky plane*). Digressing from our main problem, we note that the construction of Lobachevsky geometry outlined above gives an answer to a question which had worried mathematicians for more than two thousand years: can the fifth postulate, the parallel axiom, be deduced from the other axioms of Euclid?

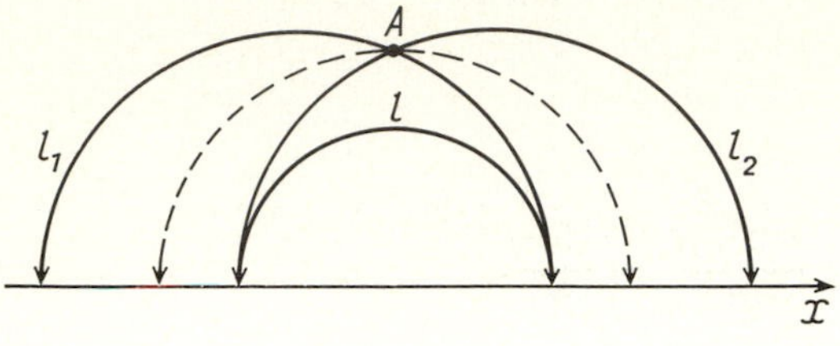

Figure 16.10

For this, note that in our construction, the Lobachevsky plane has appeared as a kind of construction within the Euclidean plane. This shows that if we take certain objects of the Euclidean plane (points of the upper half-plane; rays and semicircles orthogonal to the real line; motions of $\mathcal{G}$), and call these 'points', 'lines', 'motions', then they satisfy all the axioms of the Euclidean plane, except for the parallel axiom, which is replaced by Lobachevsky's axiom. Thus the Lobachevsky plane can be represented by objects of the Euclidean plane. This situation is described by the term *model* : we have constructed a model of the Lobachevsky plane by means of the Euclidean plane. The existence of such a model proves the impossibility of deducing the fifth postulate from the other axioms of Euclid. Indeed, if there were such a deduction, this would contradict the fact that our model satisfies Lobachevsky's axiom. Hence it would lead to a contradiction in the Euclidean plane, since the model was constructed within the Euclidean plane.

Of course, with sufficient intellectual courage one can ask the question: and why is there no contradiction in Euclidean geometry? We have actually already answered this question: the method of coordinates allows us to construct a model of the Euclidean plane in terms of the real numbers (see §6.1, Example 1a); the lines of this geometry can be defined as the set of points given by a linear equation in the coordinates x and y. Because of this, the question reduces to showing that there are no contradictions in the properties of real numbers. Defining real numbers as infinite decimal expansions, we can give a foundation for all of their properties starting from a number of intuitively obvious properties, or the axioms of the integers. Thus our investigation leads us to the problem of the non-existence of contradictions in the axioms of the integers, or the 'consistency of arithmetic'. It would seem that nothing logically simpler than the integers can be thought of, and hence the method of models is already inapplicable here. The answer however turns out to be completely different from all that has appeared so far: in mathematical logic, it is proved that it is impossible to prove the consistency of arithmetic starting from the axioms of arithmetic and logic! Thus the consistency of arithmetic is a 'postulate of logic', which we accept on the basis of faith in the whole practical and intellectual experience of mankind.

The reader can get to know other question of Lobachevsky geometry, for example from the books:

H.S.M. Coxeter, Introduction to geometry, Wiley, New York,1969;
N.V. Efimov, Higher geometry (Part 3), Mir, Moscow, 1980;
A.F. Beardon, The geometry of discrete groups, Springer, 1983;

the more experienced reader (familiar with analytic geometry and matrices) can be recommended F. Klein, Vorlesungen über nicht-euklidische Geometrie, Springer, Berlin, 1968.

Exercises

1. Let ℓ be a line of Lobachevsky geometry and A a point of ℓ; using the inequality (20) and motions in $\mathcal{G}$, prove that points M for which the distance $\|M, A\|$ is the shortest of the distances of M to points of ℓ form a line.

2. Prove that the group of motions of the Lobachevsky plane is exhausted by motions in $\mathcal{G}$ (use the preceding exercise, and §16.3, Exercise 1).

3. Prove that the sum of angles of a triangle in the Lobachevsky plane can be arbitrarily small. For this, consider an 'ideal triangle', formed by three semicircles centred on the real line and intersecting the real line in points α and β, β and γ, and α and γ; all of the angles of the ideal triangle are 0. Then consider a triangle in the Lobachevsky plane sufficiently close to the ideal triangle (this illustrates the theorem of Lobachevsky geometry that the sum of the angles of any triangle is less than π).

§17. The Lobachevsky plane, the modular group, the modular figure, and geometries on the torus

17.1. Discreteness of the modular group. In §16 we constructed a geometry, the Lobachevsky plane, for which the modular group is a group of motions, since it is a part of the group $\mathcal{G}$ of motions of this geometry.

Here we first of all complete this construction by proving that the modular group is discrete in this geometry.

We need to prove that for any point z of the Lobachevsky plane there is a radius r, depending on z such that the disc $D(z, r)$ of the Lobachevsky plane does not contain points other than z equivalent to z. We use the modular figure constructed in §14 (see Figure 14.6), which is a fundamental domain for the modular group. It is enough to verify the assertion for points z of this figure, since we can take the same radius r for equivalent points.

Notice that in the Lobachevsky plane, the distance d from a point A to a line ℓ is well-defined; if z_0 is a point of the line L given by $\operatorname{Re} z = 0$, and ℓ is the line of the Lobachevsky plane given by the semicircle centred at O (see Figure 16.9), this is exactly what is being asserted in the inequality (20) of §16: the distance from z_0 to points of this semicircle is not less than the distance from z_0 to z'. The general case reduces easily to this by motions belonging to $\mathcal{G}$.

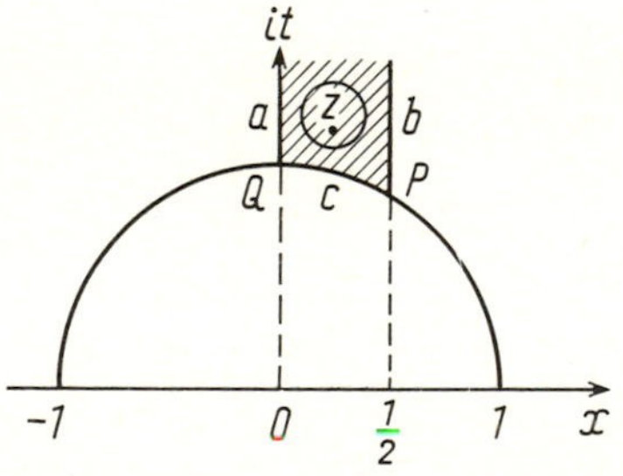

Figure 17.1

With this in mind, for a point z in the interior of the modular figure, we must obviously take r to be less than the least distance from z to the sides of the modular figure, the lines a, b and c of the Lobachevsky plane (Figure 17.1).

To deal with the other cases, we use the result of §15.2, Exercise 1, where it is shown that the modular figure I and the seven neighbouring figures II–VIII obtained from it by the motions of the modular groups listed there form a region entirely containing the modular figure and its boundary curves in its interior (Figure 17.2). Each of the figures II–VIII also does not contain distinct equivalent points, since they are obtained from I, which has this property, by motions of the modular group. After this, the arguments do not present any difficulty: we repeat word-for-word the proof of the discreteness of the group Γ of §12.4. For points z on the boundary of the modular figure we must take r such that the disc D(z, r) only

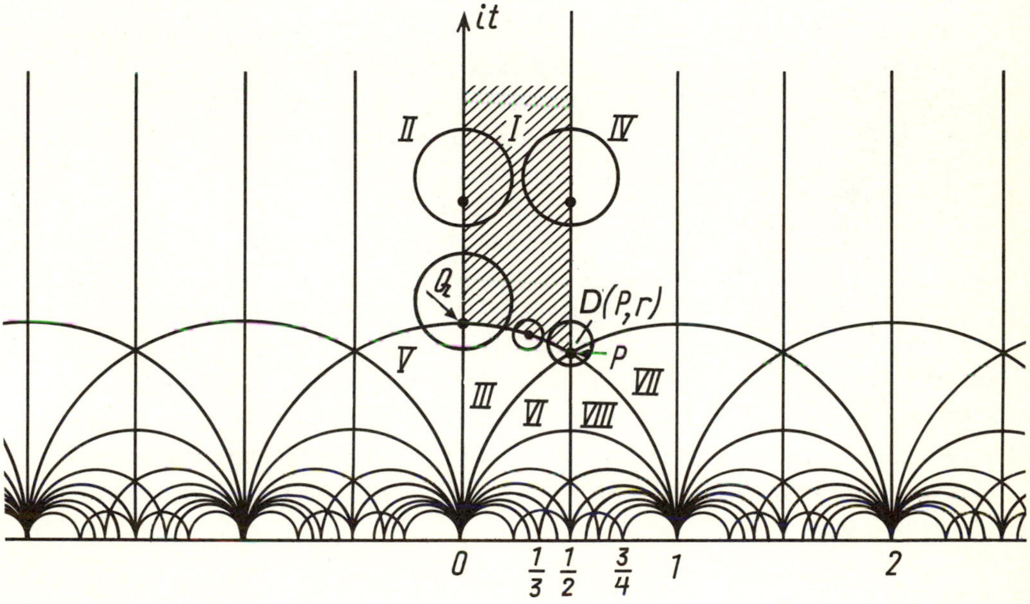

Figure 17.2

intersects those sides of the figures I–VIII that pass through z (see Figure 17.2). Indeed, for example, for the vertex P of the modular figure, the disc D(P, r) is broken up into six sectors by the figures I, III, VI, VIII, VII and IV, each of which contains P. Hence none of these sectors contain points equivalent to P, since each sector is contained in one of the figures I, III, VI, VIII, VII and IV, and these do not contain distinct equivalent points. This completes the proof that the modular groups is discrete.

One can continue by analogy with §12. Since the modular figure is a fundamental region, we can continue, as in Figure 17.2, to fill up the Lobachevsky plane with figures obtained from it by motions in the modular group, and we get a decomposition of the whole of the Lobachevsky plane into figures congruent to the modular figure (§9, Exercise 1). This can be represented without difficulty (see

Figure 17.2), since each of the figures, just like the modular figure itself, has three sides, and the figures adjacent to it along these sides are obtained from it by reflections in these sides (in the same way as figures II, III and IV are obtained from the modular figure). Obviously, the pattern in the Lobachevsky plane formed by the sides of these figures, illustrated in Figure 17.2, is taken into itself by motions of the modular group. Moreover, it is very easy to show that the modular group is just the symmetry group of this pattern (Exercise 1). Thus, as in §12, we have found another intuitively clear description of the modular group, as the symmetry group of a pattern in the Lobachevsky plane. At the same time we get an example of a pattern, or an imaginary crystal in the Lobachevsky plane. We must just remember that in Figure 17.2, we are looking at this pattern from the perspective of an inhabitant of the Euclidean plane, and we are therefore not in a position to evaluate its component parts from the point of view of size and distance betwen them, which is crucial for a pictorial representation. In Figure 17.2, it appears to our eyes that the triangles making up the pattern decrease, becoming arbitrarily small near the real line. But from the points of view of an inhabitant of the Lobachevsky plane they are all congruent, so are of the same size; we need formulas (14) and (15) of §16.3 to see the pattern from his perspective.

The pattern of Figure 17.2 looks more complicated than a pattern in the Euclidean plane. This complexity can also be expressed in more precise terms. Take the modular figure, then its neighbours II–VIII, then their neighbours, and so on. The number of regions obtained after n steps grows very fast with n, in fact as an exponential function of n. However, if we carry out this construction for the fundamental domain of a discrete group in the Euclidean plane (for example, with the patterns of Figure 12.23 and their fundamental regions), then the number of regions is a function of n which grows much slower, in fact as a polynomial of degree at most 2.

Exercise

1. Prove that the modular group is the symmetry group of the pattern of Figure 17.2. [Hint: use §16.3, Exercise 1, and §16.4, Exercise 2, to prove that the modular figure does not have any symmetries.]

17.2. The set of all geometries on the torus. As we have already said at the end of §15, the most important corollary of the discreteness we have just proved is another description of the set of all geometries on the torus (or of lattices in the plane) up to similarity, as the geometry Σ corresponding to the modular group of motions of the Lobachevsky plane.

What kind of geometry is this?

We recall the properties of the modular figure stated at the end of §15. Firstly, it is a fundamental domain for the modular group. Secondly, and especially important for our purposes, it does not contain distinct equivalent points, even on the boundary. Finally, as the reader can easily see for himself, it is convex in the

Lobachevsky plane, that is, together with any two points, it also contains the line segment of Lobachevsky geometry joining them. Hence, just like the geometries of II.3-4 and IV.3-6 in the Euclidean plane, (see Figure 12.6), it is itself the geometry we want, that is the geometry is part of the Lobachevsky plane. The corresponding arguments are entirely similar to those we gave in §12.3 for the geometry D_n.

What is this geometry, the modular figure from the point of view of Lobachevsky geometry?

It is almost a triangle, except for the fact that two of its sides are given on the upper half-plane by two parallel rays $\mathrm{Re}\, z = 0$ and $\mathrm{Re}\, z = 1/2$. As lines of Lobachevsky geometry, they are of course also parallel, that is, they do not intersect. However, they are 'special' parallels, of a kind which only appear in Lobachevsky geometry: there are points on them which are arbitrarily close, so that

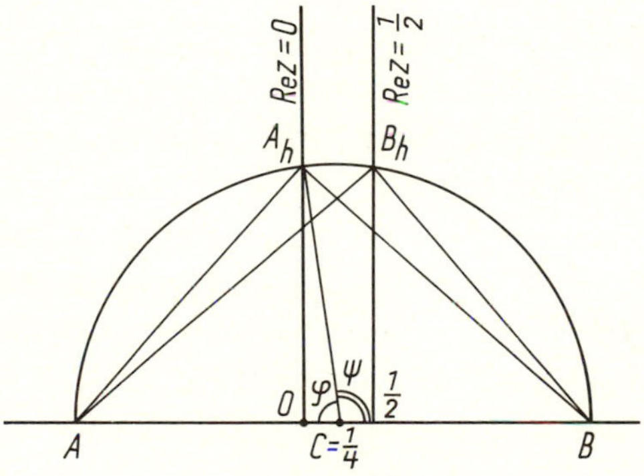

Figure 17.3

they converge together at infinity. In fact, consider the two points A_h and B_h on these lines, $A_h = ih$ and $B_h = (1/2) + ih$ (Figure 17.3). In Lobachevsky geometry, choosing 10 for the base of logarithms, and writing lg for $\log_{10}$, we have

$$\|A_h, B_h\| = \lg \frac{|A_hA|\cdot|B_hB|}{|A_hB|\cdot|B_hA|} = 2 \lg \frac{|A_hA|}{|A_hB|}.$$

From Figure 17.3, one sees that

$$|A_hA| = 2\cos(\varphi/2)\cdot|AC|, \quad |A_hB| = 2\cos(\psi/2)\cdot|BC|,$$

and $|AC| = |BC|$. Hence

$$\|A_h, B_h\| = 2\lg\frac{\cos(\varphi/2)}{\cos(\psi/2)},$$

and since as h tends to infinity, both φ and ψ obviously tend to $\pi/2$, it follows that

$$\cos(\varphi/2) \to 1/\sqrt{2}, \qquad \cos(\psi/2) \to 1/\sqrt{2},$$

hence

$$\frac{\cos(\varphi/2)}{\cos(\psi/2)} \to 1$$

and therefore

$$\|A_h, B_h\| = 2\lg\frac{\cos(\varphi/2)}{\cos(\psi/2)} \to 0 \quad \text{as} \quad h \to \infty.$$

In view of this, one speaks of the modular figure as a triangle with one vertex at infinity.

As can be seen from Figure 17.2, at the vertex P of this triangle, six angles congruent to the angle at P come together, forming a complete angle. Thus it is natural to give the angle P of our triangle the value $\pi/3$. In exactly the same way, the angle at the vertex Q should be given the value $\pi/2$. This incidentally agrees with the general definition of angle in Lobachevsky geometry, given in §16.1, Exercise 1; notice that if we use the definition of this exercise and apply it to the vertex at infinity of our triangle, then we have to give the angle at infinity the value 0.

We summarise our conclusions:

Theorem 1. The set of all geometries on the torus (or of lattices on the plane) up to similarity is itself a geometry, that of a triangle in the Lobachevsky plane with one vertex at infinity and angles of $\pi/3$ and $\pi/2$ at the two finite vertices.

A further result of our study is the fact that we have constructed an example of a discrete group of motions of the Lobachevsky plane, the modular group. It is not uniformly discontinuous in the Lobachevsky plane: indeed, its transformations have certain points as fixed points, which as we have seen, cannot happen in a uniformly discontinuous group. For example, the motion taking z into $-\bar{z}$ fixes all points of the ray $\operatorname{Re} z = 0$, and the motion taking z to $-1/z$ fixes i. Thus we run at once into a more complicated example of a discrete group; in the Euclidean plane such groups appeared only in §12 in connection with crystallographic groups in the plane. But in the Lobachevsky plane there are also uniformly discontinuous groups, which lead to 2-dimensional hyperbolic geometries, that is, geometries which are identical to Lobachevsky geometry on a sufficiently small scale. There are very many more such groups than in the Euclidean plane. Those which lead to bounded geometries are of especial interest. These geometries cannot be realised on the torus, they are realised on more complicated surfaces, such as those illustrated in Figure 17.4.

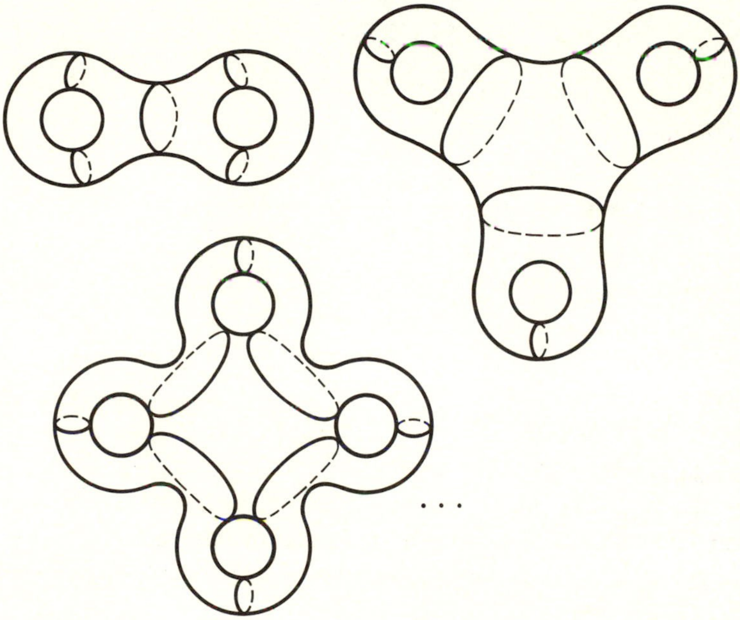

Figure 17.4

Both discrete groups of motions of the Lobachevsky plane and their corresponding geometries have an enormous number of applications in the most diverse branches of mathematics and mathematical physics.

Exercises

1. Prove that the modular group is generated by the reflections in the sides of the modular figure.

2. Prove that motions of the first kind of the modular group form a group (called the modular group of the first kind). Prove that it has as its fundamental domain the shaded area of Figure 17.5 (called the modular figure of the first kind). Which points of the boundary are equivalent? Notice that the corresponding surface is the same as that of the geometry II.7 of Figure 12.21. What are generators of the modular group of the first kind?

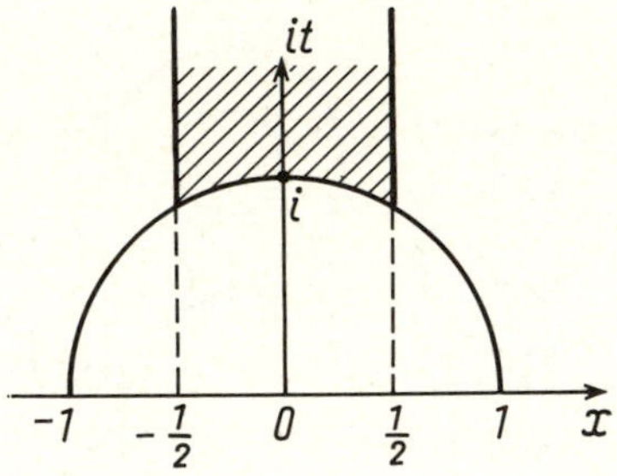

Figure 17.5

Historical remarks

The basic idea of this book is that geometry is not some rigid system of definitions and theorems, predetermined once and for all. On the contrary, geometrical intuition is capable of creating a particular geometry for each new sphere of consciousness. This point of view, which belongs among the most fertile ideas of mathematics and theoretical physics, has two sources. One of these was the investigation of the axioms lying at the basis of geometry, and more concretely, the attemps to prove the parallel axiom, which led eventually to the creation of Lobachevsky geometry. For the first time, the idea that a geometry in which the parallel axiom does not hold should exist, and should be logically as valid as Euclidean geometry, was developed in writings of N.I. Lobachevsky (1793-1856) published in 1826. Independently of Lobachevsky, the same ideas were advanced by the Hungarian mathematician J. Bolyai (1802-1860) in a work appearing in 1829. It turned out subsequently that about 10 years before this, two German mathematicians, C.F. Gauss (1777-1855) and F.K. Schweikart (1780-1857) arrived at the same conclusions, but did not publish their works, perhaps because they were not confident in them, or perhaps because they were afraid they would not be understood.

The other source was the study of the geometry of surfaces. Here Gauss, in 1817, advanced the idea that the central problem of the geometry of surfaces is not the study of the position of a surface in space, but of the intrinsic properties of the surface, viewed as a carrier of its geometry. From this point of view, as Gauss points out, a piece of a cylinder or of the plane, although they look different, must be considered to be the same.

A very general notion of geometry, reflecting and uniting these points of view, was worked out by the German mathematician G.F.B. Riemann (1826-1866) in 1854. The intuitive representation of an 'inhabitant' of the geometry was proposed to explain these ideas by the German physicist and physiologist H. von Helmholtz (1821-1894) in 1868. The most general notion of geometry, which we have used as the basis of our treatment, appeared only in the 20th century. It was prepared by the preceding development to such an extent that it is hard to single out its author.

Geometries which are locally identical with the Euclidean plane or 3-space were considered by the English mathematician W.K. Clifford (1845-1879) at the end of the 19th century; he discovered their relation with uniformly discontinuous groups. Interest in the theory of discrete groups of motions of space became especially strong when their relation with crystallography became clear. All the crystallographic groups in 3-space were classified independently by two crystallographers: a Russian, E.S. Fëdorov (1853-1919) in 1890, and a German, A. Schoenflies (1853-1928) in 1891.

The idea of a model as a proof of the consistency of Lobachevsky geometry was already advanced by Lobachevsky. It was first realised by the German mathematician F. Klein (1849-1925) and the Italian mathematician E. Beltrami (1835-1900) around 1870. The systematic study of geometry from the point of view of the underlying axioms was undertaken by D. Hilbert (1862-1943) in a book which appeared at the very end of the 19th century. He also posed the problem of the consistency of arithmetic. The theorem that the consistency of arithmetic cannot be proved on the basis of the axioms of arithmetic and logic was proved by the German mathematician K. Gödel (1906-1978) in 1931.

The model of Lobachevsky geometry in the upper half-plane which we arrived at in this book was introduced by the French mathematician H. Poincaré (1854-1912) in 1882. He was led by much the same ideas as we have used: to find a geometric interpretation of the group Γ, which appears in many question of mathematics. Discrete groups contained in Γ (but without the interpretation of their elements as motions of the Lobachevsky plane) appeared much earlier. In particular the modular group and the modular figure were introduced by Gauss, as in our book, in connection with the question of similarity of lattices. According to his diary, Gauss arrived at these ideas in 1800. Using the intuition of Lobachevsky geometry, Poincaré in a work of 1882 constructed a theory of geometries which are locally identical to the Lobachevsky plane. For bounded geometries and the corresponding uniformly discontinuous groups of motions of the Lobachevsky plane he constructed a theory which, although more complicated, was almost as exhaustive as the one we have treated in §8 for the case of Euclidean geometry. This research found a large number of applications to questions of mathematics which no-one had previously suspected had any relation at all to Lobachevsky geometry, or even to geometry at all.

Nevertheless in this theory, there remain many unsolved questions. The most important of these is the description of the set of all 2-dimensional hyperbolic geometries, that is geometries which are locally identical to the Lobachevsky plane, and are realised on surfaces of a given type (see Figure 17.4). In the results obtained so far in this direction, we can see the application of the same kind of ideas which we used in Chapter IV of our book for the description of locally Euclidean geometries on the torus. One again constructs a certain new geometry (analogous to the Lobachevsky plane), and a certain discrete group (analogous to the modular group), so that the geometries we are interested in are represented by points of the new geometry equivalent under the action of this new group.

In other words, hyperbolic geometries again determine a certain geometry. However, up to now, a description of this has not been obtained which comes close to being as intuitive and definitive as that which we have given for the set of all geometries on the torus: the triangle in the Lobachevsky plane with angles 0, $\pi/3$ and $\pi/2$.

List of notation

Σ	a geometry
Π	the plane
Γ	group of motions, p.59
Σ_Γ	geometry constructed from a group Γ, p.61
a, b, ..., x	points of the geometry
A, B, ..., X	points of the plane
A, B, ..., X	sets of equivalent points in the plane, p.61
Φ	figure
$AB^{\rightarrow}$	vector from A to B
[AB]	line segment joining A and B in the plane
[ab]	line segment joining a and b in a geometry, p.99
(AB)	line through A and B
$\|AB\|$	distance between A and B in the plane or a geometry
$\\|AB\\|$	extrinsic distance between points of sphere in 3-space, p.48
$\\|z_1, z_2\\|$	distance between z_1 and z_2 in Lobachevsky plane, p.225
f	track, p.14
$f^{\sim}$	curve covering a track f, p.15
f_i	curve in the plane
ℓ	generator of a cylinder
c	directrix of a cylinder
$\ell, \ell_0, \ell_1, m, \ldots$	lines
Π_i	strip in the plane
D	disc
D(X, r)	spherical neighbourhood of X (disc centre X, radius r), p.51
B(A, s)	ball neighbourhood of A (ball centre A, radius s), p.144
$\mathcal{G}$	group of motions of Lobachevsky plane, p.220
G^F	motion conjugate to G under F, p.188
H_O^λ	dilation with centre O and ratio of similarity λ, p.192
$T_{\mathbf{a}}$	translation in the vector **a**, p.66
$T_{OO'}$	translation in the vector $OO'^{\rightarrow}$
S_ℓ	reflection in the line ℓ, p.66
$S_\ell^{\mathbf{a}}$	glide reflection with axis ℓ and vector **a**, p.25

R_O^{φ}	rotation of plane through angle φ about centre O, p.66
$R_{\ell,\varphi}^{\mathbf{a}}$	twist, p.122
S^{Π}	reflection in a plane Π
$S_{\mathbf{a}}^{\Pi}$	glide reflection with plane Π and vector **a**, p.122
$S_{\ell,\varphi}^{\Pi}$	rotary reflection, p.123
$\varphi: \Sigma_1 \to \Sigma_2$	covering of the geometry Σ_2 by a geometry Σ_1, p.102
$< F_1, F_2, ..., F_n >$	group generated by motions $F_1, F_2, ..., F_n$, p.86
C_n	group of rotations of regular n-gon, also the geometry constructed from it, p.165
D_n	group of symmetries of regular n-gon, also the geometry constructed from it, p.165
i	unit imaginary number, with $i^2 = -1$, p.212
$z = a + bi$	complex number, p.213
$\bar{z}$	complex conjugate of z, that is $\bar{z} = a - bi$
Re z	real part of z
Im z	imaginary part of z
$\lvert z \rvert$	modulus (absolute value) of z
arg z	argument of z

Index

M. Berger

Geometry I

Translated from the French by M. Cole and S. Levy

Universitext

1986. 426 figures. XIII, 427 pages. ISBN 3-540-11658-3

M. Berger

Geometry II

Translated from the French by M. Cole and S. Levy

Universitext

1986. 364 figures. X, 407 pages. ISBN 3-540-17015-4

This two-volume textbook is the long-awaited translation of the French book "Géometrie" originally published in five volumes. It gives a detailed treatment of geometry in the classical sense.
An attractive characteristic of the book, and of Prof. Berger's writing in general, is that it appeals systematically to the reader's intuition and vision, and systematically illustrate the mathematical text with many figures (a practice which has fallen into disuse in more recent years).
For each topic the author presents a theorem that is esthetically pleasing and easily stated - even though the proof of the same theorem may be quite hard and concealed. Many open problems and references to modern literature are given.
The third principal characteristic of the book is that it provides a comprehensive and unified reference source for the field of geometry in all its subfields and ramifications, including, in particular the following topics: crystallographic groups: affine, Euclidean and non-Euclidean spherical and hyperbolic geometries, projective geometry, perspective and projective completion of an affine geometry, cross-ratio; geometry of triangles, tetrahedron circles, spheres; convex sets and convex polyhedrons, regular polyhedrons, isoperimetric inequality; conic sections and quadrics from the affine, Euclidean and projective viewpoints.
A companion volume of exercises in geometry has already been published by Springer-Verlag in its "Problem Book" series.

Springer-Verlag
Berlin Heidelberg New York
London Paris Tokyo

Springer